# Mass Spectrometry in Drug Discovery

# Mass Spectrometry in Drug Discovery

edited by

## David T. Rossi
*Pfizer Global Research and Development*
*Ann Arbor, Michigan*

## Michael W. Sinz
*Bristol-Myers Squibb*
*Pharmaceutical Research Institute*
*Wallingford, Connecticut*

MARCEL DEKKER, INC.    NEW YORK · BASEL

**ISBN: 0-8247-0607-2**

This book is printed on acid-free paper.

**Headquarters**
Marcel Dekker, Inc.
270 Madison Avenue, New York, NY 10016
tel: 212-696-9000; fax: 212-685-4540

**Eastern Hemisphere Distribution**
Marcel Dekker AG
Hutgasse 4, Postfach 812, CH-4001 Basel, Switzerland
tel: 41-61-261-8482; fax: 41-61-261-8896

**World Wide Web**
http://www.dekker.com

The publisher offers discounts on this book when ordered in bulk quantities. For more information, write to Special Sales/Professional Marketing at the headquarters address above.

Current printing (last digit):
10 9 8 7 6 5 4 3 2

**PRINTED IN THE UNITED STATES OF AMERICA**

# Preface

The recent, pervasive use of liquid chromatography/mass spectrometry (LC/MS) is providing sensitive, selective, rapid, and information-rich analytical methodology to drug discovery research. With the possible exception of combinatorial chemistry, no single technology has revolutionized and streamlined the drug discovery process to such a great extent. In the hopes of expediting the drug discovery process, major pharmaceutical companies have invested considerable financial and human resources in mass spectrometers.

Current pharmaceutical development has become more demanding, especially with the advent of high-throughput discovery programs across the industry in the areas of chemistry, pharmacology, and pharmacokinetics. In this regard, interdisciplinary rapid-screening protocols that allow for the examination of an increased number of new chemical entities (NCEs) have arisen. Many of these new screening protocols have been successful due in large part to the ability of the mass spectrometer to function as a sensitive and selective detector with minimal method development effort. Drug discovery has been the subject of many articles in peer-reviewed journals, but until now no concise treatise had been written to bring together the numerous applications of mass spectrometry and drug discovery.

*Mass Spectrometry in Drug Discovery* is intended to bring the knowledge gap that the widespread use of LC/MS has created. From a practical and applied point of view, the text will introduce readers to the basic concepts of mass spectrometry and atmospheric pressure ionization LC/MS. Although many aspects of mass spectrometry are encompassed, the book focuses more on applications associated with pharmaceutics, pharmacokinetics, and combinatorial chemistry;

it is not intended to incorporate the abundance of applications in the areas of pharmacology and toxicology. Nor is this book intended just for those in drug metabolism who have extensive experience in mass spectrometry. Rather, it is hoped that any scientist interested in applying LC/MS to drug discovery will benefit, including those who may not have received any prior training in the area. Moreover, this book should serve as a background text for those who will use the information that comes out of LC/MS experiments but may not be using LC/MS directly, including toxicokineticists, pharmacologists, and drug discovery scientists.

Parts I and II introduce the principles of mass spectrometry instrumentation and Part III illustrates some of the theoretical aspects of the LC/MS experiment. Part IV discusses the various applications of mass spectrometry, such as concepts of combinatorial chemistry and chemical libraries. Part IV also discusses pharmacokinetic drug discovery strategies and the application of mass spectrometry to discover better pharmaceutical agents, as well as give the novice some background in the disciplines of pharmacokinetics and drug metabolism. The last chapter is dedicated to future applications of mass spectrometry in both drug discovery and development that could have interesting prospects in the near future.

*Mass Spectrometry in Drug Discovery* provides a solid background in this realm of pharmaceutical science. The book is expected to be useful for the industrial scientist as well as students and academicians working in this field.

*David T. Rossi*
*Michael W. Sinz*

# Contents

v

# Contributors

**Lucinda R. H. Cohen**   Pfizer Global Research and Development, Ann Arbor, Michigan

**Scott T. Fountain**   Pfizer Global Research and Development, Ann Arbor, Michigan

**Krishna Ramalingam Iyer**   Bombay College of Pharmacy, Mumbai, India

**Erick K. Kindt**   Pfizer Global Research and Development, Ann Arbor, Michigan

**Ryan M. Krisko**   Department of Chemistry, University of Kansas, Lawrence, Kansas

**Michael J. Lovdahl**   Pfizer Global Research and Development, Ann Arbor, Michigan

**Craig E. Lunte**   Department of Chemistry, University of Kansas, Lawrence, Kansas

**Steven M. Michael**   Pfizer Global Research and Development, Ann Arbor, Michigan

**Richard A. J. O'Hair**   School of Chemistry, University of Melbourne, Victoria, Australia

**Terry Podoll**   CEDRA Corporation, Austin, Texas

**David T. Rossi**   Pfizer Global Research and Development, Ann Arbor, Michigan

**Michael W. Sinz**   Bristol-Myers Squibb, Pharmaceutical Research Institute, Wallingford, Connecticut

**Barbra H. Stewart**   Pfizer Global Research and Development, Ann Arbor, Michigan

**Narayanan Surendran**   Pfizer Global Research and Development, Ann Arbor, Michigan

**Jayesh Vora**   Chiron Corporation, Emeryville, California

**David J. Weiss**   Department of Chemistry, University of Kansas, Lawrence, Kansas

# 1

# The Demand for Drug Discovery Today

**Michael W. Sinz**
*Bristol-Myers Squibb, Pharmaceutical Research Institute,
Wallingford, Connecticut*

> I have not failed. I've just found 10,000 ways that won't work.
> —*Thomas Edison*

## I. INTRODUCTION

Drug discovery is the process of generating (compounds and data) and evaluating all of the necessary information to determine the feasibility of a new chemical entity (NCE) to become a safe and efficacious drug. The challenge of drug discovery is to determine the most efficient and accurate means of synthesizing compounds; generating, processing, and evaluating data; and ultimately combining this information into an assessment of drug safety and efficacy in humans. The most successful drug companies of the future will be those that develop and implement the most efficient and reliable methods to screen and evaluate NCEs, either through high-throughput means or rational drug design, to improve the quantity, quality, and performance of drug candidates. The demand for improved drug discovery is greater today than ever before due to the intense need for novel drugs to treat existing disease states and for the multitude of new therapeutic targets being discovered each day related to disease. By reducing the time to discover and develop new drugs, the cost to the pharmaceutical industry will decrease and ultimately reduce the cost of new medications to the consumer.

Major advances and improvements in genomics and proteomics have greatly impacted a variety of industries; however, the most effected to date is the pharmaceutical industry. The number of disease targets uncovered by genomics and proteomics has exploded, thereby increasing the number of ways in which each disease can be treated [1,2]. Other advances in the areas of combinatorial chemistry, molecular biology, automation, instrumentation, and computing (data storage/retrieval and software) have enhanced the pharmaceutical industry's ability to perform drug discovery and development. Who would have imagined that experiments performed years ago in individual test tubes could now be done in a single 1536-well plate?

An indication of the importance and fast pace of drug discovery is illustrated by the increased number of articles published dealing with discovery and the number of new journals dedicated to the topic, such as *Combinatorial Chemistry and High Throughput Screening, Drug Discovery and Development, Drug Discovery Today, Modern Drug Discovery*, and *Bioinformatics.* New industries, many of which did not exist 10 years ago, are growing and thriving due to the demand for improved techniques and methodologies, hardware with greater flexibility, and software for data analysis and prediction (Table 1). Many of these hardware, software, and process advancements are briefly described in this chapter and in much greater detail in subsequent chapters.

**Table 1**  Companies That Develop New Technologies in the Area of Drug Discovery

| Company | Area of development | Web site |
| --- | --- | --- |
| Trega Bioscience | ADME prediction | www.trega.com |
| Aurora Biosciences | Discovery technologies | www.aurorabio.com |
| Camitro Corp. | ADME prediction | www.camitro.com |
| Affymetrix | Genetic information—gene chip | www.affymetrix.com |
| Celera Genomics | Genomics and proteomics | www.celera.com |
| Cellomics Inc. | Cellular screening systems | www.cellomics.com |
| GeneLogic Inc. | Genomic databases/software | www.genelogic.com |
| Oxford Glycosystems | Proteomics | www.ogs.com |
| Large Scale Biology | Proteomics and genomics | www.lsbc.com |
| Tecan | High-throughput instrumentation | www.tecan.com |
| Xenometrix | Toxicology screening | www.xeno.com |
| Tripos Inc. | Discovery software and compound libraries | www.tripos.com |
| MDL Information Systems | Discovery informatics and databases | www.mdli.com |
| Pharsight Corp. | Clinical trial simulations | www.pharsight.com |
| Micromass | Mass spectrometry instrumentation | www.micromass.com |
| PE Sciex | Mass spectrometry instrumentation | www.pesciex.com |

## II. DISCIPLINES IN DRUG DISCOVERY

### A. Chemistry

The first step in any drug discovery process is the decision of which chemical structures to synthesize [3]. There are various means by which compounds can be proposed or actually synthesized for biological activity testing (Table 2). Rational drug design (information-based synthesis) involves preparing compounds based on a great deal of information known about the biological target, such as what structural conformation or functional groups are necessary for an NCE to bind to a particular receptor. Sometimes the exact molecular structure of the receptor is known from crystallography studies or by structural probes. On a much larger scale, compounds can be prepared by combinatorial methods with some knowledge about the biological target; however, in this approach the number of compounds generated improves the odds of finding an active compound with the appropriate physiochemical, toxicological, or biopharmaceutical properties [4]. An alternative approach to chemists actually preparing compounds in the laboratory is to screen compounds in chemical libraries (collections of synthesized compounds) or virtual libraries (software generated compounds). Chemical libraries are composed of actual compounds synthesized in the laboratory or found as natural products. These libraries range from hundreds of thousands to millions of compounds. Virtual libraries can contain trillions of compounds that could be potentially synthesized based on sequences of known reactions and reagents [5]. Whether the screening takes place with actual or virtual compounds, the chemical libraries can be searched for compounds matching particular aspects of the biological target (receptor, protein, enzyme, or organism). The number of compounds that can be screened generally depends on how much information is known about the biological target. When a great deal of information is known about a particular target or there are major constraints that limit the structural

**Table 2** Different Methods of Discovering New Chemical Entities (NCE)

| Method | Number of potential compounds | Comment |
|---|---|---|
| Rational design | 10–1,000s | Actual synthesis |
| Combinatorial synthesis | 1,000–1,000,000s | Actual synthesis |
| Chemical library | | |
|    Natural products | 1,000,000s | Many compounds still undiscovered |
|    Proprietary | 1,000,000s | Owned by individual companies |
| Virtual library | 1,000,000,000,000s | Software generated |

diversity of the NCEs, then the screening can be done on a smaller scale (less than 2 million compounds). When very little is known about a biological target, screening on a much larger scale would be appropriate (more than 2 million compounds) [6]. By screening a larger database of structures, the chemist can learn more about what defines the structure necessary for activity at a biological target and also increases the odds of finding compounds that will elicit a pharmacological response.

Once synthesized, multiple physical-chemical properties of the molecule can be determined, some by experimental means and others through the use of software. For example, physical-chemical properties, such as molecular weight, solubility, $pK_a$, LogP (LogD), chemical stability at various pH's or in different environments (light, humidity, or liquid), can be determined. Generally, these are the first parameters entered into a large database of information that will eventually contain all that is known about the properties of an individual compound.

## B.  Pharmacology

After synthesis, each NCE is tested against a single or multiple therapeutic target(s) to determine if the NCE has any potential to elicit a pharmacological response. This pharmacological testing can be done in multiple different ways depending on the number of compounds to be screened, the speed at which you need to screen, the limitations of what is known about a biological target, or the reagents available to the researcher. If there are a large number of compounds that need to be screened rapidly, a method and process must be proposed that will achieve the necessary turnaround time. Generally this can be done when the biological target is well established and in vitro methods are available to screen many compounds simultaneously, e.g., multiwell plates [7]. In addition, the biological response must be rapid and easily measurable. For example, thousands of compounds can be screened in a single day using a receptor-based assay that elicits a spectrofluormetric response [8].

State-of-the-art technology today belongs to Aurora Biosciences, who are developing an ultra-high-throughput screening system (UHTSS™) capable of screening over 100,000 compounds per day. This is accomplished by the use of an automated compound storage and retrieval module that stores over 1,000,000 compounds in solution for rapid access and testing in 3,456-well plates. This rate of testing is highlighted by the use of novel microfluidic technologies that accurately transfer small volumes (one-billionth of a liter) of liquid at rates of up to 10,000 wells/hr coupled with fluorescence detectors to measure response [9,10].

In contrast to the UHTSS, when assays using individual proteins or subcellular preparations are not appropriate to measure receptor binding or illicit a pharmacological response, more sophisticated cell-based or in vivo assays are necessary [11]. The speed at which individual compounds can be screened dramatically decreases in these systems, especially in the case of in vivo animal testing, where the number of compounds that can be simultaneously screened decreases and often the time to measure a response is increased (see Chap. 11).

## C. ADME Properties

The majority of topics covered in this book will deal with the properties of absorption, distribution, metabolism, and elimination (ADME) of potential new drugs. These are the disciplines of pharmacokinetics, pharmaceutics, drug metabolism, and drug transport, which are covered in much greater detail in subsequent chapters. Historically, due to the more laborious nature of these studies and the lack of emphasis on ADME properties in drug discovery, these aspects of a drug's development are somewhat behind the fields of chemistry and pharmacology. However, since 1990, multiple advances, such as cassette dosing [12], in vitro metabolic stability [13] and cellular permeability assays, sample preparation/ extraction [14], and atmospheric pressure ionization mass spectrometry have brought ADME properties to the forefront of discovery screening [15]. Many of the physical-chemical properties of the molecule mentioned previously do have an impact on the ADME properties of a potential new drug. Therefore, parameters such as molecular weight, solubility, and hydrophobicity (LogP) are evaluated in conjunction with or in the prediction of such phenomena as formulation, absorption, permeability, and metabolism [16].

## D. Pharmacogenomics

Pharmacogenomics (pharmacogenetics) is the grouping of a pharmacological drug response from a population based on the genetic variation of that population [17]. In some instances, different individuals in a population respond to the same drug in different ways; these variations can be due to genetic variations in the biological target (receptor), genetic differences in drug transport, or the metabolizing enzymes that are involved in drug absorption and clearance [17,18]. Therefore, pharmacogenomics is a combining of pharmacology, genetics, pharmacokinetics, and toxicology in order to give the correct drug to the patient so as to produce a safe and efficacious response. By applying pharmacogenomics early and to a greater number of drugs, pharmaceutical companies can potentially reduce the number of unwanted side effects, improve clinical trials, restrict drugs from those patients who will not respond, and tailor some drugs to individuals with the appropriate genetic makeup, resulting in greater efficacy.

## E. Toxicology

In addition to the above-mentioned properties, the potential toxicological aspects of discovery compounds should be evaluated at the earliest opportunity. Although the details are beyond the scope of this text, the reader should be aware that new techniques are currently being employed and newer, more predictive methods are underway to explore both the in vitro and in vivo toxicology of NCEs in discovery [19]. For example, the use of genomics and metabonomics or biomarkers (the relationship between genes and metabolite patterns, respectively, with disease) allow much higher throughput and earlier toxicology screening of NCEs [20–22]. Other screening tools include the use of desktop or web-based databases of known toxic compounds or structural filters that eliminate compounds with unwanted functional groups known to cause toxicity. Of all the disciplines described, toxicology screening appears to be trailing the rest, in part due to the complex nature of studying toxicological events in vitro, as these events are often multifaceted and can occur in several organs. Also, preclinical toxicology is performed in animals with the assumption that animal toxicity and safety are correctly related to human toxicity and safety, which in some instances may not be true.

Figure 1 illustrates the process of drug discovery, which involves a multidisciplinary approach with a constant influx of new technologies and output of information that results in a lower compound attrition rate and compounds with improved "druglike" characteristics. Ultimately, this entire process comes together to produce a greater number of better drugs to treat disease.

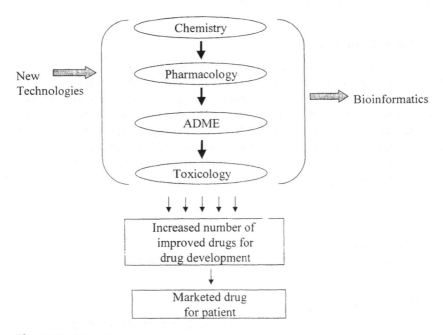

**Figure 1** Drug discovery process.

## III. BIOINFORMATICS

We are in the information business; our success is dependent on how well we use this information (data). One must always remember that the highest throughput systems will generate the largest amount of information. Databases are now the foundation of our information handling systems. The ability to store, retrieve, visualize, and learn from such databases is the challenge that effects drug discovery [23]. Databases are an essential part of drug discovery, as they allow us to organize data, find correlations (in vitro/in vivo), and predict drug properties. Unfortunately, bioinformatics is generally an afterthought when setting up a new discovery screening protocol. Issues such as determining which information needs to be stored, in which format the data should be stored, how to input or retrieve data from the system, how the data can be visualized, or how to integrate a database with data from other sources are typically dealt with at a later date. Bioinformatics needs be addressed at the outset of a new discovery screening technique to avoid delays in implementation of a new screen due to lack of an appropriate storage place (database) for the large amounts of data generated.

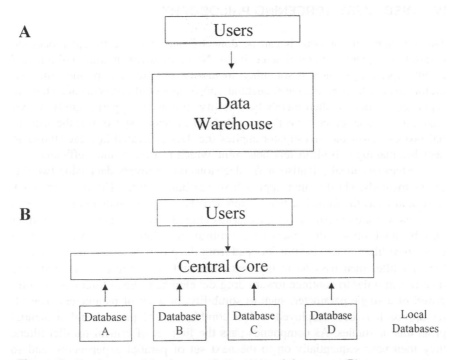

**Figure 2**   Database models: data warehouse (A) and data marts (B).

In general, there are two database models for the storage of information: data warehouses and data marts (repositories). Data warehouses are large databases that contain most or all of the information collected in discovery, much akin to a warehouse that stores raw materials or goods. This model of data storage places all of the information in one place that can be centrally administered. A disadvantage to this large single database is the difficulty in setting up and modifying the database. Having a single database implies that all information inputs and retrievals are by the same or similar mechanisms, which can be difficult, given the large variability in data output formats from discovery instrumentation. Data marts or repositories are a collection of smaller local databases that all connect to a central core for access by multiple users and software products. The advantages of data marts are that they are easier to set up and modify and have greater functionality with software products that generate or analyze data; however, they can be difficult to integrate. Figure 2 is a graphical representation of both the data warehouse and data mart models. Both models are currently in use across the industry and have advantages and disadvantages depending on the amount of data, types of data, and user access needs.

## IV.  DISCOVERY SCREENING PHILOSOPHY

The screening of compounds can be done by one of two different processes: parallel or sequential. Parallel screening of NCEs involves running multiple different experiments, such as solubility, metabolic stability, and permeability, simultaneously. In contrast, the sequential approach would perform each study in sequence: solubility, then metabolic stability, followed by permeability. Obviously, the parallel process has the advantage of increased speed over the sequential process. How the investigator applies the data generated in these studies to decision making will ultimately determine which process is most efficient.

Pharmaceutical scientists make decisions on discovery data using two different methods: (1) the filter approach or (2) buffet style. The filter approach (also known as the funnel or tiered approach) takes compounds through a series of filters that are descriptors of good drugs (Fig. 3). For example, an early filter may be solubility—a drug must have a minimum solubility in order to pass on to the next filter, otherwise it is dropped from the program. Compounds that pass the first filter then pass on to the next filter and again must show good drug qualities in order to continue toward drug development. Each filter can be composed of a single parameter, such as solubility, or a set of parameters. The filter approach is best employed with a combination of parallel and sequential programs of studies. As compounds pass the first set of critical parallel filters, they then pass sequentially on to the next set of parallel experiments, and so forth, toward drug development. Typically, larger numbers of compounds can be

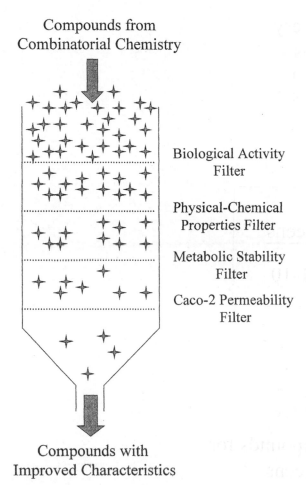

Figure 3   Filter (tiered) drug discovery.

screened by this method. One disadvantage of this approach is that it removes compounds from the system with little ability to reenter the process at a later stage.

The buffet-style approach to drug screening involves collecting all of the necessary data (i.e., running all of the experiments either in parallel or in sequence) on a compound to make decisions on whether a compound should proceed toward development (Fig. 4). In this approach, all of the necessary data (parameters) are determined in a data set (package) that the researcher(s) can utilize in total to make decisions on which are the best compounds to take forward

Set of Discovery
Compounds

Data on all compounds for
all 10 screens

**Figure 4**   Buffet-style drug discovery.

in development or which need further analysis. Unless extremely high-throughput systems are in place, this approach is not capable of screening the same number of compounds as the filter approach. However, it does offer the distinct advantage of having more data to make an informed decision on the progression of compounds toward development.

What types of screens are used and where (chronologically) they appear in the process are unique to each company. There are some basic parameters, such as physical-chemical properties and biological activity, that are early screens in nearly every program. However, the remaining screens are dependent on company philosophy or subjective factors, such as what has recently caused a compound in clinical development to fail. Other information, such as characteristics of poor compounds as identified by regulatory agencies (FDA), what makes the most sense to screen, and what screens are predictable, also define each company's discovery process. Ultimately, most all discovery programs are a combination of both of the filter and buffet-style processes, along with parallel and sequential data collection steps.

Finally, the education of data customers is essential in every discovery program. With the increased amount of data stored in databases, access by many more data users naturally occurs. The scientist who generated the data is no longer the only researcher to examine and interpret that data or otherwise draw conclusions from them. Most databases are merely an assembly of data with no information on how an experiment was performed or the conclusions drawn from the data. There is a responsibility associated with the development of discovery screens and that is the education of potential users of the information.

## V. SUMMARY

Drugs have been shown to fail in preclinical and clinical development for a variety of reasons. The major reasons that drugs fail in later development are ADME, toxicology, human efficacy, or marketing issues, in about equal proportions. The amalgamation of the above-mentioned disciplines and techniques in discovery should result in a lower compound attrition rate and the selection of higher quality compounds for drug development. By eliminating drugs with poor characteristics in discovery, the industry circumvents the expense and time of trying to ''patch up'' or fix compounds with significant issues during development. Drugs of higher quality (fewer issues) that come out of discovery programs will move through the development process at an increased rate and should have a better chance of making it all the way to patients.

The subsequent chapters describe the role of mass spectrometry in the drug discovery process in greater detail. Initial chapters deal more with instrumentation and sample preparation as they relate to mass spectrometry. The remaining chapters describe the utility of mass spectrometry in different areas of the drug discovery process, such as combinatorial chemistry, drug transport, drug metabolism, pharmacokinetics, and microdialysis.

## REFERENCES

1.  JH Wang, RM Hewick. Proteomics in drug discovery. Drug Discov Today 4:129–133, 1999.

2.  S Borman. Proteomics: Taking over where genomics leaves off. C&EN July 31:31–37, 2000.
3.  DJ Ecker, ST Crooke. Combinatorial drug discovery: which methods will produce the greatest value? Biotechnology 13:351–360, 1995.
4.  S Borman. Combinatorial chemistry. C&EN April 6:47–67, 1998.
5.  J Boguslavsky. Lead discovery optimized. Drug Discov Dev March:67, 2000.
6.  G Karet. Are you screening too many compounds? Drug Discov Dev Nov/Dec:39–40, 1999.
7.  G Karet. Factory automation speeds screening. Drug Discov Dev Jun/Jul:34–38, 2000.
8.  U Haupts, M Rudiger, AJ Pope. Macroscopic versus microscopic fluorescence techniques in (ultra)-high-throughput screening. Drug Discov Today 1:3–9, 2000.
9.  www.aurorabio.com.
10. I Gibbons. Microfluidic arrays for high-throughput submicroliter assays using capillary electrophoresis. Drug Discov Today 1:33–37, 2000.
11. M Stevens, DB Roark. Liquid handler facilitates miniaturizing cell-based assays. Drug Discov Dev Jun/Jul:79–80. 2000.
12. MK Bayliss, LW Frick. High-throughput pharmacokinetics: cassette dosing. Curr Opin Drug Discov Dev 2:20–25, 1999.
13. PJ Eddershaw, M Dickins. Advances in in vitro drug metabolism screening. Pharm Sci Technol Today 2:13–19, 1999.
14. AP Watt, D Morrison, DC Evans. Approaches to higher-throughput pharmacokinetics (HTPK) in drug discovery. Drug Discov Today 5:17–24, 2000.
15. DA Smith, H Van de Waterbeemd. Pharmacokinetics and metabolism is early drug discovery. Curr Opin Chem Biol 3:373–378, 1999.
16. R Borchardt, CA Lipinski, C Pidgeon, R Evers, H Van de Waterbeemd, DD Christ, AD Rodriguez, TJ Raub, Q Wang. Screening and profiling compounds for biopharmaceutical properties. Proceedings of American Cancer Society 217th National Meeting, Anaheim, California, 1999, pp 59–61.
17. www.celera.com.
18. R Pettipher. Pharmacogenetics aims to improve drug efficacy. Drug Discov Dev May:66–68, 2000.
19. DE Johnson, GHI Wolfgang. Predicting human safety: screening and computational approaches. Drug Discov Today 5:445–454, 2000.
20. DG Robertson, SJ Bulera. High-throughput toxicology: practical considerations. Curr Opin Drug Discov Dev 3:42–47, 2000.
21. C Sansom. Pharmacokinetic and toxicology screening. Drug Discov Today 4:199–201, 1999.
22. DE Johnson, RE Braeckman, GH Wolfgang. Practical aspects of assessing toxicokinetics and toxicodynamics. Curr Opin Drug Discov Dev 2:49–57, 1999.
23. BR Roberts. Screening informatics: adding value with meta-data structures and visualization tools. Drug Discov Today 1:10–14, 2000.

## APPENDIX

The editors realize that, like with any new field, there is an increase in new terminology used to define that discipline. Drug discovery is no different and is even more convoluted because it is an accumulation of multiple scientific disciplines. Therefore, the following brief list of terms has been compiled to help the reader. Some references adopted from Ref. 17.

**ADME**—absorption, distribution, metabolism, and elimination. These are the defining pharmacokinetic characteristics of how a drug is handled by the body.

**Bioinformatics**—the collection, organization, visualization, and analysis of large amounts of biological data, using computers and databases. Bioinformatics also includes the integration and "mining" (detailed searching) of databases.

**Combinatorial chemistry**—the use of a small set of chemical building blocks, combined together in multiple ways using standard chemistries, to create large libraries of compounds that may be screened for potential new drugs.

**Genomics**—establishes the relationship between gene activity and particular diseases.

**Metabonomics**—a technique of using proton-NMR to compare changes in the production or elimination of standard biochemicals found in blood or urine due to the in vivo administration of new chemical entities.

**In cerebro**—defined as using ones own mind (knowledge) to think through a problem, interpret data, or make conclusions.

**In silico**—the use of software to analyze data or make prediction (estimations) through some form of logic or knowledge (learning rules or database).

**In vitro**—"in glass." Studies that are performed in test tubes, flasks, or multiwell plates or any biological preparation that is not considered a living organism.

**In vivo**—"in life." Studies that are done in living microorganisms, animals, or humans.

**Pharmacokinetics**—the study of the kinetics associated with drug absorption, distribution, metabolism, and elimination (ADME).

**Pharmacodynamics**—the study of the effect of a drug and its mechanism of action.

**Proteomics**—establishes the relationship between a protein or enzyme and a particular disease state.

**Rational drug design (structure-based drug design)**—the systematic design of new drug molecules using the three-dimensional structure of the drug target as a starting point. Rational drug design (also known as structure-based drug design) uses the high-resolution (atomic) structure of the target molecule and of molecules that bind to it.

# 2

# The Impact of Atmospheric Pressure Ionization

**David T. Rossi**
*Pfizer Global Research and Development, Ann Arbor, Michigan*

> The laws of physics haven't changed. They've just been patented and marketed.
>
> *—mass spectrometry adage*

The current pace of drug discovery is rapid and still accelerating. This growth is possible only because of the set of strongly interdependent technologies that are available to do the necessary tasks. A better understanding of the human genome will continue to lead to better molecular targets for drug molecules. These targets will be used as focal points for new and highly automated synthetic techniques, such as those used in combinatorial chemistry [1]. Although these techniques are capable of churning out many thousands of possible drug candidates in short order, most of this effort would be useless without accurate screening approaches for understanding and predicting the effectiveness of the drug candidates in humans. Better surrogate biological models and screening approaches are continually being sought with the hope that they will allow drug candidates with good biopharmaceutical properties and true therapeutic potential to be found more quickly [2].

The pyramid model for discovering and evaluating viable drug candidates, shown in Fig. 1, assumes that there is a strong interdependence between the several components of drug discovery and that discovery is only as effective as the weakest of the building blocks. The Analytical Technology and Information Systems components of drug discovery are especially interesting in that they

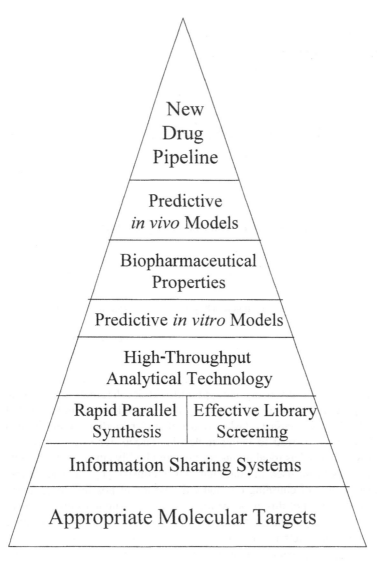

**Figure 1**  The pyramid model suggests that the structure of successful drug discovery is built on several interrelated technologies. No single technology is more important than any other.

impact most of the other components of the pyramid structure. It is only recently that these systems have become full participants in the discovery process.

For scientists involved in the Analytical Technology component of drug discovery, the primary ongoing challenges are speed and effectiveness of assay methods (qualitative and quantitative) that are used in the screening of these drug candidates. Of course, the speed at which an assay can be completed is important. To facilitate quicker decision making, pharmaceutical scientists need to shorten the cycle time between when an experiment is run and when the results are available. Although more compounds are now being synthesized, many of these do not have appropriate pharmaceutical properties. They are tested once and never examined again. Therefore, the speed at which new analytical methods can be developed is as important as the speed at which an established assay method can be run.

Assay effectiveness is critical for all analytical work as well. Assay *performance* is judged by precision, accuracy, sensitivity, and so on, but assay *effectiveness* is judged by the ability of the analytical method to answer the question being posed. What is the structure of this newly synthesized compound? How quickly is this drug candidate metabolized? To what extent is this compound biologically available after an oral dose? What is the biological half-life? In the drug discovery phase, these types of questions are often asked on a relative basis; that is, drug candidates are rank ordered relative to one another. If the analytical method can quickly answer the pertinent question and allow the discovery/development process to proceed, it is effective.

So why is it that mass spectrometry (MS) has recently emerged as a leading technology in drug discovery? This is an interesting question, in light of the fact that, in one form or another, mass spectrometry has been gainfully used for more than 80 years, but has only recently attained a dominant role in pharmaceutical analysis [4,5]. The answer lies in the continuously evolving capabilities of modern analytical mass spectrometry and how these capabilities are now meeting the demands of modern drug discovery. Figure 2 represents a historical time line of pharmaceutical discovery and mass spectrometry development and gives us some insight as to how the two paths have recently become interwoven.

For most of its history, mass spectrometry had been a narrowly focused set of gas-phase analytical techniques. A single sample was introduced by means of a vacuum lock and ionized in vacuo by bombardment with energized particles, such as electrons (electron impact ionization), neutral atoms (fast atom bombardment), or other ions (secondary ion mass spectrometry), or through energy transfer (laser desorption, electric field ionization, or spark discharge) [2]. Gas-phase ionization was hard, in the mass spectrometry sense of the word, meaning that it usually caused extensive fragmentation of the analyte ion [3]. The chances of observing the intact molecular ion was small and, because of the complexity of many real world samples, trace determination of drugs in matrix was not often

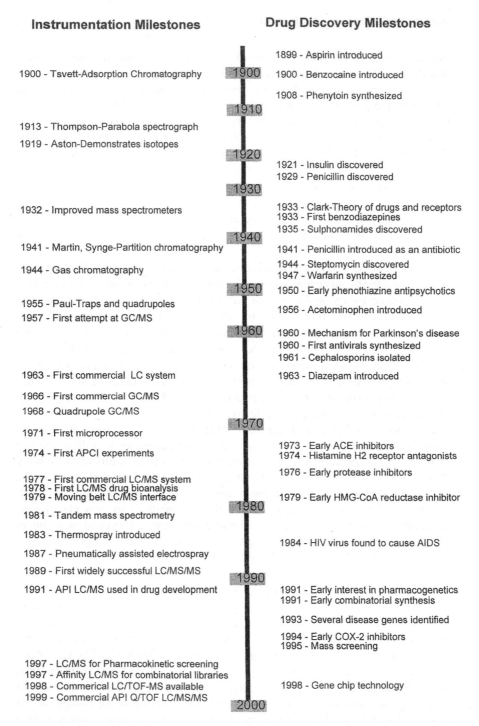

**Instrumentation Milestones**

1900 - Tsvett-Adsorption Chromatography

1913 - Thompson-Parabola spectrograph
1919 - Aston-Demonstrates isotopes

1932 - Improved mass spectrometers

1941 - Martin, Synge-Partition chromatography

1944 - Gas chromatography

1955 - Paul-Traps and quadrupoles
1957 - First attempt at GC/MS

1963 - First commercial LC system

1966 - First commercial GC/MS
1968 - Quadrupole GC/MS

1971 - First microprocessor

1974 - First APCI experiments

1977 - First commercial LC/MS system
1978 - First LC/MS drug bioanalysis
1979 - Moving belt LC/MS interface
1981 - Tandem mass spectrometry

1983 - Thermospray introduced

1987 - Pneumatically assisted electrospray

1989 - First widely successful LC/MS/MS

1991 - API LC/MS used in drug development

1997 - LC/MS for Pharmacokinetic screening
1997 - Affinity LC/MS for combinatorial libraries
1998 - Commerical LC/TOF-MS available
1999 - Commercial API Q/TOF LC/MS/MS

**Drug Discovery Milestones**

1899 - Aspirin introduced

1900 - Benzocaine introduced

1908 - Phenytoin synthesized

1921 - Insulin discovered
1929 - Penicillin discovered

1933 - Clark-Theory of drugs and receptors
1933 - First benzodiazepines
1935 - Sulphonamides discovered

1941 - Penicillin introduced as an antibiotic

1944 - Steptomycin discovered
1947 - Warfarin synthesized
1950 - Early phenothiazine antipsychotics

1956 - Acetominophen introduced

1960 - Mechanism for Parkinson's disease
1960 - First antivirals synthesized
1961 - Cephalosporins isolated

1963 - Diazepam introduced

1973 - Early ACE inhibitors
1974 - Histamine H2 receptor antagonists

1976 - Early protease inhibitors

1979 - Early HMG-CoA reductase inhibitor

1984 - HIV virus found to cause AIDS

1991 - Early interest in pharmacogenetics
1991 - Early combinatorial synthesis

1993 - Several disease genes identified

1994 - Early COX-2 inhibitors
1995 - Mass screening

1998 - Gene chip technology

1900 · 1910 · 1920 · 1930 · 1940 · 1950 · 1960 · 1970 · 1980 · 1990 · 2000

**Figure 2**  A time line depicts the major events in pharmaceutical discovery and LC/MS development.

achievable. The interference from the matrix was too great to make the approach widely practical. To some, the problem seemed to be selectivity and was added more by developing tandem-mass spectrometry capabilities [6]. Of course, tandem mass spectrometry and the use of molecular fragmentation through collision induced dissociation was also useful for structural elucidation, studying gas-phase chemistry and novel experiments. Nevertheless, in terms of discriminating against a complicated matrix, tandem mass spectrometry by itself was not enough, and adding a chemical separation prior to the mass spectrometry step seemed to be a good idea.

Gas chromatography/mass spectrometry was a different story. Because molecules that had been separated by gas chromatography were already in the gas phase, they could be readily ionized and manipulated. Well into its 5th decade, gas chromatography continues to be a successful analytical technique, especially in environmental science. The difficulty with gas chromatography was that it only worked for a minor (~20) percentage of the molecules of interest [7], and it worked better for nonionic molecules than for ionic molecules. Because of this, gas chromatography has been of limited utility in pharmaceutical science, where the majority of drugs and metabolites are polar or ionic.

Analytical scientists had struggled for years to find ways of interfacing the high vacuum domain of mass spectrometry with the condensed-phase domain of analytical separations such as liquid chromatography (LC), and for a long time nothing worked well enough [8]. Three fundamental compatibility problems had to be solved: the volume of solvent eluting from the separation (1 mL/min), the volatility of the mobile phase, and the polar nature of the analytes. This last problem included such issues as analyte volatility, stability, and ionizability. Most early efforts focused on direct liquid introduction and on various types of moving belt interfaces.

Direct liquid introduction was an obvious approach to coupling LC and MS [9,10], generally involving a capillary separation or other low liquid flow arrangement. Because of solvent and analyte volatilization dynamics, nonvolatile components tended to collect at the outlet of the capillary and block the flow, giving the approach a very narrow range of applicability. The premise of the moving belt interface was to continually deposit drops of mobile phase effluent from a liquid chromatography experiment onto a miniature conveyer belt. After solvent evaporation, the residue was conveyed into the ion source and traditional ionization, such as electron impact, was conducted. This approach had a number of difficult technical hurdles, including carryover, loss of chromatographic resolution, and poor response time. Although the object of widespread study for more than a decade, moving belts never became very successful, technically or commercially [11]. Other interfacing approaches that were evaluated and have fallen into disuse include particle beam [12], pneumatic nebulizer [13], and continuous-flow fast atom bombardment (FAB) [14]. Although these techniques represented incremental improvements in the state of the art, each has difficulties, such as need for excessive heat that caused thermal degradation of analytes (particle beam) or an inability to handle reasonable flow rates (flow FAB), restricting the applications.

Even the most useful of these interfaces was never more than moderately successful, yet an important lesson was learned: it was more productive to eliminate the solvent and ionize nonvolatile compounds at atmospheric pressure, not in vacuo. From this liquid-based ionization strategy, atmospheric pressure ionization (API) mass spectrometry was born. After this breakthrough, analytical chemists could interface their separation module (e.g., liquid chromatography and capillary electrophoresis) to a mass spectrometer using atmospheric pressure ionization. A small difference in implementation between thermospray (strong heating and much thermal degradation) [15] and electrospray (high voltage, mild heating, and very soft ionization) [16,17] led to a large difference in performance and applicability. Now a tool that was applicable to most compounds of pharmaceutical interest and yielded soft ionization was available. Atmospheric pressure chemical ionization (APCI), an ionization technique similar to electrospray, became available and popular because it could be operated at the higher flow rates that chromatographers were accustomed to [18,19]. As described in Chapters 3 and 5, this ionization approach complements electrospray in several ways.

Since the invention of API, desktop computers became plentiful, powerful, and cheap. Software development caught up with the hardware, and graphical–user interfaces helped make instrument control and acquisition software easy to use. Mass spectrometers were reengineered and optimized for routine use. The few remaining lens, gas flow, and temperature controls became readily adjustable and controlled by the software. The hardware became easy to set up and use. For example, electrospray and APCI source changeover is a minor operation that any trained technician could perform in a few minutes.

After coupling API tandem mass spectrometry to liquid chromatography, scientists in the pharmaceutical industry were impressed by the tremendous selectivity, qualitative and quantitative information, good-to-excellent sensitivity, and applicability to many small molecules of pharmaceutical interest. Whereas in the past they had needed weeks to develop quantitative bioanalytical methods for small drug molecules, now these methods could be developed in a day or 2 with LC/MS. The resulting quantitation limits were comparable to and often better than other types of detection. Method selectivity was no longer the great stumbling block it once had been and this encouraged widespread multicomponent quantitation, where 20 or more trace drug components could be simultaneously quantified in a single biological sample. Because of the gainful evolution of technologies, mass spectrometry, particularly API LC/MS, became a tool that could effectively contribute to the rapid pace of modern drug discovery. Once considered an exclusive field, exercised solely by technicians dedicated to the art, mass spectrometry had evolved and matured into a commonly used tool for many scientists, regardless of formal training. For analytical chemists in the pharmaceutical industry, not routinely using mass spectrometry now puts them at a significant competitive disadvantage.

Atmospheric pressure ionization mass spectrometry has now developed to the point where it can bring significant problem-solving power to bear upon quantitative and qualitative problems in many of the disciplines needed for drug discovery. For example, it is often necessary to determine or confirm the molecular weight of an entire plate of freshly synthesized compounds. This process can be automated with the aid of API/MS and simple flow injection sample introduction. It is of interest to rank order the in vitro permeability of a collection of compounds in a single class or to compare classes of compounds. This can be accomplished in a few days, using quantitative API LC/MS. The oral bioavailability of compounds can also be an important criterion in drug candidate selection. Using a simple cassette-dosing paradigm, this parameter can be simultaneously assessed for a number of test compounds in any common animal model, again within a few days.

Figures 3 and 4 demonstrate the dramatic growth of API, APCI, and electrospray, both within and external to pharmaceutical science, by graphing the

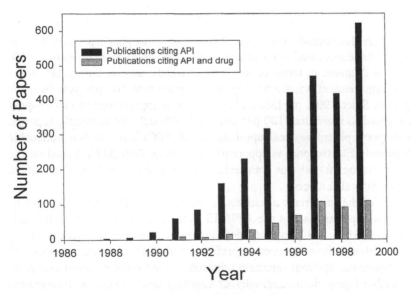

**Figure 3** Bar chart representing the annual number of articles involving atmospheric pressure ionization (API) mass spectrometry and atmospheric pressure ionization mass spectrometry in the pharmaceutical sciences.

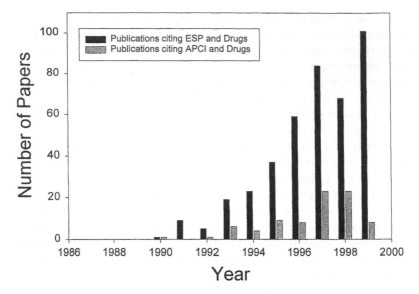

**Figure 4**  Bar chart representing the annual number of articles involving electrospray and atmospheric pressure chemical ionization (APCI) mass spectrometry in relationship to pharmaceutical science.

numbers of articles published per year. Articles on atmospheric pressure ionization (Fig. 3) have increased from zero in 1986–1987 to more than 600 in 1999. While not as dramatic in terms of numbers or trends, articles citing API in the field of pharmaceutical science have grown to more than 100 per year over the same period. Since 1990, published pharmaceutical applications of electrospray have increased to more than 100 per year and, although not as widely reported as electrospray, pharmaceutical applications of APCI have shown noticeable growth as well. Electrospray is apparently used more than APCI, based on the number of citations found, but both techniques continue to gain applications in the pharmaceutical sciences.

Many of the important questions associated with drug discovery can be addressed by experiments that use API/MS as the analytical approach. It is one of the most flexible and powerful tools currently available to pharmaceutical scientists. Although flow injection is preferred for qualitative determinations of major components, up-front separation approaches are often required and commonly involve liquid chromatography or capillary electrophoresis. Information obtained can be qualitative (molecular weight or detailed structural information) or quantitative (concentration in biological or in vitro samples). The time scale required to develop and conduct the assay procedure or to screen compounds has decreased drastically and this has helped remove analytical bottlenecks in the drug discovery process. How this is done is, in principle and application, the subject for the rest of this book.

## REFERENCES

1. DJ Ecker, ST Crooke. Combinatorial drug discovery: which methods will produce the greatest value? Biotechnology 13:351–360, 1995.
2. S Borman. Reducing time to drug discovery. Chem Eng News March 8:33–48, 1999.
3. GW Ewing. Analytical Instrumentation Handbook, New York: Marcel Dekker, 1997, chapter 16.
4. G Siuzdak. The emergence of mass spectrometry in biochemical research. Proc Natl Acad Sci USA 91:11290–11297, 1994.
5. MS Lee, EH Kerns. LC/MS applications in drug development. Mass Spectrom Rev 18:187–279, 1999.
6. FW McLafferty. Tandem mass spectrometry. Science 214:280–287, 1981.
7. LR Snyder, JJ Kirkland. Introduction to Modern Liquid Chromatography. New York: Wiley, 1991.
8. WMA Niessen. Liquid Chromatography/Mass Spectrometry. New York: Marcel Dekker, 1999, chapter 4.
9. WMA Niessen. Review of direct liquid introduction interfacing for LC/MS part I: instrumental aspects. Chromatographia, 21:277, 1986.
10. WMA Niessen. Review of direct liquid introduction interfacing for LC/MS part II: mass spectrometry and applications. Chromatographia, 21:342, 1986.
11. WH McFadden, DC Bradford, G Eglinton, SK Hajbrahim, N Nicolaides. Application of combined liquid chromatography/mass spectrometry (LC/MS): analysis of petro-porphyrins and meibomian gland waxes. J Chromatogr Sci 17:518–522, 1979.
12. MA Baldwin, FW McLafferty. Liquid chromatography mass spectrometry interface—I: the direct introduction of liquid solutions into a chemical ionization mass spectrometer. Org Mass Spectrom 7:1111–1112, 1973.
13. AP Bruins. Developments in interfacing microbore HPLC with mass spectrometry (a review). J Chromatogr 323:99–111, 1985.
14. M Barber, RS Bordoll, GS Elliott, RD Sedgewick, AN Tyler. Fast atom bombardment mass spectrometry. Anal Chem 54:645A–657A, 1982.
15. ML Vestal, GJ Fergusson. Thermospray liquid chromatograph mass spectrometer interface with direct electrical heating of the capillary. Anal Chem 57:2373–2378, 1985.
16. M Dole, LL Mach, RL Hines, RC Mobley, LP Ferguson, MB Alice. Molecular beams of macroions. J Chem Phys 49:2240–2249, 1968.
17. CM Whitehouse, RN Dreyer, M Yamashita, JB Fenn. Electrospray interface for liquid chromatographs and mass spectrometers. Anal Chem 57:675–679, 1985.
18. EC Hornung, DI Carroll, F Dzidic, KD Haegle, MG Hornung, RN Stillwell. Atmo-

spheric pressure ionization mass spectrometry. Solvent-medicated ionization of samples introducted in solution and in a liquid chromatograph effluent stream. J Chromatogr Sci 12:725, 1974.

19. AP Bruins, TR Covey, JD Henion. Ion spray interface for combined liquid chromatography/atmospheric pressure ionization mass spectrometry. Anal Chem 59: 2642–2646, 1987.

# 3
# A Mass Spectrometry Primer

**Scott T. Fountain**
*Pfizer Global Research and Development, Ann Arbor, Michigan*

> If it wasn't for mass spectrometry, I'd probably be in the dry-cleaning business.
>
> —*With apologies to Joseph Heller.*

## I. WHAT IS A MASS SPECTROMETER?

A *mass spectrometer* is an instrument that is capable of forming, separating, and detecting ions, either atomic or molecular, based on their mass-to-charge ratio. The mass-to-charge ratio of an ion is often abbreviated as *m/z*, and this nomenclature is used consistently throughout this chapter.

## A. Components of a Mass Spectrometer

Any mass spectrometer can be generically described as a set of functional modular components as shown in Fig. 1. These components include (a) a sample inlet, (b) an ionization source, (c) a mass analyzer, (d) a detector, and (e) a data recorder/processor [1]. For each of these modular components, a number of unique technologies exist, distinguished mechanically by their mode of operation and practically by their advantages and disadvantages for a particular analytical application. By linking individual components together, diverse mass spectrometer systems have been configured, typically with specific analytical requirements in mind. Table 1 gives an overview of some common mass spectrometer components, and although not all individual components are universally compatible, to a large extent the analytical diversity of mass spectrometry, as exploited by ven-

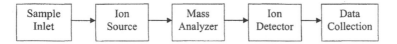

**Figure 1** Modular components of a modern mass spectrometer system.

dors and academicians, has relied heavily on the mix-and-match capabilities of these individual components [2].

## 1. Sample Inlet System

Because a mass spectrometer separates and distinguishes ions based on some translational characteristic associated with their mass-to-charge ratios, it is necessary that mass analysis occur in a vacuum so that an ion's mean free path (that is, the average distance an ion travels before it collides with another particle) is greater than the travel distance from ion source to ion detector. Much of the engineering behind a mass spectrometer, therefore, is centered on its pumping design and subsequent vacuum interlock. There are two options in the design of a sample inlet system. One option is to introduce the sample into the mass spectrometer as a neutral species through a controlled vacuum leak, followed by ionization in the vacuum chamber. The other option is to create ions at atmospheric pressure and then introduce the ions into the mass spectrometer through a controlled vacuum leak with the aid of electrostatic and/or electromagnetic focusing lenses. This process is called *atmospheric pressure ionization* (API) and provides the best option to date when coupling a dynamic liquid system, such as a liquid chromatograph, to a mass spectrometer. Atmospheric pressure ionization is largely responsible for the dramatic growth of mass spectrometry and the pharmaceutical industry is no exception. For this reason, API is discussed in detail this chapter.

## 2. Ionization Sources, Mass Analyzers, and Ion Detectors

The choice of an ionization source must be consistent with the mass spectrometer inlet system, as described above, and the ionization characteristics of the analyte. Likewise, the appropriate choice of mass analyzer is highly application-, compound-, and matrix-dependent, and each mass analyzer requires certain characteristics from its ion detector. The variety of individual ionization, mass analyzer, and ion detector systems that can be integrated creates a myriad of unique mass spectrometer systems, each with certain advantages and disadvantages. To add even more complexity to the situation, modular configuration of these individual components (ionization source, mass analyzer, and detector) is not limited

**Table 1** Common Components of Modern Mass Spectrometers, Including Inlet Systems, Ionization Sources, Mass Analyzers, and Ion Detectors

| Sample inlet systems | Ionization sources | | Mass analyzers, single stage | Ion detectors | |
|---|---|---|---|---|---|
| | Atmospheric pressure | Vacuum | | Point | Array |
| Gas chromatography | *Electrospray ionization* | *Electron impact ionization (EI)* | *Quadrupole mass filter* | *Photomultiplier* | *Multichannel plate detector* |
| Liquid chromatography | *Atmospheric pressure chemical ionization (APCI)* | *Chemical ionization (CI)* | *Time-of-flight* | *Electron multiplier* | Photographic film |
| Direct insertion probe | | *Matrix-assisted laser desorption (MALDI)* | *Magnetic sector* | | |
| Particle beam | | Fast atom bombardment (FAB) | *Fourier transform* | | |
| | | Secondary ion mass spectrometry (SIMS) | *Quadrupole ion trap (QUISTOR)* | | |

Note: Those components which are discussed in some detail are italicized within the table.

to a single choice from each category listed in Table 1. In fact, one of the greatest strengths of the technology has been the creation of system hybridization, particularly with mass analyzers, in which multiple components within a given category in Table 1 are linked together, providing tandem mass spectrometry (MS/MS) capabilities. A common example of tandem mass spectrometry is the triple quadrupole instrument, in which two single-stage quadrupole mass analyzers are linked via an ion reaction cell, providing a number of unique and powerful applications.

## B.  Isotopes and Molecular Weight

Mass spectrometers determine atomic and molecular isotope ratios. Table 2 lists the relative isotopic abundance of elements commonly encountered in pharmaceutical analysis [3,4]. The values in Table 2 have been empirically determined and refinements in the values are necessary as atomic mass measurements improve, but for this discussion any inaccuracies in the table are insignificant. For some elements there are only two naturally occurring isotopes. For example, if you were to randomly sample carbon atoms in nature, 99% of the time you would find $^{12}$C, and roughly 1% of the time a $^{13}$C would turn up. Other elements, such as chlorine and bromine, have elemental isotope ratios that are not as heavily

**Table 2**   Common Elements with Their Naturally Occurring Isotopic Relative Abundance

| Element | Most abundant isotope | Other isotopes | |
|---|---|---|---|
| | | Mass | Relative % |
| Hydrogen, H | $^{1}$H | $^{2}$H | 0.015 |
| Carbon, C | $^{12}$C | $^{13}$C | 1.08 |
| Nitrogen, N | $^{14}$N | $^{15}$N | 0.37 |
| Oxygen, O | $^{16}$O | $^{17}$O | 0.04 |
| | | $^{18}$O | 0.20 |
| Fluorine, F | $^{19}$F | | |
| Silicon, Si | $^{28}$Si | $^{29}$Si | 5.1 |
| | | $^{30}$Si | 3.4 |
| Phosphorous, P | $^{31}$P | | |
| Sulfur, S | $^{32}$S | $^{33}$S | 0.80 |
| | | $^{34}$S | 4.40 |
| Chlorine, Cl | $^{35}$Cl | $^{37}$Cl | 32.5 |
| Bromine, Br | $^{79}$Br | $^{81}$Br | 98.0 |
| Iodine, I | $^{127}$I | | |

*Source*: From Refs. 3 and 4.

biased to one isotopic form. Because a molecule is nothing more than a weighted combination of a set of elements, the theoretical isotope ratio of a molecule can be predicted mathematically, based on the percentages in Table 2, which determine the weighing factors for isotope distribution.

A compound will have a distribution of individual isotopic molecular weights based on the relative abundance of its constituent ions, and it is important when expressing the molecular weight of a compound to understand how its weight is calculated. A common way of defining the molecular weight of a compound is by expressing it as a weighted average of its constituent isotopes. This is typically called the *average molecular weight* and is a convenient way to express a molecular weight when physically handling a compound, such as when weighing on a balance or calculating molar equivalents. For stoichiometric calculations it is practical to deal with average molecular weights. Because a mass spectrometer is typically capable of resolving the constituent isotopes of a compound, it becomes convenient to define molecular weight not by average mass, but rather by some other isotopic characterization. This is done by expressing the molecular weight based on a single isotope, most commonly the lowest molecular weight isotope of the compound. This is sometimes called the *monoisotopic molecular weight* and is a convenient measure when dealing with mass spectra because that is the empirical observation. For example, Fig. 2a shows the theoretical isotope pattern of diphenhydramine ($C_{17}H_{21}NO$). The average molecular weight for diphenhydramine is 255.4 Da, which results from a weighted average of the individual isotopes shown. The monoisotopic molecular weight, however, is 255.2 Da, corresponding to the lowest, and in this case most abundant, molecular isotope.

It is interesting to note that the difference between the monoisotopic and average molecular weight of a compound increases with molecular weight. At relatively low molecular weights the monoisotopic molecular weight contributes the most to the mass. Such is the case with the diphenhydramine distribution shown in Fig. 2a. As molecular weight increases, isotopic contribution to the average mass becomes more significant, and ultimately a point will be reached where the monoisotopic mass is no longer the most intense mass spectral peak. This trend is shown in Fig. 2b, showing the isotope pattern for thiostrepton ($C_{72}H_{85}N_{19}O_{18}S_5$), which has an average molecular weight of 1664.9 Da and a monoisotopic molecular weight of 1663.5 Da. Considering this from a statistical viewpoint, at relatively low molecular weight a compound consisting of carbon, hydrogen, oxygen, and nitrogen is more likely to contain only ions of the most abundant atomic weight because 98.92% of carbon atoms are $^{12}C$, 99.985% of hydrogen atoms are $^{1}H$, 99.63% of nitrogen atoms are $^{15}N$, and so forth. As molecular weight increases a point is reached where, statistically, it is more likely that *at least one* atom in the molecule is not the most abundant atomic weight. For example, if a molecule contained >108 carbon atoms, statistically it is more

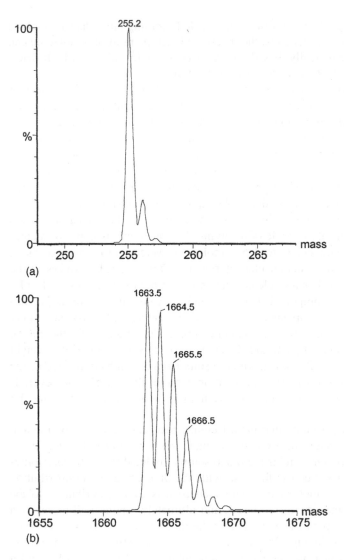

**Figure 2** Theoretical isotope distributions of (a) diphenhydramine and (b) thiostrepton. In general, the difference between the monoisotopic and average molecular weight of a molecule increases as molecular weight increases (isotope distributions were generated using Micromass MassLynx 3.1 software).

likely you would find at least one $^{13}C$ atom. As a rule of thumb, this conversion of isotopic base peak roughly occurs around 2000 Da.

## C. Mass Resolution

The *resolution* of a mass spectrometer is qualitatively defined as its ability to discriminate between adjacent ions in a spectrum [5–12]. Resolution is often defined as a function of molecular weight and is given by the following equation:

$$\frac{m}{\Delta m} \tag{1}$$

where $m$ is the molecular weight of the ion and $\Delta m$ is the peak width of the ion at half-height, defined at a given percentage valley between the adjacent ions. As peak width decreases for a given $m$, or as $m$ increases without a concurrent increase in peak width, resolution improves. Resolution, like sensitivity, is commonly used as a performance specification for an instrument and varies greatly by mass spectrometer analyzer, sample ionization mechanism, and the detailed design components of a particular instrument. Most ion source/mass spectrometer configurations are capable of providing unit mass resolution, in which adjacent singly charged ions can be distinguished with 10–30% valley separation, within the $m/z$ range of the instrument. Although mass accuracy and mass resolution are not necessarily related, mass spectrometers that provide exceptional resolution, such as ion resonance cyclotron instruments, tend to generate spectra with high accuracy.

## D. Dynamic Range

The *dynamic range* of a mass spectrometer is defined as the range over which a linear response is observed for an analyte as a function of analyte concentration. It is a critical instrument performance parameter, particularly for quantitative applications, because it defines the concentration range over which analytes can be determined without sample dilution or preconcentration, which effects the accuracy and precision of an analytical method. Dynamic range is limited by physiochemical processes, such as sample preparation and ionization, and instrumental design, such as the type of mass analyzer used and the ion detection scheme.

## E. Mass Accuracy

The accuracy to which a mass spectrometer can determine the molecular weight of an ion is, like sensitivity and dynamic range, a common performance benchmark [5–12]. Depending on configuration and choice of mass analyzer, mass spectrometers can be designed to provide mass accuracy anywhere from ±0.5 to ±0.0001 $m/z$, with cost typically proportional to performance. Instrument accuracy to ±0.5 $m/z$, sometimes called a *nominal mass measurement*, is often sufficient to provide the level of accuracy required for an analytical application. Instruments with exceptional mass accuracy can provide an additional level of mass selectivity, particularly for the discrimination of isobaric molecules, which have the same nominal mass (±0.5 $m/z$) but differ in their accurate mass (±0.001 $m/z$). The term *accurate mass* has been reserved for instruments or applications that provide mass measurements to within ±5 ppm of the theoretical molecular weight. For most analytical applications, however, mass accuracy in excess of ±0.1 $m/z$ is not necessary, and sufficient analytical performance can be obtained without the additional cost of an accurate mass measurement. A common example would be triple-quadrupole mass spectrometry, in which the mass accuracy of the analyzer typically does not exceed ±0.1 $m/z$, where lack of selectivity from mass accuracy is compensated by the increase in selectivity associated with the tandem mass spectrometric capabilities of the instrument.

## II. ION SOURCES: HOW IONS ARE FORMED

Ionization can proceed by either of two fundamental processes, the loss/gain of an electron or the loss/gain of a charged particle (such as a proton), generating odd- or even-electron ions respectively [4]. Each ionization mechanism has fundamental characteristics which can effect analytical methodology and both are discussed. An odd-electron ion is generated by the gain or, more commonly, loss of an electron in vacuum, generating an ionic species of identical nominal molecular weight (differing only by the mass of an electron) to the neutral species from which it was generated. Classic electron impact ionization takes advantage of this mechanism. In comparison, an even-electron ion is generated by the adduction or loss of an even-electron species from a molecule, in the most simple case a proton, to generate an ion with a molecular weight dissimilar to the molecule from which it was generated. Molecular ions generated by adduction or ionic loss are sometimes referred to as *pseudomolecular ions* because their mass (independent of charge) is different from that in its un-ionized state. The most commonly used even-electron ionization sources are electrospray and atmospheric pressure chemical ionization; these sources fall under the general heading of atmospheric pressure ionization to distinguish them from ionization processes which take place in a vacuum.

Once an ion is generated in a mass spectrometer, its mass-to-charge ratio ($m/z$) can then be determined based on the trajectory an ion takes in the mass analyzer. Quite often the charge of a molecule is simply +1 or −1 and a direct measurement of the $m/z$ ratio reveals the mass of the molecular or pseudomolecular ion. In cases where ions are multiply charged ($z \geq 1$), various mathematical methods can be applied to determine the charge state of the ion and thus its molecular weight. Techniques to determine charge state are discussed later in this chapter, specifically in the section describing atmospheric pressure ionization.

Another convenient way to classify ionization sources, rather than from the perspective of odd- or even-electron ion generation, is in relation to where the ions are created relative to the vacuum system; that is, either generated at atmospheric pressure or in a vacuum. The two most common atmospheric pressure ionization sources, electrospray and atmospheric pressure chemical ionization, are arguably the most common ionization techniques applied in quantitative mass spectrometry today. However, discussion of earlier ionization sources is useful, as many of these techniques are still commonplace and their understanding provides a framework for appreciation of atmospheric pressure ionization technology and what it has to offer the pharmaceutical industry.

## A. Electron Impact Ionization

*Electron impact* (EI) ionization is a technique in which a neutral molecule, M, is ionized following bombardment by an accelerated electron beam creating a radical cation, $M^{\cdot+}$, and an additional free electron [5–12] as follows:

$$M + e^- \rightarrow M^{\cdot+} + 2e^- \tag{2}$$

A metal filament is heated in a vacuum in order to generate free electrons, and these electrons are accelerated within a restricted source, into which an unionized analyte has been introduced. Sample introduction can be through a number of mechanisms, including gas chromatography, liquid chromatography, and direct insertion probe. Although EI ionization sources may contain a variable potential for ion acceleration, a 70-eV potential is typically applied. At this voltage, ionization and fragmentation become very reproducible among different source designs, making possible the creation of standardized mass spectral databases. In fact, it is this property, i.e., its ability to produce highly reproducible "fingerprint" spectra, both in ion fragmentation and relative abundance, that has afforded EI ionization a steady market share in the diverse world of mass spectrometric ionization sources. Qualitative analysis of compounds amenable to EI ionization can be streamlined by library searching in which spectra of unknowns are compared to standard libraries [13–15]. Searching algorithms generate statistical matches relative to the spectral library and list possible hits in order of descending probability. Most software applications also allow the creation of customized libraries that contain selected or proprietary compounds.

Electron impact ionization has a number of advantages and limitations as an ionization source. It is a nonspecific ionization technique that will ionize most organic molecules. For relatively small, nonpolar, and volatile compounds EI is a good choice for ionization, while techniques such as atmospheric pressure ionization require some minimal polar functionality to produce an ion. While such polarity is not a requirement for electron impact ionization, this ionization technique also introduces a significant amount of internal energy into an ion, which can induce excessive fragmentation. The technique generates relatively unstable odd-electron radical cations, and while EI fragmentation tends to be reproducible and generates mass spectral "fingerprints" for molecules, this fragmentation also limits applications of EI for the characterization of large molecules, such as peptides or proteins, and labile molecules. Because EI requires ionization in a vacuum, liquid chromatographic interfaces such as particle beam, which do not take advantage of differential pumping (see Sec. II.D), have poor sensitivity performance relative to atmospheric pressure ionization.

## B.  Chemical Ionization

To improve the stability of ions formed in a vacuum, and thereby limit ion fragmentation, an alternative to electron impact called *chemical ionization* was developed. Chemical ionization generates an even-electron ion rather than an odd-electron ion and although ions are still generated directly in a vacuum with chemical ionization, the even-electron configuration of the analyte improves the stability of molecular ion significantly such that molecules previously not amenable to vacuum ionization with EI can be detected [5–12,16]. Unlike *hard ionization* techniques, such as EI, where ion fragmentation is commonly observed, chemical ionization (CI) is a *soft ionization* technique in that molecular ions can be generated without significant fragmentation. The ionization mechanism of CI relies on EI ionization for the initial ionization event, but within the source is added a chemical ionization gas, such as methane, isobutane, or ammonia, that becomes ionized through charge transfer processes and ultimately creates a reactive species that can ionize the analyte compound. For example, using ammonia as the CI gas, creation of the ammonia radical cation can be induced by electron impact ionization as follows:

$$NH_3 + e^- \rightarrow NH_3^{\cdot+} + 2e^- \tag{3}$$

The ammonia radical cation can then form the reactive species, $NH_4^+$, through further reaction with ammonia in the ionization source as follows:

$$NH_3^{\cdot+} + NH_3 \rightarrow NH_4^+ + NH_2^{\cdot} \tag{4}$$

The $NH_4^+$ species can then react with a basic amine analyte compound, $RNH_2$, to create an ionized species with an $m/z$ ratio 1 Da in excess of the parent compound's molecular weight as follows:

$$NH_4^+ + RNH_2 \rightarrow NH_3 + RNH_3^+ \qquad (5)$$

Charge transfer to an analyte molecule can be improved by judicious choice of CI gas, as the gas-phase acidity of the chemical ionization gas influences the efficiency of charge transfer.

## C. Matrix-Assisted Laser Desorption/Ionization

Electron impact ionization and chemical ionization are two vacuum ionization techniques that have been applied to relatively low-molecular-weight molecules (roughly <1000 Da). In order to produce gas-phase ions of oligopeptides, peptides, and large biomolecules, alternative ionization methods are required. One method, described as *matrix-assisted laser desorption/ionization* (MALDI), has since become a powerful analytical technique for the characterization of biopolymers [6,17,18]. This method has primarily been interfaced with time-of-flight mass spectrometry because this mass analyzer configuration has an inherently high $m/z$ range and is easily coupled with pulsed or modulated ionization sources.

Matrix-assisted laser desorption/ionization produces stable high-mass ions through an energy transfer process rather than by direct laser ionization. A molecule of interest (such as a protein) is mixed in a suitable organic matrix, typically at a 1:1000 molar ratio of analyte to matrix, and the analyte–matrix mixture is allowed to crystallize on a metallic or ceramic plate or other suitable surface. Following crystallization, the plate is introduced into the mass spectrometer through a vacuum interlock. To ionize the sample, pulsed laser light is focused onto the analyte–matrix mixture [17,18]. With appropriate choice of matrix and laser wavelength, the MALDI process is designed such that the matrix rather than the analyte absorbs the photons [19–20]. Absorption of photons results in a nonequilibrium phase change in which the matrix crystals and embedded analyte ions are volatilized. Molecules previously embedded in the matrix are then ionized through charge transfer with the expanding matrix plume. Common MALDI matrices include nicotinic, sinapic, cinnamic, and gentisic acid (2,5-dihydroxybenzoic acid) for UV laser desorption in addition to urea and succinic acid for IR desorption.

A significant characteristic of MALDI is its ability to produce gas-phase ions of large biopolymers with little or no molecular fragmentation and typically with a single charge [17,18]. This is extremely useful for molecular weight determination but has limitations when structural information is required. There are two major causes for the soft ionization of MALDI. First, desorption produces a phase transition in which the matrix, rather than the analyte, primarily absorbs the energetic photons. As a result, little internal energy is transferred to the solvated analyte molecules. Second, it appears that the phase transition is analogous to a supersonic jet expansion, with the analyte ion "seeded" in an expanding plume of matrix "carrier gas." Because the gas-phase ion has very little internal energy, unimolecular fragmentation is typically not observed on the time scale of source extraction (see Sec. III.F.2).

Although MALDI applications for quantitative analysis of biopolymers have been shown recently, most applications involve qualitative studies [21–23]. Limitations in the quantitative addition of matrix to analyte and on-line configurations of MALDI, compared to gas chromatography/MS and liquid chromatography (LC)/MS systems, have kept the primary focus on qualitative applications.

## D. Atmospheric Pressure Ionization

The need for an ionization source that provided both softer ionization, i.e., less fragmentation of the molecular ion, and a convenient interface with liquid chromatography (at ambient pressure) to mass spectrometry (at high vacuum) helped spur the creation of atmospheric pressure ionization. Two techniques fall under the heading of API, electrospray and atmospheric pressure chemical ionization (APCI), and the technical aspects of each are discussed individually. However, many of the fundamental principals that describe these ionization mechanisms can be applied to both electrospray and APCI sources.

### 1. Differential Pumping

Atmospheric pressure ionization, whether electrospray or atmospheric pressure chemical ionization, takes advantage of ionization outside of the vacuum system of the mass spectrometer. This provides a number of advantages over ionization in vacuum. For example, ion enrichment relative to ambient neutral background molecules is possible with API, creating a vacuum interface with improved sensitivity and efficiency over vacuum ionization techniques. In an API source, analyte ionization competes with background matrix from ambient air, chromatographic effluent and buffer additives, sample matrix, or impurities. Some selectivity is possible with judicious choice of ionization mechanism, but ionization is never totally efficient and a mixture of ions and neutrals is produced. When ions are generated at atmospheric pressure, it is necessary to allow a controlled leak of the ions into the mass spectrometer vacuum region, and the most efficient process to date is through *differential pumping* (see Fig. 3) [6,24,25]. In a generic differential pumping system, ions are generated at atmospheric pressure in a bath of neutral background molecules. Typically a sampling cone, face-plate orifice, or steel capillary provides the first interface where the API and differential pumping regions meet. In Fig. 3 the vacuum interlock is represented by a face-plate orifice. The interface introduces both neutral molecules and ions into the first pumping stage of the mass spectrometer. Once ions and neutral molecules are introduced into the first pumping stage, ion enrichment is possible because the paths of the neutral molecules and the ions are differentially influenced by the pumping dy-

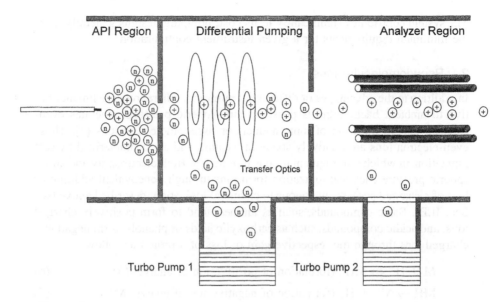

**Figure 3** Schematic of a generic atmospheric pressure ionization source showing a differential pumping system.

namics and the electronic fields of the tuning lenses, respectively. Separating this first pumping stage from a second pumping stage is a larger vacuum orifice, and the tuning lenses are configured to focus ions into the next stage. Subsequent pumping stages are similar to the first, except that pumping capacity decreases and orifice diameters increase with each additional pumping stage. With each stage of pumping, a subsequent lowering of the operating pressure is obtained such that the mass analyzer is capable of maintaining a vacuum pressure amenable to ion detection (sufficient mean free path) even with a considerable solvent load at the source (such as a 1 mL/min solvent flow from an analytical scale liquid chromatography column). Where further vacuum pumping is required additional differential pumping stages can be added.

A number of lens systems, both electrostatic- and RF-based, have been employed in differential pumping applications. Most commercial API/MS systems available today rely on RF-only quadrupole, hexapole, or octapole lens designs, largely due to their automated tuning capabilities within a quadrupole mass spectrometer detection system and their efficiency at relatively high vacuum pressure. The ultimate vacuum pressure required for mass analysis is dependent on the analyzer and detection system configuration. Quadrupole mass analyzers can operate at relatively high pressures ($10^{-5}$ Torr) relative to time-of-flight analyzers ($10^{-7}$ Torr), and because the vacuum pumping system design of an integrated

MS system contributes significantly to its cost, pumping designs generally meet the minimum requirements for a given instrument configuration.

## 2.  Pseudomolecular Ions

In addition to the advantages of differential pumping, where ion enrichment relative to ambient background is possible, atmospheric pressure generates even-electron ions in the form of ionic adducts or pseudomolecular ions [4]. These even-electron ions are relatively stable in vacuum and are characterized by soft ionization, in which ion fragmentation is minimized. Both electrospray and atmospheric pressure chemical ionization form ions through noncovalent addition or loss of a charged species, analogous to chemical ionization, described above (see Sec. II.B). Basic compounds, such as amines, tend to form positively charged ions, and acidic compounds, such as carboxylic acids or phenols, form negatively charged ions through the respective gain or loss of a proton as follows:

$$M + H^+ \leftrightarrow MH^+ \text{ (formation of positive ion of analyte M)} \qquad (6)$$

$$MH \leftrightarrow M^- + H^+ \text{ (formation of negative ion of analyte M)} \qquad (7)$$

Though the charged species is commonly a proton, thereby generating an ion that has a molecular weight 1 Da higher or lower than the parent molecule, ionization is also possible through addition or loss of other charged ions, such as sodium, potassium, or ammonium ions. The molecular weight of the ion changes accordingly depending on the mass of the charged species lost or gained. Because the process of API does not perturb the electron configuration of the molecule upon ionization, molecules charged by this method are typically very stable and can be detected without significant fragmentation, especially when compared to other vacuum ionization techniques such as electron impact ionization.

Although both electrospray and APCI can be described generically as atmospheric pressure ionization techniques, each has its own mode of operation and a wide realm of applications. To a large extent, either technique can be applied to a given analytical method, especially for low-molecular-weight analytes of roughly less than 1000 Da. In fact, it is commonly observed that an industrial analytical laboratory will tend to favor one technique over the other out of convenience rather than analytical rigor.

## 3.  Electrospray Ionization

*Electrospray ionization* is an atmospheric pressure ionization technique in which ions are generated in solution phase, the carrier solvent is evaporated, and a gas-phase ion is produced (see Fig. 4) [26]. An appropriate solvent (typically the effluent from a liquid chromatography system) is passed through a metal capillary to which is applied a static DC voltage. This voltage induces ionization of the

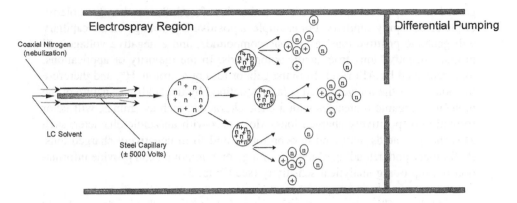

**Figure 4**   Schematic of a generic electrospray ionization source.

effluent, creating charged droplets of solvent as the liquid emerges from the end of the capillary. As the solvent evaporates, charge density increases, creating coulombic repulsion and subsequent droplet disassociation. Further evaporation of the droplets creates an environment in which charge transfer from the solvent to the analyte occurs, generating an even-electron gas-phase ion [27]. Capillary voltages and solvent flow rates are dependent on instrument design, but some generalization can be made. Typically a voltage of ±2.5 to 5.0 kV is applied to the capillary to induce ionization, and the polarity of the voltage determines the charge of the ions generated. Early prototype electrospray ionization sources were limited to low rates (<10 µL/min) to allow sufficient desolvation. Most designs today rely on coaxial gas flow, typically nitrogen, which improves desolvation and allows higher flow rates to be used. This is particularly true for LC/MS applications. Electrospray ionization sources with coaxial gas flows are sometimes called nebulization-assisted electrospray. Flow rates for most commercial nebulization-assisted electrospray ionization sources range from 1 to 1000 µL/min, which is compatible with standard microbore and analytical liquid chromatography (1.0, 2.1, and 4.6 mm I.D.). Efficient desolvation is critical to ionization, and commercial electrospray sources are designed with pneumatically assisted nebulization and heated source design to aid in droplet evaporation. In cases where droplet desolvation is not complete, because either nebulization or heating is insufficient for the aqueous percentage or solvent flow rate, ionization efficiency is reduced. This low ionization efficiency can sometimes be observed as solvent clustering or excessive analyte adduct formation.

*Ionization Polarity.*   As discussed above, either a positively or negatively charged ion can be generated by atmospheric pressure ionization [see Eqs. (6) and (7)]. The type of ion generated is dependent on the charge of the adduct gained or moiety lost upon ionization and is therefore influenced on the polarity of the electrospray capillary. For example, a positive DC voltage on the capillary will generate positive ions from basic compounds, and a negative voltage will produce negative ions from acidic compounds. In the majority of applications, ions generated by API result from the gain or loss of a proton, $H^+$, and therefore the polarity of the ion generated is largely defined by the acid/base nature of the molecule. Organic molecules with a basic character, such as amines, will favor formation of positively charged ions. Molecules with an acidic character, such as carboxylic acids, will tend to deprotonate and form negatively charged ions. Preliminary understanding of the $pK_a$ of a given compound can provide information on improving analytical sensitivity (see Chap. 5).

*Multiple Charging.* A characteristic of electrospray that is often exploited for the analysis of high-molecular-weight molecules such as peptides and proteins is the generation of multiply charged ions [28–30]. Mass spectrometry requires sample ionization, and regardless of the specific type of mass analyzer used, the mass-to-charge ratio, $m/z$, of the ion is measured. The $m/z$ range that can be measured by a mass spectrometer is analyzer-dependent. For example, a quadrupole mass analyzer can have an upper limit of 2,000–5,000 $m/z$. A time-of-flight mass analyzer, on the other hand, can have an upper mass limit of $>100,000$ $m/z$. Discussion to this point has assumed that ionization is limited to one charge, i.e., $z = 1$. In cases where this is true, the analytical $m/z$ limitation of a mass spectrometer is nominally limited to the ion's molecular weight, as the *charge state* is considered unity (charge state refers to the number of charges on an ion). As a result, for a typical commercial quadrupole mass analyzer, which can have an upper limit of 4,000–5,000 $m/z$, detection of single-charged ions in excess of 5,000 Da is not possible. Quadrupole mass analyzers have been applied to protein characterization, however, and are capable of measuring $m/z$ ratios in excess of 5,000 $m/z$. This is possible through multiple charging, in particular with electrospray ionization. As the molecular weight of a given compound increases, the statistical probability of multiple acidic or basic functional groups increases. In a solvent system that is not charged-limited, such as a pH buffered solution, this often results in multiply charged ions, and an equilibrium will be formed in which a distribution of ions is generated over multiple charge states.

Electrospray ionization of a protein in an acidic solution will create the potential for multiply charged ionization, as the excess of $H^+$ in solution provides a distribution of charge states. Because a mass spectrometer measures the mass-to-charge ratio of an ion, the effective mass limit of an instrument can be extended beyond its single-charged $m/z$ range. Because the mass spectrum contains multiple measurements of the molecular weight of a compound (after compensation for charge state and mass difference due to proton adduction), multiple mass measurements for the protein can be taken, thereby allowing for improved mass

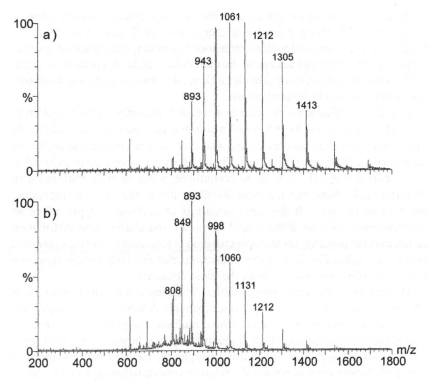

**Figure 5** Multiple charging of a horse heart myoglobin sample analyzed by a quadrupole MS instrument with electrospray ionization collected at (a) pH 6.0 and (b) pH 3.5.

accuracy through statistical averaging. For example, Fig. 5b shows the electrospray spectrum of pure horse heart myoglobin obtained by flow injection in 50:50 acetonitrile: water (0.1% formic acid). The average molecular weight of horse heart myoglobin is 16,954 Da, and the ions observed in the spectrum correspond to myoglobin with a varying number of protons attached, the distribution of which is determined by a number of chemical and physical parameters. The ion at 893 $m/z$, for example, corresponds to horse heart myoglobin with 19 $H^+$ attached as follows:

$$m/z = \frac{(M + nH)}{nH} = \frac{16,954 + 19H}{19H} = 893. \tag{8}$$

Adjacent ions in the spectrum correspond to horse heart myoglobin at different charge states (for example, 998 is charge state 17; 849 $m/z$ is charge state 20; and 808 is charge state 21). The charge state of an ion is defined as the number of charges an ion carries such that a singly charged ion has a charge state of 1, a doubly charged ion has a charge state of 2, and so on. With the multiple charging associated with electrospray ionization, a commercial quadrupole mass spectrometer can determine the molecular weight of a protein to within $\pm 0.01\%$. This level of accuracy provides a clear advantage over gel electrophoresis for molecular weight determinations.

One factor influencing the charge state of a multiply charged protein is solution pH (see Fig. 5). As the pH of a solution increases, the availability of free protons for ionization decreases, and a shift of the charge state distribution of a multiply charged protein is observed. Figure 5b shows a horse heart myoglobin spectrum obtained at pH 3.5 from an acetonitrile:water solution buffered with 0.1% formic acid. Note that the most abundant ion is 893 $m/z$, corresponding to the 19th charge state. If the same sample is characterized at pH 6.0 in an acetonitrile:water solution without acid (Fig. 5a), the charge state distribution shifts because the potential for multiple charging is regulated by the concentration of protons available for electrospray adduction, and the 16th charge state, observed as the 1061 $m/z$ ion, becomes the most abundant.

Depending on the application, multiple charging can be either deleterious or advantageous to the assay. The previous example describing the characterization of proteins illustrates an example where multiple charging is exploited. For the quantitative determination of small pharmaceuticals, however, singly charged ions are preferable not only for intuitive simplicity, but also because multiple charging of the analyte will reduce the relative concentration of a given charge state. Because quantitative mass spectrometry using tandem mass spectral methods typically relies on selective detection of a single charge state, assay detection limits could suffer as a result. Quantitation of a multiply charged ion can also detract from assay linearity, or quantitative dynamic range. Charge state is not only effected by the concentration of protons, but also by the analyte concentration itself, and when an equilibrium exists between multiple charge states, ionization efficiency can decrease dramatically with increased analyte concentration. The result is a quantitation curve that tends to flatten at higher concentration if the multiply charged ion is monitored and raises if the singly charged ion is monitored. This nonlinearity can be difficult to reproduce, causing potential problems with assay validation or reproducibility, and for this reason particular care is needed when analyzing compounds with multiple charge states.

One way to determine if an ion is multiply charged is to investigate the ion's isotope pattern. Each element has a natural abundance of isotopes, and for any given charged molecule these isotopes are observed in a mass spectrum as multiple peaks, the distribution and intensity of which are dependent on the combined relative abundance of each individual ion. Figure 6a shows the theoretical isotope pattern for singly charged gramicidin-S ($C_{60}H_{92}N_{12}O_{10}$), a cyclic decapeptide. The first and largest peak corresponds to the singly charged ion, $(M + H)^{1+}$,

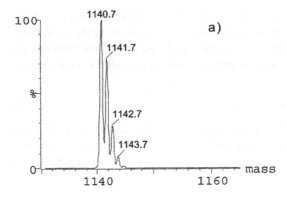

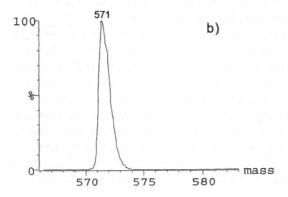

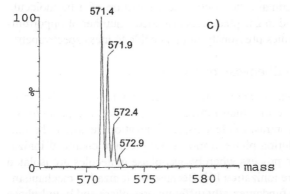

**Figure 6** Theoretical isotope distributions of single- and double-charged gramacidin-S. (a) The theoretical distribution of the single-charged molecular ion with nominal mass resolution. (b) The double-charged molecular ion with nominal mass resolution. (c) Discrimination of the double-charged isotopes is possible with higher instrument resolution.

in which each atom in the ion is represented by the lightest isotope, thereby generating the lowest possible molecular weight measurement for the compound. For a singly charged ion, where $z = 1$, the observed difference between the monoisotopic ion ($m_{m.o.}$, all $^{12}C$ atoms) and an ion with a single $^{13}C$ atom, $m_{m.o.+1}$, is as follows:

$$\Delta m/z = (m_{m.o.} - m_{m.o.+1})/1 = (1)/1 = 1. \tag{9}$$

Therefore, in the case of a singly charged ion, the observed difference in $m/z$ between the monoisotopic weight and the next heavier isotope is one $m/z$. If the charge state of the ion is 2, where $z = 2$, then the equation changes as follows:

$$\Delta m/z = (m_{m.o.} - m_{m.o.+1})/2 = (1)/2 = 0.5. \tag{10}$$

When an ion carries two charges, therefore, the difference between the monoisotopic $m/z$ and the next heavier isotope becomes one-half $m/z$. This causes the isotope distribution to compress together, creating a distinctive isotopic pattern indicative of charge state. An example of this is shown in Figure 6b, which is the theoretical doubly charged gramicidin-S ion, $(M+2H)^{2+}$. If the resolution of the mass spectrometer were increased, it would be possible to distinguish subsequent charge states (Fig. 6c). Monitoring isotope patterns can be used empirically to determine the charge state of any ion, but there are practical limitations on the resolving power of mass spectrometers, severely limiting this technique as charge state increases. However, for most small pharmaceutical candidates (<800 Da) this technique is sufficient to determine charge states from one to three charges and should be used as the primary indication of multiple charging.

*Noncovalent Interactions.* An additional characteristic of electrospray ionization is its ability to monitor noncovalent interactions [31]. The ability of electrospray ionization to maintain the noncovalent structure of a biomolecule in its transition from the liquid to gas phase has provided a number of opportunities to study molecular dynamics previously not amenable to mass spectrometry.

## 4. Atmospheric Pressure Chemical Ionization

Atmospheric pressure chemical ionization, like electrospray ionization, is a mass spectrometer ionization source in which ionization occurs not in a vacuum but at atmospheric pressure. In contrast to electrospray ionization, in which the ionization process occurs in solution phase, atmospheric pressure chemical ionization is a gas-phase ionization process whereby gas-phase molecules are isolated from the carrier solvent before ionization [6]. Because the ionization mechanisms of APCI and electrospray are fundamentally different (gas-phase and liquid-phase ionization, respectively) the two methods have the potential to provide complimentary analyte characterization. To generalize, electrospray ionization is more applicable to high-molecular-weight compounds, as electrospray ionization requires less heat and has the potential to produce multiply charged ions. Less polar molecules, such as steroids, generally ionize better by APCI. However, for the majority of pharmaceutical drug candidates, compounds with some acidic or basic characteristics and with relatively low molecular weight, either electrospray ionization or APCI is applicable. To a large extent empirical investigation is required to determine which API ionization method is most appropriate for a given compound or sample matrix environment [32]. It should be noted that once gas-phase ions are generated, the process of introducing the ions into the mass spectrometer system via differential pumping, and their subsequent separation and detection, are identical. Because ion sampling into the mass spectrometer is independent of the API method, most commercial API instruments are designed to handle both ionization techniques with minor hardware switchover.

A generic APCI source consists of a capillary interface for liquid introduction, a heated nebulization system, and a high-voltage corona discharge needle (see Fig. 7). In contrast to electrospray ionization, where flow rates less than 1 µL/min are possible, APCI requires comparably high liquid flow rates from roughly 200 to 2000 µL/min. Although specific design specifications differ among commercial instrument manufacturers, the APCI source capillary is grounded and does not induce ionization. Nebulization of the effluent is assisted by heated gas flow (typically nitrogen) generating gas-phase neutral molecules of both the effluent and analyte as well as any solvent modifiers or matrix impurities. Between the nebulization probe and the first vacuum orifice is placed a metal needle, upon which is applied a ±2- to 5-kV DC voltage. This static voltage generates a charged plasma from the ambient source atmosphere and evaporated

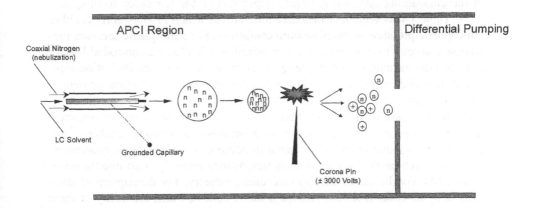

**Figure 7**  Schematic of a generic atmospheric pressure chemical ionization source.

solvent, and neutral molecules passing through this plasma are ionized through chemical charge transfer similar to that of chemical ionization.

Because APCI is not as closely coupled to the solvent environment as electrospray ionization, certain advantages of APCI over electrospray are observed, particularly in applications involving relatively nonpolar analytes or chromatographic effluent. Any effective API process requires some degree of polarity in both the analyte and the carrier solvent, and to a large extent electrospray ionization requires relatively more polar characteristic in the ionization mechanism. Atmospheric pressure chemical ionization is, however, often used in normal-phase chromatographic systems because solvents such as hexane, chloroform, and methylene chloride are generally nonpolar and pH neutral. In cases where normal-phase chromatography is used, addition of alcohol at 5–10% relative volume is useful in adding some proton-donating character to the solution. Likewise, for weakly polar analytes, such as steroids, APCI is often the ionization source of choice, as improved ionization efficiency is observed.

## III.  MASS ANALYZERS: HOW IONS ARE SEPARATED

### A.  Relationship Between Ionization Source and Mass Analyzer

The ionization source and mass analyzer of a mass spectrometer are inextricably linked, for the mass analyzer requires a particle be charged in order to separate it from other ions or neutral components. Any mass analyzer contains some electric or magnetic field, or combination of the two, which is capable of manipulating the trajectory of the ion in a vacuum chamber. Integration of an ionization source and mass analyzer must maintain the integrity of the vacuum, and the complexity of the engineering involved is largely dependent on the ionization technique.

Table 1 lists a number of ionization sources which produce ions at either atmospheric pressure or under vacuum conditions. For atmospheric pressure ionization sources a suitable interface is required which allows a controlled leak of ions into the vacuum region of the mass spectrometer. Vacuum ionization techniques likewise require a controlled leak, or mechanical introduction, of neutral molecules into the vacuum chamber, followed by ionization.

An impressive diversity of mass analyzers are utilized in modern analytical instrumentation. An overview of the common mass spectrometer analyzers follows, with particular emphasis on linear quadrupole mass analyzers, quadrupole ion traps, and time-of-flight mass analyzers, as they arguably constitute the quantitative MS workhorses of the pharmaceutical industry. The description of alternate analyzer systems should provide a framework in which the utility of these three particular systems provides the most cost-effective analytical mass spectrometer systems for pharmaceutical analysis.

## B. Magnetic Sector Mass Analyzers

A charged particle passing through a magnetic field assumes a radius of curvature, normal to the magnetic field, which is a function of the magnetic field strength and the mass, charge, and kinetic energy of the charged particle. The equation which characterizes the radius of curvature as a function of the $m/z$ ratio is given by the following:

$$m/z = \frac{r^2 B^2}{2V},\qquad (11)$$

where $r$ is the radius of curvature, $B$ is the magnetic field strength, and $V$ is the acceleration voltage of the ion source. By measuring the radius of curvature for a given magnetic field strength and acceleration voltage (5–10 kV) it is possible to determine the $m/z$ ratio of an ion.

A *magnetic sector* mass spectrometer measures the radius of curvature of an ion in a magnetic field to determine its $m/z$ ratio [4–8]. This can be accomplished by either simultaneously measuring the ion current for all $m/z$ ions with an array detector or by selectively detecting a single radius of curvature using a point detector (see Sec. V). With an array detector, the magnetic field strength and ion acceleration voltage can be held constant and the $m/z$ determined by the $r$ value for a given ion. When using a point detector, which corresponds to a central radius of curvature, either the acceleration voltage or the magnetic field strength must be scanned to generate a mass spectrum. Because the radius of curvature is a function of the energy of the ion, and any ionization source imparts a distribution of energies into an ion, magnetic sector instruments are often configured with an electrostatic filter. The electrostatic filter reduces the kinetic energy distribution of the ions, improving mass resolution and precision.

Magnetic sector mass analyzers are high-performance instruments, providing exceptional mass resolution and tandem mass spectrometry capabilities. Their relatively high cost compared to quadrupole mass filters and quadrupole ions traps have made them less desirable for many routine analytical applications. This increase in cost is due to the magnets and vacuum pumping requirements of magnetic sector instruments, which require lower operating pressures to $10^{-7}$ Torr compared to quadrupole mass filters, which can function at $10^{-5}$–$10^{-4}$ Torr. Because magnetic sector instruments rely on relatively high acceleration voltages for ion extraction, however, high-energy collisions can be studied, and where both resolution and MS/MS capabilities are required, they provide a viable alternative.

## C. Quadrupole Mass Analyzers

Quadrupole mass analyzers have developed into the analytical workhorses of today's pharmaceutical research and coupled with liquid chromatography they have essentially replaced LC/UV for routine quantitative and qualitative bioanal-

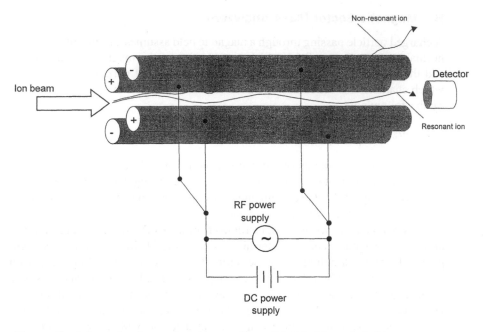

**Figure 8**    Schematic of a quadrupole mass filter. (Modified from Ref. 3.)

ysis. The inherent selectivity of this mass spectrometric method provides a power-ful tool for quantitative bioanalysis, particularly when a tandem mass spectrome-ter configuration is employed.

Quadrupole mass analyzers, sometimes called quadrupole mass filters, con-sist of two pairs of electrically conducting rods or electrodes, onto which have been applied a concurrent radio frequency (RF) and direct current (DC) voltage (see Fig. 8) [4–8,33,34]. Quadrupole mass analyzers consist of four conducting metal rods, composed of molybdenum or a gold-plated ceramic, and individual rods can have a diameter of 1 cm and can be up to 30 cm long. Opposing quad-rupole rod pairs are connected electrically such that they carry the identical RF/DC ratio, resulting in only two independent RF/DC voltages being applied to the quadrupole rod set at any time. For a given RF/DC voltage ratio on the quadrupole mass filter, only ions within a selected mass/charge ratio have a stable trajectory as they pass from one end of the quadrupole rod set to the other. The physics governing quadruple mass filters are described by the Mathieu equations [34,47]. Ions are given their initial incentive for a translational shift from one end of the quadrupole to the other through supplemental DC voltage offsets.

One reason for the broad applicability of quadrupole mass spectrometers in analytical applications is their ability to provide either selective ion detection or mass spectrum generation. If the RF/DC is held constant, only a narrow $m/z$ range will traverse the quadrupole, and in this configuration the quadrupole mass spectrometer acts as a mass filter. The resolution of this stable mass range can be modified, but in general a quadrupole mass filter is configured to provide selectivity to discriminate against ions differing by more than one $m/z$ unit. It is this configuration that is exploited for quantitation because selective detection of a single $m/z$ ion is possible in the presence of a milieu of ions. Selective ion detection is sometimes called single-ion recording (SIR) or selected-ion monitoring (SIM). Quadrupole mass analyzers can also generate mass spectral data (called a full-scan spectrum) in which a finite $m/z$ range is plotted as a function of ion intensity. By applying an RF/DC voltage gradient to the quadrupole rod set, cycling once per second for example, a mass spectrum can be created. In order to create an RF/DC voltage gradient, either the frequency or the peak-to-peak voltage of the RF voltage must be varied, and following a complete $m/z$ scan, the RF/DC voltage must reset back to the start point before initiating a subsequent scan. Because a quadrupole mass spectrometer must scan to create a spectrum, the speed at which it can generate spectra is limited by a number of factors, including the mass/charge range, the speed and stability of the system electronics and power supplies, and the time necessary to reset the RF/DC voltage for the next scan (sometimes called the *interscan delay*). The mass range of the instrument is limited by the quadrupole rod dimensions and the RF/DC power supply voltage. Commercially available quadrupole mass spectrometers typically have upper limits of 1500–4500 $m/z$, with cost proportional to performance.

Quadrupole mass spectrometers configured in tandem, commonly called *triple-quadrupole mass spectrometers*, increase the capabilities of the instrumentation exponentially. An analytical mass spectrometer with a single-quadrupole mass analyzer is capable of either SIM detection or full-scan acquisition. By adding an additional quadrupole mass analyzer in tandem, coupling the two analyzers with a gas reaction cell (*collision cell*), at least four additional analytical experiments can be conducted. Table 3 outlines these experiments and summarizes their capabilities. The collision cell may not be a true quadrupole configuration; rather it may have a hexapole or octopole configuration, and it does not provide mass discrimination, rather ion collision and focusing, hence, the triple quadrupole gets its name from its *quadrupole analyzer–collision cell–quadrupole analyzer* configuration.

It is useful to remember that a triple-quadrupole mass spectrometer can be configured to perform any single-quadrupole experiment (namely full-scan and SIM acquisition), but the remaining experiments (product ion scanning, precursor ion scanning, neutral loss/gain, and multiple reaction monitoring) require a tandem mass spectrometer configuration. Mass spectrometers or experiments that only employ a single mass analyzer are sometimes called single stage, as opposed to tandem configurations, which have more than one mass analyzer.

**Table 3** Single-Stage and Tandem MS Experiments of the Quadrupole Mass Analyzer Summarizing the Qualitative and Quantitative Applications of Each Configuration

| MS experiment | Quadrupole 1 configuration | Quadrupole 2 configuration | Collision cell | Application | System requirements |
|---|---|---|---|---|---|
| Full scan spectrum | Scanning analyzer | N/A | Not required | Qualitative analysis: Nonselective detection. Mass spectrum obtained. | Single- or triple-quadrupole |
| SIR | Selective mass filter | N/A | Not required | Quantitative analysis: Improved detection limit for a target ion. No mass spectrum obtained. | Single- or triple-quadrupole |
| Product (daughter) ion scan | Selective mass filter | Scanning analyzer | Yes | Qualitative analysis: Selective fragmentation and detection of an isolated ion. | Triple-quadrupole only |
| Precursor (parent) ion scan | Scanning analyzer | Selective mass filter | Yes | Qualitative analysis: Determine possible precursor ions of a given ion fragment. | Triple-quadrupole only |
| Neutral loss (linked MS/MS scanning) | Scanning analyzer (Q1/Q3 linked) | Scanning analyzer (Q1/Q3 linked) | Yes | Qualitative analysis: Determine neutral molecules lost following ion fragmentation. | Triple-quadrupole only |
| Multiple reaction monitoring (MRM) | Selective mass filter | Selective mass filter | Yes | Quantitative analysis: Selective detection of target ions. Optimum ion detection mode. No mass spectrum obtained. | Triple-quadrupole only |

Note: The triple-quadrupole mass spectrometer can be configured for single-stage operation.

## 1.  Full-Scan Spectra

The single-stage full-scan experiment is the fundamental qualitative configuration of the quadrupole mass spectrometer (see Fig. 9). Full-scan spectra (sometimes called an MS1 or Q1 scan) are obtained by scanning the RF/DC voltages on the quadrupole mass filter such that the stable $m/z$ ion trajectory sequentially changes from low to high $m/z$ (for example, one $m/z$ scan per second would be typical). This allows ions of different $m/z$ (represented by triangles, squares, and circles) to be sequentially detected (see Fig. 9a). A mass spectrum is generated in which ion intensity is plotted as a function of the range of $m/z$ ions (see Fig. 9b).

Full-scan spectra can be obtained on either single-stage or tandem quadrupole systems (in the latter configuration, the collision cell is evacuated and one quadrupole scans the RF/DC voltage while the other operates in an RF-only mode to provide uninhibited ion transmission). Because a quadrupole mass analyzer generates a mass spectrum by scanning the RF/DC voltage, the $m/z$ range has a stable trajectory which is transient such that only a small percentage of the total ions produced in the ionization source are detected at any given moment. As an

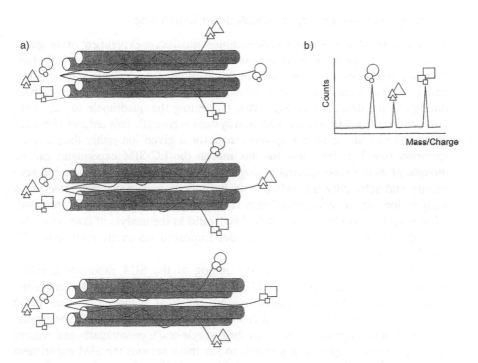

**Figure 9**  Full-scan acquisition mode of the quadrupole mass filter.

example, assume that the resolution of a particular quadrupole is 1 $m/z$ unit, and the instrument is scanning from 100 to 2000 $m/z$. At any given moment only 1 of 1900 $m/z$ windows has a stable trajectory and is detected. Under these instrument conditions the $m/z$ window transmitted through the mass filter relative to the total $m/z$ window of interest can be roughly estimated to be $(1/1900) \times 100$ = 0.05%. The ratio of ions transmitted relative to the total window of ions is called the *duty cycle* and is a measure of an instrument's transmission efficiency. This simple example illustrates two things. First, the quadrupole mass filter is relatively inefficient at detecting ions in full-scan mode. Second, the detection limit of a quadrupole analyzer in full-scan mode is a function of the $m/z$ range scanned (as the $m/z$ range is increased, sensitivity tends to decrease for a given $m/z$ ion). It is therefore prudent when acquiring full-scan spectra that the $m/z$ window be minimized to allow improved detection limits. Duty-cycle limitations can be addressed with time-of-flight and quadrupole ion trap mass spectrometers; however, current time-of-flight technology is not capable of efficient ion selection, while quadrupole ion traps do not provide quantitative performance comparable to that from triple-quadrupole instruments. For these reasons quadrupole mass analyzers are still the first-tier MS technology for quantitation.

## 2.   Single-Ion Recording or Selected-Ion Monitoring

Just as the full-scan experiment is the simplest qualitative experiment of the quadrupole mass analyzer, *single-ion recording*, or *selected-ion monitoring*, is its fundamental quantitative configuration. In this configuration, a single-stage quadrupole is used as a static mass filter, allowing only a selected $m/z$ window to pass through to the detector (see Fig. 10a). By setting the quadrupole to pass only ions of interest, a selective ion detection system is created. This reduces chemical noise and provides a time–response curve for a given ion rather than a mass spectrum (see Fig. 10b), and for this reason the LC/SIM experiment can be thought of as the mass spectral analog of LC/UV detection. Mass spectral sensitivity and selectivity still suffer from limitations from ion source mechanics such as ion suppression and relative response to ionization. Interference with other ions that have the same nominal $m/z$ ratio as the analyte of interest is also possible, but in general LC/SIM provides improved selectivity relative to LC/UV detection.

The tandem mass spectrometry analog to the SIM experiment, called multiple-reaction monitoring (MRM), provides additional selectivity and is the configuration of choice for quantitative bioanalytical methodology. The MRM experiment requires more financial capital to initiate, as tandem mass spectrometers are clearly more expensive than their single-stage counterparts and require more operational expertise in general, so for these reasons the SIM experiment is still widely used as a primary quantitative tool. Note that tandem mass spec-

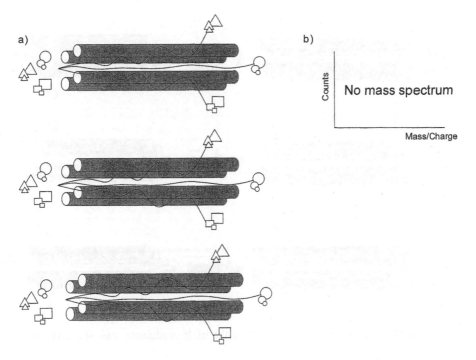

**Figure 10** Single-ion recording (SIR) mode of the quadrupole mass filter.

trometry systems can be configured for SIM acquisition by simply configuring either the first or second quadrupole mass analyzer as a bandpass filter, deactivating the collision cell, and using the remaining quadrupole as a discriminating mass filter. Because of the almost universal improvement in selectivity with MRM, SIM is used on tandem MS systems only in rare cases, such as the inability to produce sufficient product ion fragmentation in the collision cell to allow an MRM transition to be monitored.

## 3. Product (or Daughter) Ion Scanning

The two quadrupole mass analyzer configurations discussed above (full-scan and SIM) are amenable to either single- or multiple-stage mass spectrometers. With discussion of product ion scanning we begin the investigation of MS/MS technology and its applications. In the product ion scanning configuration, the first quadrupole (Q1) acts as a mass filter to selectively isolate a single mass/charge ion from the ion source. The selected ions are sent to the collision cell, ion fragmentation is induced, and the fragment ions are focused into the second quadrupole

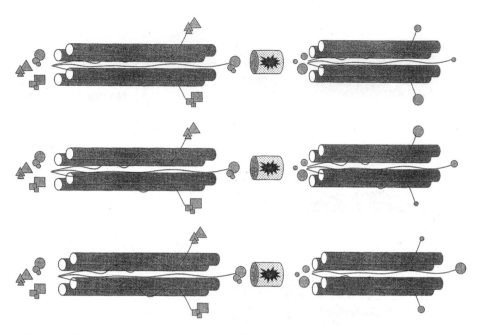

**Figure 11**   Product ion scanning mode of the triple-quadrupole mass spectrometer.

mass analyzer (see Fig. 11). The second analyzer (Q3) operates in scanning mode, producing a mass spectrum of the fragments. It is therefore possible to generate a mass spectrum of a single compound from a complex mixture free of background interference without chromatographic separation, so long as no compounds in the mixture have the same molecular weight. This is the impetus of tandem mass spectrometry. In cases where a single-stage quadrupole mass filter is used, coeluting peaks in a chromatographic separation produce ambiguous full-scan spectra, as it is difficult to assign fragment ions to a particular precursor ion. With tandem mass spectrometry, discrete isolation and correct assignment of fragmentation spectra for each analyte precursor ion is possible, even for coeluting peaks (again assuming that the ions have a different mass/charge). This significantly reduces the benchmark for ''sufficient'' chromatographic separation, speeding analytical method development and shortening chromatographic run times.

Product ion scanning has applications in qualitative analysis when unambiguous determination of the fragmentation pattern of an ion is required. Product ion scanning is also an efficient way to determine specific ion fragmentation pathways, or ion transitions, of a given compound when optimizing quantitative MRM experimentation.

## 4. Single- and Multiple-Reaction Monitoring

Multiple-reaction monitoring, as implemented on triple-quadrupole mass spectrometers, is currently the preferred technique for quantitative mass spectrometry [35]. In simple terms, this technique can be considered a back-to-back SIR experiment in which a precursor ion is selectively filtered by the first quadrupole and fragmented in the collision cell, and a unique fragment ion, possibly a structural fingerprint, is selectively filtered by the second quadrupole (see Fig. 12). Multiple reaction monitoring provides minimal qualitative information, but it is an excellent technique for quantitative bioanalytical applications. If only a single ion transition is monitored by the triple-quadrupole MS system, this is sometimes described as a single-reaction monitoring (SRM) experiment. However, the power supplies, electronics, and software of commercially available quadrupole MS/MS systems allow multiple mass transitions to be monitored sequentially and on a chromatographic time scale. When multiple ion transitions (multiple-product precursor sets) are monitored simultaneously, the experimental configuration is defined as multiple-reaction monitoring. This semantic distinction between SRM

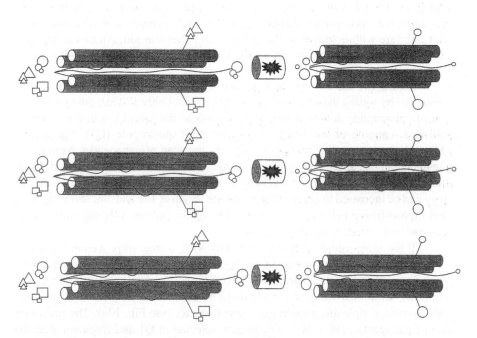

**Figure 12** Multiple-reaction monitoring (MRM) mode of the triple-quadrupole mass spectrometer.

and MRM is not always made, and it is common that the term MRM is used to define a reaction monitoring experiment with either one or multiple ion transitions.

Figure 13 shows a sensitivity and selectivity comparison between a number of quadrupole MS and MS/MS techniques. Each of the chromatograms in Fig. 13 is from a 1000-ng/mL standard of the amino acid gabapentin in artificial cerebral spinal fluid. Figure 13a shows the full-scan total ion chromatogram (TIC) of gabapentin [monoisotopic MW = 171; positive ion electrospray molecular ion $(M + H)^{1+} = 172$ $m/z$]. A total ion chromatogram is a plot of the sum total of mass spectral ion signal as a function of time. In this example, the scan range was from 20 to 300 $m/z$. With a TIC plot, the ion signal from a specific analyte, in this case gabapentin, is typically reduced by background chemical noise. The chromatographic retention time for gabapentin is approximately 0.95 min, and the inherent background of the total ion chromatogram severely limits detection of the gabapentin chromatographic peak. If only the molecular ion of gabapentin is plotted over time, a marked improvement in $S/N$ is observed (see Fig. 13b). This type of chromatogram showing a single ion plotted over time and extracted from a set of full scan data is often called an *extracted mass chromatogram.* Fig. 14a shows the full-scan mass spectrum, averaged across the peak, obtained from the extracted mass chromatogram. Note the large number of background ions that interfere with detection of the gabapentin molecular ion. An extracted mass chromatogram does not effect spectral data, but is simply a display option of the total ion chromatogram.

Figure 13c shows a product ion chromatogram of the gabapentin sample generated by setting the first quadrupole (Q1) to pass only 172 $m/z$ (the molecular ion of gabapentin), followed by fragmentation of the parent ion in the collision cell and scanning of the fragments by the third quadrupole (Q3). Figure 14b shows the corresponding product (daughter) ion scan of gabapentin. Notice that the chemical background observed in the Q1 spectrum has been eliminated. Once the gabapentin molecular ion has been isolated (Fig. 14b), the collision cell energy can be increased to induce fragmentation. Figures 14c and 14d show medium and high collision cell energy respectively (relative to Fig. 14b) and subsequent generation of product (daughter) ions.

If the same sample is characterized by SIM, a clear improvement in selectivity and sensitivity is observed (see Fig. 13d). All spectral information on the peak is sacrificed as the instrument is configured to monitor only a single $m/z$ window. Additional selectivity is obtained by monitoring an MRM transition, in this case on a triple quadrupole mass spectrometer (see Fig. 13e). The precursor ion of gabapentin, $(M + H)^{1+}$ 172 $m/z$, is selected in Q1 and fragmented in the collision cell, and the product ion, $(M + H)^{1+}$ 154 $m/z$, is monitored by Q3. This technique provides the highest combination of selectivity and signal to noise (S/N) with the sacrifice of qualitative information.

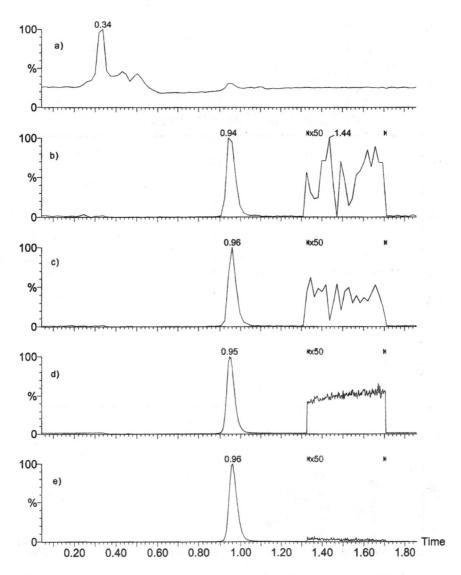

**Figure 13** Comparative example of chromatograms obtained from the API analysis of gabapentin from artificial cerebrospinal fluid. (a) Total ion chromatogram from a full scan acquisition; (b) extracted mass chromatogram from a full scan acquisition; (c) daughter ion chromatogram; (d) single-ion monitoring (SIM) chromatogram; and (e) a multiple-reaction monitoring (MRM) chromatogram. (Data courtesy of Ms. Xiaochun Lou.)

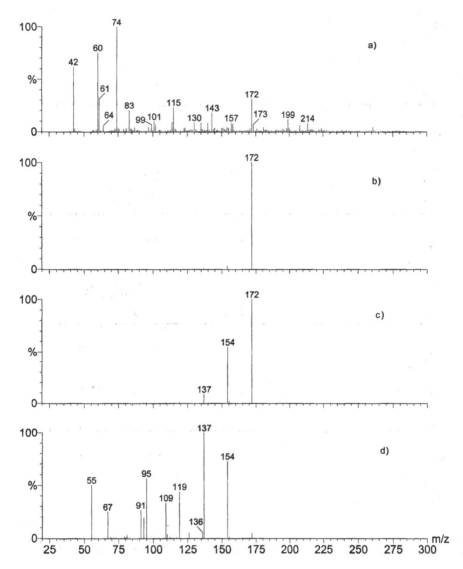

**Figure 14** Full-scan and daughter ion scan of gabapentin (172 *m/z*). (a) Full-scan spectra of gabapentin with API background ions and product ion scan of gabapentin at (b) low, (c) medium, and (d) high collision cell voltage. (Data courtesy of Ms. Xiaochun Lou.)

*Collision Cell Cross Talk.* The selectivity of the MRM experiment is not universal and a number of experimental processes can provide misleading information. One potential source of error in the MRM experiment is through collision cell cross talk, which can occur when multiple MRM transitions are being monitored in an MRM experiment. Cross talk is observed when the collision cell still holds fragment ions from the previous MRM channel, and in cases where two analyte ions have similar fragmentation patterns, the holdover ions from the cell contribute to the signal associated with the following ion. Where chromatographic separation does not exist between compounds that have a potential for cross talk, quantitation errors are likely, as discrimination between the compounds is not possible. Examples of collision cell cross talk are given in Chapter 5.

To minimize the potential for collision-cell cross talk, it is prudent to find product ions of a given compound that are unique to its structure. A practical example would be that when using stable label isotopes as internal standards, use a product ion which contains the stable label moiety in order to insure distinct product ion masses for the analyte and internal standard. Where it is not possible to find unique MRM product ions for components, it may be necessary to increase the time delay between monitored ion transitions (sometimes called the interchannel delay) to allow the collision cell to reequilibrate and empty itself of the prior ion load. For example, if two ion transitions are monitored in an MRM experiment, and both monitor the identical product $m/z$ ion, the interchannel delay may need to be increased from 100 to 300 msec. This additional time allows the instrument to clear the collision cell of holdover ions. Another option to minimize cross talk is to insert additional ''dummy'' MRM channels that monitor a different product ion $m/z$ into the instrument acquisition method, allowing the collision cell to reequilibrate.

*In-Source Fragmentation.* Another potential source of quantitation error with MRM methodology is through in-source fragmentation. Discussions up to this point have assumed that isolation of a given precursor ion through Q1 is consistent with its original mass/charge upon ionization. This assumption is not always valid, and in cases where conjugation or adduction of a precursor compound is observed, in-source fragmentation is possible. This effect can occur when a metabolic conjugate is ionized in the API source. In-source fragmentation occurs when a labile group cleaves off within the ionization source as a neutral and the precursor chemical structure is regenerated, producing an MRM transition indistinguishable from the nonconjugated precursor ion.

A textbook example of in-source fragmentation is shown in Figure 15, in which a sugar conjugate of a parent compound produces a detectable ion signal in the parent ion MRM channel. These chromatograms were obtained from an MRM experiment in which a solution containing both guanine (Fig. 15b; molecu-

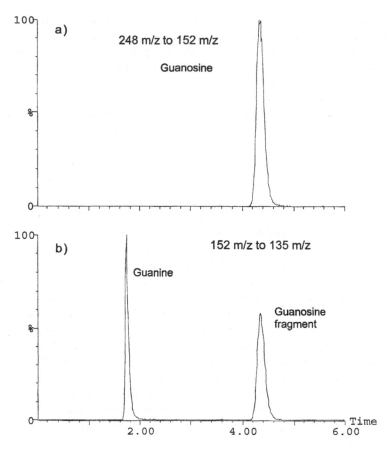

**Figure 15** Example of in-source fragmentation observed in a MRM experiment in which concurrent analysis of guanine and guanosine is required. (a) MRM chromatogram of a guanosine MRM channel. (b) MRM chromatogram of a guanine channel.

lar weight = 151 Da; MRM ion transition = 152 to 135 $m/z$) and guanosine (Fig. 15a; molecular weight = 283 Da; MRM ion transition = 284 to 152 $m/z$) were simultaneously characterized by LC/MS/MS. When ionized in the API source, the labile nature of the sugar moiety of guanosine results in its cleavage and loss as a neutral species. This process is not 100% efficient such that both the conjugated compound (guanosine) and parent compound (guanine) exist simultaneously prior to separation by Q1. Guanosine is selected by the appropriate MRM channel (284 to 152 $m/z$) and is unambiguously detected. The fragmented guanosine, however, is detected by the guanine MRM channel (152 to 135 $m/z$). The result is an additional chromatographic peak in the guanine MRM channel at

4.4 min. If sufficient chromatographic separation between guanosine and guanine exists, in-source fragmentation produces little more than an inconvenience. If chromatographic separation of conjugates from a parent compound is not possible, the experimentalist must be aware of the possibility of in-source fragmentation contributing to the integrated area counts of the parent ion peak. This is particularly acute in high-throughput chromatography, especially in discovery research, where characterization of chromatographic separations is typically not comprehensive.

*Dwell Time.* When operating a quadrupole mass analyzer in full-scan mode, the RF/DC voltage scans between two extremes, thereby generating a transient window of $m/z$ stability that ultimately generates a mass spectrum. The rate at which the RF/DC voltage scans determines the rate at which spectra are obtained. If a quadrupole mass analyzer is configured to scan 100–500 $m/z$ once per second and a chromatographic peak elutes in 15 sec, a total of 15 full-scan spectra are obtained that characterize that peak (as measured baseline to baseline). If the same quadrupole mass analyzer is configured for selective ion detection, whether single-ion monitoring or multiple-reaction monitoring, the RF/DC voltage is set to a constant ratio and scanning is not required. To produce discrete data points across the chromatographic peak, a setting analogous to scan time is required. For SIM and MRM experiments, the discrete collection time per data point is called the *dwell time* and is simply the length of time in which each discrete $m/z$ ion signal is integrated or the time interval before another point is collected. Increasing the dwell time will improve detector ion-counting statistics by averaging signal for a longer time period, but will result in fewer data points taken per chromatographic peak. Likewise, as the dwell time is decreased, the chromatographic sampling rate will increase. A common rule of thumb is to balance dwell time such that at least 20 data points are collected for any chromatographic peak. This will make any error attributed to finite chromatographic sampling negligible. Depending on the number of $m/z$ transitions being monitored, typical dwell time settings range from 0.05 to 0.30 sec. For studies in which a few ($\leq 3$) analytes are monitored dwell time is typically not a limitation and a sufficient number of data points can be taken for each chromatographic peak. Limitations do occur, however, in cassette dosing experiments (Chap. 10) where the number of $m/z$ transitions monitored can increase significantly ($\geq 8$). As the number of ions monitored increases, the dwell time for each $m/z$ transition must be decreased in order to provide sufficient chromatographic peak characterization. However, as dwell time decreases, the integration time for any given ion decreases as well, and this limitation in detector ion-counting statistics will decrease the sensitivity for all ion channels.

Figure 16 shows an example of dwell time on sensitivity and chromatographic response. Each chromatogram represents a single-channel MRM acquisi-

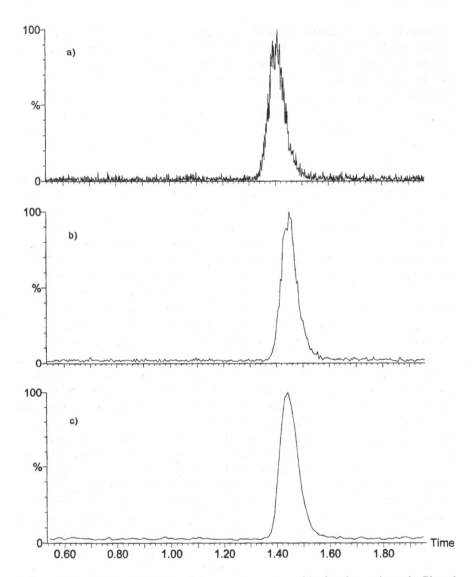

**Figure 16** Effect of MRM dwell time on chromatographic signal-to-noise ratio. Phenyl-butazone MRM chromatogram with dwell times of (a) 50 msec, (b) 250 msec, and (c) 750 msec.

tion of a phenylbutazone solution in mobile phase and analyzed by LC/MS/MS. In the first chromatogram (Fig. 16a), the dwell time for MRM acquisition is 50 msec and the baseline-to-baseline phenylbutazone chromatographic peak (10-sec peak width) is characterized by approximately 140 individual data points. As the dwell time is increased by a factor of 5 to 250 msec (Fig. 16b), the relative noise of the chromatogram diminishes significantly because the ion-counting statistics for each data point are improved. This comes at the cost of fewer data points defining the chromatographic peak, which in this case is now approximately 40. In the final example (Fig. 16c), the dwell time has been increased to 750 msec. Baseline noise and chromatographic characterization are reduced further and the chromatographic peak is still adequately sampled, with 15 data points over 10 sec. Appropriate dwell time settings balance the improvement of signal-to-noise ratio with sufficient chromatographic sampling to characterize the chromatographic peaks. The latter becomes especially important in multicomponent MRM analysis, such as used for cassette dosing, because the instrument must monitor more than one ion transition.

## 5. Precursor (or Parent) Ion Scanning

All MS/MS techniques rely on the fragmentation of a parent ion in the mass spectrometer. How this process is exploited determines the exact MS/MS method employed. For any mass spectrometric method to analyze and detect an atomic or molecular species, it must be charged and therefore manipulated under the influence of electromagnetic potential. Assuming that an ion only has one charge, upon fragmentation in the collision cell both a neutral fragment and an ion of lower $m/z$ will be generated.

For qualitative analysis in which the experimentalist is interested in product ion fragmentation, design of the MS/MS experiment requires some consideration as to whether it is the product ion or the neutral fragment that is of interest. In this way, *precursor ion scanning* and *neutral loss scanning* (discussed below) are related, as the former provides information from the perspective of the fragment ion and the latter from the neutral, or uncharged, fragment. For example, given four compounds (AW, AX, AY, and AZ) of differing molecular weights, each of which has a common structural moiety (A), it is possible that upon fragmentation the charge will remain with either the shared structural moiety, A, or with the remainder of the molecule, -W, -X, -Y, or -Z respectively (see Table 4). For discussion let us assume that an experimentalist is interested in developing an assay which investigates a series of compounds for a particular functional group, A. For the situation summarized in column I of Table 4, upon ionization and fragmentation of a precursor ion, the charge remains with the functional group of interest, namely A. Therefore, by configuring Q3 of the triple-quadrupole MS system as a static mass filter for ion $A^+$, only precursor ions

**Table 4** Possible Fragmentation Pathways of a Theoretical Homologous Series of Four Compounds.

| I. Common product ion formed from each compound | | | II. Unique product ion formed from each compound | | |
|---|---|---|---|---|---|
| Precursor ion | Product ion | Neutral | Precursor ion | Product ion | Neutral |
| $AW^+$ | $A^+$ | W | $AW^+$ | $W^+$ | A |
| $AX^+$ | $A^+$ | X | $AX^+$ | $X^+$ | A |
| $AY^+$ | $A^+$ | Y | $AY^+$ | $Y^+$ | A |
| $AZ^+$ | $A^+$ | Z | $AZ^+$ | $Z^+$ | A |

Note: Homology is represented by the common chemical moiety "A." In case I, fragmentation of the homologous series produces an identical product ion, $A^+$, requiring a precursor ion scan to distinguish the series. In case II, fragmentation produces a unique product ion for each compound. Neutral loss scanning (linked by the mass of A) will provide discrimination of the series.

which produce $A^+$ upon fragmentation in the collision cell will be detected. By scanning Q1 through a given $m/z$ range, it is therefore possible to qualitatively determine what precursor ions (e.g., $AW^+$, $AX^+$, $AY^+$, and $AZ^+$) in a mixture contain A as a common functional group. This technique, called product ion scanning, could be useful, for example, in determining which components in a mixture have a common ionic moiety, as could be the case with a drug and related substances in a bulk drug sample.

## 6. Neutral Loss/Gain Scanning

The other possibility assumes that upon fragmentation the functional group of interest, A, is lost as a neutral molecule. As the molecular weights of W, X, Y, and Z are unknown, another approach, called a neutral loss scan, is required. During a neutral loss scan, both Q1 and Q3 of the mass spectrometer act as scanning mass analyzers. The quadrupole analyzers are linked and offset by the molecular weight of molecule A. For example, as Q1 is scanning 500 $m/z$, Q3 is scanning $(500 - A)$ $m/z$, and as Q1 is scanning 501 $m/z$, Q3 is scanning $(501 - A)$ $m/z$, and so forth. As a result, only precursor ions that fragment to lose a neutral species of mass A are detected. Neutral loss scanning is useful when investigating a complex mixture for compounds that are structurally similar and have the potential to lose a common neutral, such a $CO_2$ or $NO_2$. Neutral gain scanning works on the same principle, but is applicable in situations where adduction of neutral species is being investigated.

The precursor ion scan and neutral loss scan provide the same qualitative information on the structure of an ion, with specific applicability determined by the characteristic of ion fragmentation or, more specifically, depending on what part of the ion retains the charge after fragmentation. If the fragment of interest retains the charge in the collision cell, parent ion scanning is the appropriate method for characterization. If a neutral fragment of interest is generated in the collision cell, the linked-scan neutral loss experiment will provide a tandem mass spectrometric method for characterization.

## D. Ion Cyclotron Resonance and Fourier Transform Mass Analyzers

From first principles of physics it is known that the motion of a charged particle can be manipulated as it travels through a magnetic field. This concept was introduced above in the section discussing magnetic sector mass analyzers. Another class of mass spectrometer, referred to as *ion cyclotron resonance* (ICR), takes advantage of this principal not only to manipulate the motion of the ion through a transient trajectory, but to trap it for an extended period. This creates a time-dependent mass spectrometric method (analogous to the quadrupole ion trap described below). In both quadrupole ion traps and ion cyclotron resonance mass analyzers, ion storage provides the conduit for $m/z$ analysis and tandem mass spectrometric capabilities. Fourier transform mass spectrometry (FTMS) can be thought of as a subtype of ICR mass spectrometer and is a mathematical transform approach for ICR data analysis.

Ion cyclotron resonance mass spectrometers determine the $m/z$ ratio of an ion from the rotational frequency of the ion within the magnetic field [36–38]. The relationship between an ion's $m/z$ ratio and it rotational frequency is given by the following:

$$m/z = \frac{B}{2f_c} \tag{12}$$

where $B$ is the strength of the magnetic field and $f_c$ is the frequency of the trapped ion. Therefore, by determining the frequency of the ion and knowing the magnetic field strength of the instrument, it is possible to determine the mass of a trapped ion.

To determine the $f_c$ of an ion trapped in an ICR instrument requires application of a resonant frequency to the system. If we consider the ICR mass analyzer to be an evacuated box, the sides of which can independently conduct an electrical current, the box becomes a magnetic field and ions of various $m/z$ ratios can be stored in it using a distinct rotational frequency for each. By applying a variable frequency to the ion storage compartment that is resonant with $f_c$, the radius of a given window of ions can be made to increase as the ions absorb the resonant energy. The measured decrease in energy of the resonant frequency relates to an energy absorption of the ion and can be used to determine $f_c$. Alternatively, the intensity of the resonant waveform can be increased such that it causes ejection, and detection, of the ion from the ICR cell.

Fourier transform mass spectrometry is a more sophisticated method of determining the $f_c$ values of ions trapped in a magnetic field and can be considered an alternative approach to the ICR methodologies described above [37,38]. Considering the same trapping compartment and magnetic field, as a trapped ion approaches the sides of the compartment it produces a measurable current through induction. If a broadband frequency pulse is applied to the compartment (through the range of the trapped ions' resonant $f_c$ frequencies) the trajectories of all the ions in the cell can be made to approach the compartment walls. Each $m/z$ ion then induces an electrical impulse, called an *image current*, at the cell wall which is a function of its $f_c$ value. This image current, which is a time-domain function, is measured as it is allowed to decay following application of the broadband frequency and is then converted to the frequency domain through a Fourier transform process. Once the frequency-domain picture of the cell is produced, the $m/z$ of the trapped ions can be determined based on the fundamental relationship described in Eq. (12).

A distinct advantage of ICR and Fourier transform mass spectrometry is the superior mass resolution obtained by these instruments [39–40]. Distinction of nominally isobaric compounds (e.g., having the same mass to within 0.1 Da) is possible using ICR and Fourier transform mass spectrometry. Magnetic trapping instruments also provide excellent sensitivity and the potential for extended ion trapping and isolation. This latter characteristic has found FTMS applications for a number of fundamental quantitative and gas-phase ion reactions [41,42]. Because Fourier transform methodology relies on unimolecular ion decay to accurately characterize the image current, these systems require exceptionally high vacuums. Coupled with hardware requirements, such as magnet technology, FTMS instrumentation is relatively expensive and typically requires expert technical support for operation and analytical applications. For these reasons, FTMS is not as widely used as quadrupole ion trap instrumentation.

## E.  Quadrupole Ion Trap Mass Analyzers

Quadrupole ion trap mass analyzers merge the trapping characteristics of the ICR with the physical principles of the linear quadrupole mass analyzer. *Quadrupole ion traps* produce time-dependent spectra with excellent sensitivity and tandem mass spectrometry capabilities, but unlike the ICR they provide these ion trapping characteristics with physically smaller and considerably less expensive instrumentation, giving them a reputation as a powerful and accessible tool for both qualitative and quantitative mass spectrometry [43–47]. The capability of quadrupole ion traps to be configured with either internal and external ionization sources has expanded their utility for modern analytical applications [48–52].

Quadrupole ion traps consists of three electrodes: a single donut-shaped electrode (called the *ring electrode*) and two disklike electrodes (called *endcap electrodes*), each of which has a hyperbolic cross section. The quadrupole ion trap is constructed such that the endcap electrodes, with the ring electrode between, are brought to within approximately 2.5 mm of one another and act as "caps" to the ring electrode (see Fig. 17). This provides the mechanical boundaries for a potential well that is capable of storing a finite amount of ions. To create the potential well, an RF voltage of approximately 1 MHz and ≤15,000 $V_{pp}$ is applied to the ring electrode while the endcap electrodes are grounded and a *buffer gas*, typically helium, is added at approximately $10^{-3}$ Torr [53]. The RF voltage alternately accelerates and decelerates ions within the trapping volume, and depending on the frequency and voltage of the trapping potential, a range of $m/z$ ions is stored. The buffer gas acts to thermodynamically cool the ions, increasing mass spectral resolution and improving trapping efficiency. Each $m/z$ ion has its own characteristic trapping frequency, called the *secular frequency*, which describes its macroscopic motion within the ion trap. Ions whose secular frequency is not compatible with a particular RF trapping voltage are destabilized and are not stored.

## 1.  Operational Modes of the Quadrupole Ion Trap

A number of operation modes have been developed for the quadrupole ion trap that provide the foundation for its use as a mass spectrometer. These would in-

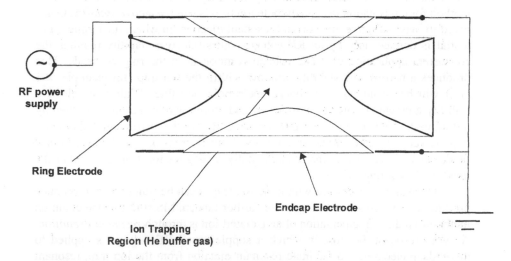

**Figure 17**  Schematic of a quadrupole ion trap mass spectrometer (QUISTOR).

clude the ability to detect trapped ions in a systematic fashion (for instance, to collect a mass spectrum) and to isolate and fragment ions within the trap (for tandem MS operation).

*Mass-Selective and Resonant Ejection.*   In principle, either the frequency or peak-to-peak voltage ($V_{pp}$) of the RF trapping potential, or a combination of the two, could be used to define the $m/z$ stability range of the quadrupole ion trap. In most system configurations, however, the RF frequency remains constant such that a given $m/z$ stability window is defined by $V_{pp}$. A variable $V_{pp}$ defines not only the initial $m/z$ storage range but can also sequentially eject ions from the trap based on their $m/z$ ratio, thereby producing a means of generating a mass spectrum. A systematic increase in the RF trapping voltage to induce controlled ion ejection is called *mass-selective ejection*. Ions are commonly ejected through a hole in one of the endcap electrodes and are detected with an electron multiplier or ion detector. Early quadrupole ion trap designs that relied on mass-selective ejection were relatively $m/z$-limited ($\leq 650$ $m/z$). By adding a supplementary RF voltage to the endcap electrodes coincident with the secular frequency of an ejected ion, it is possible to extend the $m/z$ range of the quadrupole ion trap as the trapped ions can be ejected with a lower RF $V_{pp}$. This mode of operation is called *resonant ejection* and is largely responsible for the mass-range extension of the quadrupole ion trap ($\geq 10,000$ $m/z$) [54].

*Mass-Selective Stability and Resonant Excitation.*   A defining characteristic of the quadrupole ion trap is its ability to provide multistep tandem mass spectrometry. As with any tandem mass spectrometry application, a precursor ion must first be isolated, followed by ion fragmentation and detection. One method for ion isolation in a quadrupole ion trap is *mass-selective stability*. Compared to mass-selective ejection and resonant ejection, in which the trapping potential is defined only by an RF voltage, mass-selective stability requires the concurrent application of a DC voltage component to the ring electrode. This produces a narrow $m/z$ stability window within the ion trap (for example, $\leq 1$ $m/z$), which is analogous to the linear quadrupole mass filter. With mass-selective stability a precursor ion can be selected and stored within the ion trap, thereby initiating the tandem mass spectrometry capability of the instrument. Other techniques, such as stored-waveform inverse Fourier transform and filtered-noise field, exist which provide the fundamentally analogous ion isolation within the quadrupole ion trap [55–57].

Once an ion is isolated within the ion trap it can be fragmented to produce product ions which act as the basis for further tandem MS studies. One common method to induce fragmentation of an isolated ion is through *resonant excitation*. As with resonant ejection, in which a supplemental RF voltage is applied to the endcap electrodes to facilitate resonant ejection from the ion trap, resonant excitation results when the resonant voltage induces additional translational energy into the ion without ejecting it from the trap. The resonant excitation voltage, or *tickle voltage*, can be tuned to specific $m/z$ secular frequencies, allowing selective excitation and fragmentation of ions within the trap. Resonant ejection of these fragment ions generates an MS/MS spectrum analogous to product ion scanning of the triple quadrupole mass spectrometer.

## 2. Time-Dependent Mass Spectrometry

Quadrupole ion traps offer some distinct advantages over quadrupole mass filters, particularly due to the fact that ion traps produce time-dependent mass spectrometric analysis. Ion traps can accumulate and store ions for extended periods relative to nontrapping instruments, providing a means to amplify weak ion currents and thereby, in many applications, improving detection limits. The time-dependent tandem mass spectrometric characteristics of ion traps also allow multiple ion fragmentation steps to be investigated. For instance, a triple-quadrupole mass spectrometer is capable of isolating one precursor ion, fragmenting it, and monitoring the product ion(s) that result (refer to Fig. 11). A quadrupole ion trap, however, is capable of isolating and fragmenting product ions through multiple steps, generating a selective cascade of precursor/product ions characteristic of a chemical species. The capability of producing multiple tandem MS investigations of a single precursor ion, sometimes termed $MS^n$, can provide unequivocal insight into ion fragmentation mechanisms and ion–molecular reactions [58].

Quadrupole ion trap mass spectrometers operate under some unique limitations as well. Under conditions of relatively high ion current within the trap, coulombic repulsion, or *space-charging* effects, can have deleterious effects on mass resolution and dynamic range. Some commercial quadrupole ion trap mass spectrometers, in fact, are configured with *automatic gain control*, which dynamically inhibits trap saturation, to limit space-charging effects [53]. Experiments such as precursor ion scanning and neutral loss/gain scanning are not amenable to quadrupole ion traps, and because of the low $m/z$ cutoff region associated with the technology, certain precursor/product ion pairs cannot be detected [59]. If a product ion is $\leq 30\%$ of the precursor ion mass, direct tandem MS monitoring is not possible because a stable trapping potential for both the precursor and product ion cannot be concurrently applied. For these reasons, the quadrupole ion trap has not supplanted the tandem quadrupole mass filter as the analytical workhorse for quantitative mass spectrometry.

## F. Time-of-Flight Mass Analyzers

One limitation of the quadrupole mass analyzer is that when configured for optimal quantitative performance, e.g., SIM or MRM, virtually no qualitative information can be obtained from an assay (aside from some limited applications, such as in-source fragmentation of a metabolic conjugate). Quantitative analysis

is possible with full-scan quadrupole mass spectrometry by extracting and integrating selected ion chromatograms (extracted mass chromatograms) from the
TIC, but the higher limit of detection makes this realistic only for assays that
are not sensitivity limited. For this reason SIM and MRM are almost exclusively
used for quantitative applications. With quadrupole mass analyzers, the loss of
sensitivity between selective and full-scan mass configurations stems from the
inherent time required to ramp the RF/DC voltage, giving rise to a scanning MS
method with a relatively low duty cycle (see Sec. III.C.1). Time-of-flight mass
spectrometers have the potential to provide a complimentary approach in such
instances.

Time-of-flight mass spectrometers (TOF/MS) are useful because of their
ability to generate full-scan spectral data without instrument scanning. This feature greatly improves the instrument duty cycle during full-scan acquisition. For
a given packet of ions of differing $m/z$, a TOF/MS analyzer can detect each
$m/z$ ion in turn without selectively filtering out other ions. This allows full-scan
spectra to be generated with improved sensitivity compared to most scanning
instruments, such as quadrupole mass analyzers. In addition, ion detection is not
limited by the mass range of the analyzer (although practical $m/z$ detection limits
do exist due to electronic or detector limitations). This is of particular utility for
the characterization of large molecules such as peptides and proteins. Essentially,
a time-of-flight mass spectrometer consists of an evacuated manifold with an ion
extraction region at one end and an ion detector at the other (see Fig. 18) [60,61].
A pulsed DC voltage is applied to the ions in the extraction region, thereby accelerating them into the vacuum manifold. The vacuum manifold is a grounded
field-free region sometimes called the *drift tube* or *flight tube*. All ions are acceler-

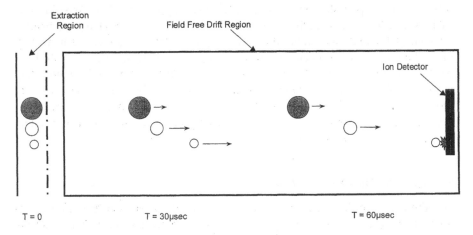

**Figure 18**  Schematic of a time-of-flight mass spectrometer (TOF/MS).

ated from the extractor region with the same nominal kinetic energy, as determined by the charge, $z$, on the ion and the voltage of the extraction pulse, $P$. Because

$$KE = zP = \frac{1}{2}mv^2 \tag{13}$$

the velocity, $v$, of an ion of mass $m$ can be expressed as follows:

$$v = \sqrt{\frac{2zP}{m}}. \tag{14}$$

Therefore, the flight time of an ion in the drift tube is

$$time = t = \sqrt{\frac{m}{2zP}} = k\sqrt{\frac{m}{z}}, \tag{15}$$

which shows that ions can be separated in time by the $\sqrt{m/z}$ ratio of an ion. Furthermore, the spectrum can be internally calibrated to determine the linear equation describing the relationship between flight time and $m/z$ as follows:

$$t = a\sqrt{\frac{m}{z}} + b, \tag{16}$$

where $a$ and $b$ are the slope and intercept respectively.

One limitation of single-stage TOF/MS instrumentation is the loss of tandem mass spectrometry capabilities. Designs allowing tandem mass spectrometry for a single-stage TOF drift tube are being developed but are not yet a viable alternative to conventional MS/MS techniques such as quadrupole ion trap, FTMS, and triple-quadrupole mass spectrometry. One solution to this problem is to couple a separate mass-selective analyzer in front of the TOF/MS analyzer. Commercial instruments combining quadrupoles and flight tubes that are designed around a quadrupole/time-of-flight configuration are now available [2].

## 1.  TOF Mass Analyzers and Mass Resolution

Historically, a common limitation of TOF/MS instrumentation was its poor mass resolution, and for this reason its implementation in many analytical applications has been slow to develop. To understand the primary mass resolution limitation it is useful to focus on the ion optics of a typical TOF/MS instrument (see Fig. 19a). The ion extraction region of a TOF/MS system can be represented by a field gradient between two distinct electrostatic lenses. The voltage gradient between these two plates is typically quite significant, say $10^3$–$10^4$ V/cm. Because the kinetic energy of an ion characterizes its velocity, and therefore its $m/z$ measurement, three ions of the same $m/z$ ratio simultaneously extracted from different

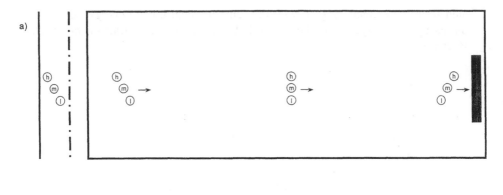

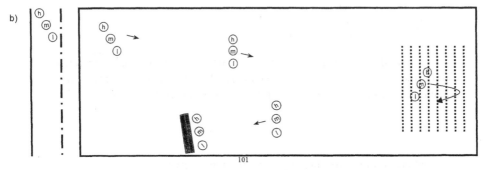

**Figure 19**  Resolution in a time-of-flight mass spectrometer (a) with and (b) without an ion reflector.

points in the extraction region will have different kinetic energies (represented by l = low, m = medium, and h = high) and therefore different velocities. This velocity difference will have a negative impact on mass resolution because slight differences in flight time will result.

Advances in TOF/MS technology, such as reflectron ion optics, delayed ion extraction, and orthogonal ion acceleration, have given rise to bench-top instrumentation with resolution capabilities in excess of that obtained by quadrupole MS systems. Each of these devices limits or corrects the kinetic energy spread of ions with identical $m/z$ ratios.

*Reflectron Ion Optics.*  The *reflectron*, or *ion reflector*, compensates for changes in flight time associated with ions of the same $m/z$ ratio that have different kinetic energies [60–62]. This correction is made after ion extraction. A reflectron consists of an array of electrostatic lenses, at the opposite end of the drift tube, to which is applied a DC voltage gradient of the same polarity as the extraction voltage (see Fig. 19b). A reflectron can be thought of as a voltage divider upon which each lens carries a DC potential of increasing voltage. As ions enter the reflectron, they are decelerated, and because the voltage magnitude of the reflectron is set higher than that of the ion extractor, ions are stopped and turned back toward the source. Ion residence time in the reflectron is proportional to the energy of the ion because ions with greater kinetic energy push further into the reflectron, and this compensates for flight time differences. Ions are focused in time and this improves spectral resolution.

*Delayed Extraction.* *Delayed extraction*, or *time lag focusing*, is a resolution-enhancing technique applied to TOF/MS in which a time delay (on the order of nanoseconds) is set between ion formation and ion extraction [60,61,63–65]. As an illustration take the ionization of three identical molecules (same molecular weight) within a TOF/MS ionization source (see Fig. 20a). These three ions, A, B, and C, may have different transient energies and different velocities such that prior to ionization A and C are moving away from and toward the drift tube respectively and ion B is stationary. If ion extraction occurs immediately upon ionization, the extraction potential applied to the source must overcome the inherent transient energy of ion A in order to change its velocity. This will delay the extraction of ion A by some unit time, called the turnaround time. Ion B will be extracted from the source at a rate proportional only to the source extraction voltage, and ion C will be extracted sooner than either ion A or B, as its inherent transient motion will be additive to the extraction potential. The resolution of the TOF/MS analyzer will suffer because these ions, though having identical molecular weights, each have a different residence time in the ion source.

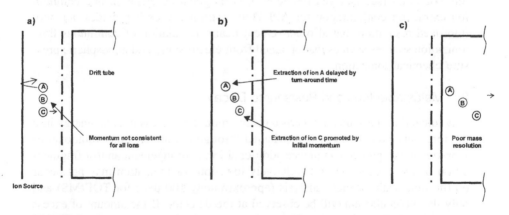

**Figure 20** Delayed extraction applied to a TOF/MS ionization source. (a) Initial positions of ions A, B, and C. (b) Position of ions A, B, and C after translational equilibration based on initial ion momentum. (c) Effect of ion momentum on flight time.

In the example of ions A, B, and C above, by delaying the extraction voltage relative to ion formation, the ions are allowed to expand, as a function of their inherent velocities, within the source under field-free conditions. Ion A therefore moves away from the extraction lens, ion C moves toward the extraction lens, and ion B remains stationary. When the extraction voltage is applied, ions A and C will have moved to regions of higher and lower extraction potential respectively, relative to their initial positions within the source (see Fig. 20b). Because of this, the turnaround time for ion A will decrease, and the additional voltage push for ion C will be minimized. This ultimately improves mass resolution because the ion source residence time differentials between A, B, and C are minimized.

Time lag focusing presumes that (a) ionization occurs within the source and (b) the ionization mechanism imparts upon the ions a distribution of transient energies and velocity vectors. An ion source such as that used in MALDI can greatly benefit from time lag focusing, and this technology is becoming standard on commercial MALDI/TOF instruments.

*Orthogonal Acceleration.* For vacuum ionization techniques, such as MALDI, delayed extraction provides resolution enhancement by compensating for ion energy spread within the extraction region. Atmospheric pressure ionization, when coupled to TOF/MS, can likewise benefit from a source design in which ion energy spread is minimized. A convenient way to do this is through orthogonal acceleration. *Orthogonal acceleration* is an ion extraction technique in which ions are introduced into the TOF/MS extraction region perpendicular to the drift tube [66]. By introducing the ion beam orthogonal to the extraction region, differences in extraction potential as a function of position within the voltage gradient are minimized, ultimately improving mass resolution following ion extraction (see Delayed Extraction). Orthogonal acceleration is a common ion extraction configuration for API/TOF instruments, as it provides not only improved resolution, but also convenient beam modulation of continuous flow ionization sources such as that obtained from electrospray and atmospheric pressure chemical ionization.

## 2. Metastable Ions and Postsource Decay

Any ionization mechanism imparts a given amount of excess internal energy into the ion. This excess energy is distributed through the ion as either rotational or vibrational energy that can induce additional ion rearrangement and/or fragmentation. If the excess energy is relatively low, ion fragmentation may not occur on the time scale of mass analysis (approximately 100 μsec for TOF/MS) and only the molecular ion will be observed at the detector. If the amount of excess energy is relatively high, ion fragmentation can be instantaneous by TOF/MS standards ($<1$ μsec) and both the molecular ion and additional fragment ions will be extracted and detected. For ions with borderline internal energies, called *metastable ions*, ion fragmentation can occur during mass analysis (1–100 μsec) rather than prior to ion extraction [67]. For TOF/MS methodology this results in ion fragmentation in the drift tube, after ion extraction, and for this reason metastable ion fragmentation is often called *postsource decay* [68].

Metastable ions provide an avenue for enhanced structural elucidation, particularly for large biomolecules ionized by relatively soft ionization techniques such as MALDI. Metastable ions in TOF/MS have identical velocities as their precursor ions but have a lower kinetic energy proportional to the mass of the fragment. Linear TOF/MS systems that do not contain a reflection cannot discriminate between metastable ions and their precursors, as both have the same velocity and strike the detector at the same time. Reflectron TOF/MS systems, however, can discriminate between precursor and metastable ions, as the reflectron can act as a variable-energy ion reflector [69]. By changing the voltage potential of the reflectron, different ion stability ranges can be created in which ions of a given kinetic energy window, and therefore metastable ion mass range, are reflected back to the detector. Higher end commercial MALDI/TOF instruments automatically switch the reflectron voltage to compensate for metastable ions with different kinetic energies, i.e., different *m/z* ratios. Individual mass spectra associated with the variable reflectron voltage steps are then patched into a single spectrum, representative of the metastable ion decay associated with the compound. This technique to characterize postsource decay is critical in obtaining fragmentation spectra in MALDI, as this is a relatively soft ionization technique and is commonly used for protein sequencing applications.

## 3. Accurate Mass Measurements

Orthogonal extraction, improvements in TOF/MS resolution, and improvements in electronics have created a new application for TOF/MS in accurate mass measurements [70,71]. Accurate mass applications have traditionally been associated with high-resolution magnetic sector and FTMS systems. The ability to provide accurate mass measurements (±5 ppm) with more cost-effective instrumentation has created potential opportunities in the pharmaceutical industry. Such applications can offer improved selectivity for single-stage instrumentation and metabolite or impurity/degradation product identification, particularly when standard compounds are not available, such as in early drug discovery.

## IV. DETECTOR SYSTEMS: HOW IONS ARE DETECTED

Detector schemes for mass spectrometers either measure the ion current directly, with possible signal amplification, or use ion conversion followed by solid-state amplification. Direct measurement of ion current has the advantage of providing precise measurement of signal intensity that is independent of the mass or kinetic energy of the ion. Examples include photographic film, on which impinging ions create a photographic response, and the Faraday cup, which produces an amplified current as ions are discharged to ground on a conducting surface. Though these detectors are generally precise, they also tend to have slow response times, and in some cases, such as photographic film, have obvious practical limitations. These detectors, while having historical significance, have been replaced with modern electronic signal detection devices. Solid-state ion detectors, such as *electron multipliers*, *multichannel plate detectors*, and *photomultipliers*, are by far the most common detection systems used [5–12]. Because solid-state detectors utilize dynodes to convert ion signal into secondary particles prior to amplification, signal response is a function of ion mass and kinetic energy. This must be considered when detecting large molecules such as proteins. Many ion detector configurations introduce an additional ion acceleration step called *postacceleration* prior to detection. This step increases ion velocity and improves sensitivity. The response time of solid-state detectors, and their obvious synergy with digital electronics, far outweigh their response limitations.

## A. Electron Multipliers

Electron multipliers provide signal detection through ion conversion and subsequent amplification. Ions impact on a conversion dynode, which converts the ions to secondary particles, such as electrons, and the secondary particles are accelerated into an electron multiplier, on which is applied a DC voltage. As the secondary particle strikes the surface of the electron multiplier, it causes the ejection of one or more electrons, which are subsequently accelerated further into the multiplier, ultimately creating an amplification cascade of electrons. The conversion dynode can be configured to respond to either polarity, depending on whether positive or negative ions are being detected. Electron multipliers are generally curved to optimize electron amplification and to limit the possibility electron feedback toward the conversion dynode and can produce a gain up to $10^7$. The active lifetime of an electron multiplier is a function of surface deactivation, such as caused by the absorption of water or contaminants to the multiplier surface, and can range from 1 to 2 years. As an electron multiplier ages, an increase in detector voltage is required to maintain comparable detector gain, and an exponential decrease in signal gain typically follows. For quadrupole mass analyzers and quadrupole ion traps, the electron multiplier is a detector mainstay, and they are quite common on commercial instrumentation.

## B. Photomultipliers

The conversion dynode of a photomultiplier detector generates electrons that impinge on a phosphor, which subsequently generates photons that are detected and amplified by a photomultiplier [72]. Photomultipliers have the advantage of relatively long lifetimes (10 years), as the photomultiplier itself is encased in glass and is therefore not as susceptible to the absorption of ambient water or gaseous contaminants. Photomultipliers are, however, sensitive to light background, and as such mass spectrometers that utilize photomultipliers must be isolated from ambient light. The addition of a phosphor adds an additional service element as well. Photomultipliers, like electron multipliers, are commonly used ion detection designs in modern mass spectrometers, particularly with quadrupole mass analyzers.

## C. Microchannel Plate Detectors

An electron multiplier can be thought of as a point detector in that a single conversion dynode and multiplier are configured to detect the ion signal. Ions to be detected must first be maneuvered to a single precise position. An alternative possibility, in which numerous electron multipliers are configured together to provide an array [73] requires miniaturization and juxtaposition of the individual electron multipliers into a continuous detector.

The microchannel plate detector (MCP) is one type of solid-state array detector commonly used in mass spectrometry. Microchannel (or multichannel) plates are flat waferlike detectors of leaded glass consisting of an array of $10^4$–$10^7$ electron multipliers in parallel, each channel of which is 10–100 µm in diameter. The length-to-diameter ratio of each channel is crucial for its performance and typically ranges from 40 to 100, and each channel is aligned 8–15° from normal relative to the surface in order to minimize positive ion feedback caused by ionization of background gases. A negative voltage bias (1 kV maximum) is applied across one side of the MCP relative to the other and this produces a potential gradient along each channel. As ions impact on the front of MPC, they discharge the individual channels, creating an miniaturized electron cascade. The electrons are then collected at an anode at the back of the detector. A number of MCP detector configurations are possible. A dual MCP consisting of two individual juxtaposed plates (aligned to maximize the pitch between the plate channels) is one configuration. This configuration, called a Chevron detector, provides a gain of $10^6$–$10^7$ with <1-nsec response time. Ions with sufficient velocity that strike the front of the charged MCP cause an electron cascade, and these electrons are then collected at the second MCP for further amplification.

Microchannel plate detectors are particularly useful in time-of-flight mass spectrometry, as they are flat, minimizing time spread and subsequent mass resolution of homologous ion packets. In addition, they have reasonable gain ($10^4$–$10^7$ per plate) and fast response time (100-psec time resolution). The major limitation of multichannel plate detectors is the recovery time needed for the detector to recharge. When a channel is discharged, a recovery time on the order of 10 nsec is typical. This becomes problematic if an ion follows another into a particular channel within a time frame less than the recharge time of the channel. The first ion that impacts on the channel will create an electron cascade and will be detected, but the following ion, hitting the detector before the channel has had time to recharge, will not create a cascade and will, therefore, not be detected. The time between channel discharge and recharge is typically on the order of nanoseconds and is sometimes called the detector dead time. Because of their inherent dead time, multichannel plate detectors saturate easily and this can severely limit their dynamic range relative to other detection devices. Multichannel plate detectors also have limited lifetimes, as the detector surface becomes deactivated and because it is necessary to operate the MCP at relatively high voltage across as short space, detector arcing is possible if operated at too high a pressure ($\geq 10^{-7}$ Torr).

## D. Saturation Effects

Practical understanding and appreciation of detector saturation limits are critical when using any of the aforementioned detectors. The saturation limit can be defined as the point at which a nonlinear detector response is observed with an increase in ion concentration. The saturation characteristics of a detector can be proportional to concentration up to a critical point, at which no additional signal can be obtained, or can be nonlinear, where detector response changes with concentration but in a nonlinear way.

    Detector saturation can effect both quantitative and qualitative data analysis, and each of these effects should be appreciated. The effect on sample quantitation is intuitive, where for instance a twofold increase in sample concentration produces a less than twofold increase in response. This will cause a flattening of calibration curves at higher concentrations. For API techniques, source saturation (or *ion suppression*) is another source of response saturation independent of detector saturation. Detector saturation can also effect qualitative measurements such as mass accuracy and isotope ratio calculations. In the former, when a mass spectral peak that has some finite resolution starts to saturate the detector the peak-top calculations that provide the *m/z* measurement of the peak will become ambiguous. Likewise, it is possible that as one isotope of an ion starts to saturate the detector, adjacent isotopes in the distribution will still provide a linear response. The result of this is that incorrect isotope ratios will be obtained. Changes in relative isotope ratios of individual spectra across a chromatographic peak is an indicator of possible detector saturation.

## V. DATA COLLECTION SYSTEMS AND INSTRUMENT CONTROL

Modern mass spectrometer systems rely heavily on integrated computer control systems for instrument control, data collection, and data processing. Whereas early mass spectrometers were characterized by manual instrument control and analog data collection systems, such as oscilloscopes or chart recorders, modern systems increasingly provide comprehensive, integrated control of both the instrument hardware and data collection components.

## A. Specialized Software

Most systems utilize personal computers running instrument control software locally, and this software often contains specialized applications and postprocessing capabilities. Examples include specialized software for quantitative analysis (peak detection, integration, standard calibration, unknown quantitation, and data reporting), protein characterization and proteomics, analysis of combinatorial chemistry libraries, automated chromatographic peak fraction collection, metabolite identification, and automated instrument optimization. As the utility of mass spectrometers to solve an ever-increasing array of scientific questions continues to grow, specialized software will provide advantages for ease of use and efficiency, particularly as the tools move from the hands of mass spectrometry specialists to scientists outside the field of analytical chemistry. Some novel applications include the development of both automated processing and reporting of data via e-mail [74–76] or the construction of customized mass spectral databases [77,78]. First-generation semiautomated method development software for quantitative (MRM) applications has recently become available as have qualitative applications such as metabolite structural elucidation [79,80]. These types of applications will undoubtedly expand as the bottleneck of decision making in drug discovery continues to become data reduction rather than sample processing and data collection.

## B. System Networking

In today's multiuser, open access environment, the computer connected to the back end of the mass spectrometer has become so heavily utilized in instrument control and data collection that it is rarely available for postcollection data reduction. Computer networking has become a tool to remedy this situation. Individual mass spectrometer systems are controlled by dedicated desktop computers running control software locally. Data acquired from individual instruments can be stored directly in a network server through a network link, with each data system collecting to a unique signature directory. Once stored to a centralized server, retrieval of archived information is possible through any networked computer system running compatible software. This configuration provides a number of advantages, including efficient data archiving and backup storage, consistent data collection from numerous sources, instantaneous retrieval of archived data from stand-alone computer systems in remote site locations, and data retrieval concurrent with data collection. Back-end computers can then be dedicated to instrument control and data collection, and better instrument utilization and efficiency is thereby achieved.

## REFERENCES

1. American Society for Mass Spectrometry. What Is Mass Spectrometry? 3rd ed. J Am Soc Mass Spectrom, 1998.
2. SA Lammert. 1999 directory of mass spectrometry manufacturers and suppliers. Rapid Commun Mass Spectrom 13:831–844, 1999.
3. DA Skoog. Principles of Instrumental Analysis. 3rd ed. Philadelphia: Saunders College, 1985, pp 544–546.
4. FW McLafferty, F Turcek. Interpretation of Mass Spectra. 4th Ed. Mill Valley, CA: University Science Books, 1993.
5. JT Watson. Introduction to Mass Spectrometry. 3rd ed. Philadelphia: Lippincott–Raven, 1997.
6. R Willoughby, E Sheehan, S Mitrovich. A Global View of LC/MS. 1st ed. Pittsburgh, PA: Global View, 1998.
7. KL Busch, GL Glish, SA McLuckey. Mass Spectrometry: Techniques and Applications of Tandem Mass Spectrometry. New York: VCH, 1988.
8. E DeHoffmann, J Charette, V Straoobant. Chemical ionization. In: Mass Spectrometry: Principles and Applications. Paris: Wiley, 1996.
9. AM Lawson, ed. Mass Spectrometry. Berlin: de Gruyter, 1989.
10. DH Russell. Experimental Mass Spectrometry. New York: Plenum Press, 1994.
11. DM Desiderio. Mass Spectrometry. New York: Plenum, 1992.
12. AL Burlingame, SA Carr. Mass Spectrometry in the Biological Sciences. Totowa, NJ: Humana Press, 1996.
13. SR Heller, A McCormick, T Sargent. Compound identification by computer matching mass spectra. In: GR Waller, OC Dermer, ed. Biochemical Applications of Mass Spectrometry. New York: Wiley, 1980, pp 103–124.
14. P Ausloos, CL Clifton, SG Lias, A Mikaya, SE Stein, DV Tchekhovskoi, OD Sparkman, V Zaikin, D Zhu. The critical evaluation of a mass spectral library. J Am Soc Mass Spectrom 10:287–299, 1999.
15. FW McLafferty, DA Stauffer, SY Loh, C Wesdemoitis. Unknown identification using reference mass spectra: Quality evaluation of databases. J Am Soc Mass Spectrom 10:1229–1240, 1999.
16. AG Harrison. Chemical Ionization Mass Spectrometry. 2nd ed. Boca Raton, FL: CRC Press, 1992.
17. RC Beavis, BT Chait. Matrix-assisted laser desorption ionization mass spectrometry of proteins. Methods Enzymol 270:519–551, 1996.
18. F Hillenkamp, M Karas, RC Beavis, BT Chait. Matrix-assisted laser desorption/ionization mass spectrometry of biopolymers. Anal Chem 63:1193A–1203A, 1991.
19. P Chaurand, M Stoeckli, RM Caprioli. Direct profiling of proteins in biological tissue sections by MALDI mass spectrometry. Anal Chem 71: 5263–5270, 1999.

20. R Zenobi, R Knochenmuss. Ion formation in MALDI mass spectrometry. Mass Spectrom Rev 17:337–366, 1999.
21. E Sugiyama, A Hara, K-I Uemura. A quantitative analysis of serum sulfatide by matrix-assisted laser desorption ionization time-of-flight mass spectrometry with delayed ion extraction. Anal Biochem 274:90–97, 1999.
22. Y-C Ling, L Lin, Y-T Chen. Quantitative analysis of antibiotics by matrix-assisted laser desorption/ionization time-of-flight mass spectrometry. Rapid Commun Mass Spectrom 12:317–327, 1998.
23. AJ Nicola, AI Gusev, DM Hercules. Direct quantitative analysis from thin-layer chromatography plates using matrix-assisted laser desorption/ionization time-of-flight mass spectrometry. Appl Spectrosc 50:1479–1482, 1996.
24. H Wollnik. Ion optics in mass spectrometers. J Mass Spectrom 34:991–1006, 1999.
25. WMA Niessen. Liquid Chromatography—Mass Spectrometry. 2nd ed. New York: Marcel Dekker, Inc., 1999.
26. RB Cole. Electrospray Ionization Mass Spectrometry. New York: Wiley, 1997.
27. P Kebarle, L Tang. From ions in solution to ions in the gas phase: The mechanism of electrospray mass spectrometry. Anal Chem 65:972A–986A, 1993.
28. DK Bohme, TD Mark. ed. Special issue: Multiply-charged ions. Int J Mass Spectrom 192:1–451, 1999.
29. M Scalf, MS Westphall, J Krause, SL Kaufman, LM Smith. Controlling charge states of large ions. Science 283:194–197, 1999.
30. SA McLuckey, JL Stephenson. Ion/ion chemistry of high-mass multiply charged ions. Mass Spectrom Rev 17:369–407, 1999.
31. JA Loo. Studying noncovalent protein complexes by electrospray ionization mass spectrometry. Mass Spectrom Rev 16:1–23, 1997.
32. J Sunner, G Nichol, P Kebarle. Factors effecting the relative sensitivity of analytes in positive mode APCI. Anal Chem 60:1300–1307, 1988.
33. RE Finnigan. Quadrupole mass spectrometers: From development to commercialization. Anal Chem 66:969A–975A, 1994.
34. PH Dawson. Quadrupole Mass Spectrometry and Its Applications. Amsterdam: Elsevier, 1976.
35. K Heinig, J Henion. Fast liquid chromatographic-mass spectrometric determination of pharmaceutical compounds. J Chrom B Biomed Sci Appl 732:445–458, 1999.
36. AG Marshall. Fourier transform ion cyclotron resonance mass spectrometry. Acc Chem Res 18:316–322, 1985.
37. AG Marshall, FR Verdun. Fourier Transforms in NMR, Optical, and Mass Spectrometry. Amsterdam: Elsevier, 1990.
38. J-LM Abboud, R Notario. FT ICR: Basic principles and some representative applications. NATO Sci Ser Ser C 535:281–302, 1999.
39. JJ Drader, SD-H Shi, GT Blakney, CL Hendrickson, DA Laude, AG Marshall. Digital quadrature heterodyne detection for high-resolution Fourier transform ion cyclotron resonance mass spectrometry. Anal Chem 71:4758–4763, 1999.
40. SA Lorenz, EP Maziarz, TD Wood. Electrospray ionization Fourier transform mass spectrometry of macromolecules: the first decade. Appl Spectrosc 53:18A–36A, 1999.
41. S Lorenz, M Moy, AR Dolan, TD Wood. Electrospray ionization Fourier transform

mass spectrometry quantification of enkaphalin using an internal standard. Rapid Commun Mass Spectrom 13:2098–2102, 1999.

42. H Budzikiewicz, T Cvitas, S Kazazic, L Klasinc, D Srzic. Gas phase reaction rate measurements in Fourier transform mass spectrometry. Rapid Commun Mass Spectrom 13:1109–1111, 1999.

43. JFJ Todd, RE March. A retrospective review of the development and application of the quadrupole ion trap prior to the appearance of commercial instruments. Int J Mass Spectrom 190/191:9–35, 1999.

44. RE March. Advances in quadrupole ion trap mass spectrometry: instrument development and applications. Adv Mass Spectrom 14:241–278, 1998.

45. RG Cooks, RE Kaiser. Quadrupole ion trap mass spectrometry. Acc Chem Res 23: 213–219, 1990.

46. RG Cooks, GL Glish, SA McLuckey, RE Kaiser. Ion trap mass spectrometry. Chem Eng News March 25:26–41, 1991.

47. JFJ Todd. Ion trap mass spectrometer—past, present, and future. Mass Spec Rev 10:3–52, 1991.

48. DM Chambers, DE Goeringer, SA McLuckey, GL Glish. Matrix-assisted laser desorption of biological molecules in the quadrupole ion trap mass spectrometer. Anal Chem 65:14–20, 1993.

49. DE Goeringer, GL Glish, SA McLuckey. Fixed-wavelength laser ionization/tandem mass spectrometry for mixture analysis in the quadrupole ion trap. Anal Chem 63: 1186–1192, 1991.

50. SA McLuckey, GJ Van Berkel, DE Goeringer, GL Glish. Ion trap mass spectrometry of externally generated ions. Anal Chem 66:689A–696A, 1994.

51. ST Quarmby, RA Yost. Fundamental studies of ion injection and trapping of electrosprayed ions on a quadrupole ion trap. Int J Mass Spectrom 190/191:81–102, 1999.

52. GJ Van Berkel, GL Glish, SA McLuckey. Electrospray ionization combined with ion trap mass spectrometry. Anal Chem 62:1284–1295, 1990.

53. JFJ Todd, AD Penman. The recent evolution of the quadrupole ion trap mass spectrometer—an overview. Int J Mass Spectrom Ion Processes 106:1–20, 1991.

54. GC Stafford, PE Kelley, JEP Syka, WE Reynolds, JFJ Todd. Recent improvements in and analytical applications of advanced ion trap technology. Int J Mass Spectrom Ion Processes 60:85–98, 1984.

55. X Jin, J Kim, S Parus, DM Lubman, R Zand. On-line capillary electrophoresis/microelectrospray ionization tandem mass spectrometry using an ion trap storage/time-of-flight mass spectrometer with SWIFT technology. Anal Chem 71:3591–3597, 1999.

56. DE Goeringer, KG Asamo, SA McLuckey, D Hoekman, SW Stiller. Filtered noise field signals for mass-selective accumulation of externally-formed ions in a quadrupole ion trap. Anal Chem 66:313–318, 1994.

57. RW Vachet, SW McElvany. Application of external customized waveforms to a commercial quadrupole ion trap. J Am Soc Mass Spectrom 10:355–359, 1999.

58. SA McLuckey, GL Glish, GJ Van Berkel. Multiple stages of mass spectrometry in a quadrupole ion trap mass spectrometer: prerequisites. Int J Mass Spectrom Ion Processes 106:213–235, 1991.

59. DB Gordon, MD Woods. Enlarging the stability domain for ion trap tandem mass spectrometry (MS/MS). Rapid Comm Mass Spec 7:215–218, 1993.

60. RJ Cotter. The new time-of-flight mass spectrometry. Anal Chem 71:445A–451A, 1999.

61. RJ Cotter. Time-of-Flight Mass Spectrometry: Instrumentation and Applications in Biological Research. Washington, DC: American Chemical Society, 1997.

62. VM Doroshenko, RJ Cotter. Ideal velocity focusing in a reflectron time-of-flight mass spectrometer. J Am Soc Mass Spectrom 10:992–999, 1999.

63. LL Haney, DE Riederer. Delayed extraction for improved resolution of ion/surface collision products by time-of-flight mass spectrometry. Anal Chim Acta 397:225–233, 1999.

64. E Gleitsmann, M Karas. Computer simulation and practical performance of DE MALDI-TOF mass spectrometer. Adv Mass Spectrom 14:B073480/1–B073480/10, 1998.

65. BD Gardner, JF Holland. Nonlinear ion acceleration for improved space focusing in time-of-flight mass spectrometry. J Am Soc Mass Spectrom 10:1067–1073, 1999.

66. IV Chernushevich, W Ens, KG Standing. Orthogonal-injection TOFMS for analyzing biomolecules. Anal Chem 71:452A–461A, 1999.

67. JH Beynon, RM Caprioli. Metastable ions as an aid in the interpretation of mass spectra. In: GR Waller, OC Dermer, eds. Biochemical Applications of Mass Spectrometry. New York: Wiley, 1980, pp 89–102.

68. T Pfeifer, M Drewello, A Schierhorn. Using a matrix-assisted laser desorption/ionization time-of-flight mass spectrometer for combined in-source decay/post-source decay experiments. J Mass Spectrom 34:644–650, 1999.

69. ST Fountain, DM Lubman. Wavelength-specific resonance enhanced multiphoton ionization for isomer discrimination via fragmentation and metastable analysis. Anal Chem 65:1257–1266, 1993.

70. P Michelsen, AA Karlsson. Accurate mass determination of a metabolite of a potential diagnostic imaging drug candidate by high performance liquid chromatography with time-of-flight mass spectrometry. Rapid Commun Mass Spectrom 13:2146–2150, 1999.

71. N Zhang, ST Fountain, H Bi, DT Rossi. Quantification and rapid metabolite identification in drug discovery using atmospheric pressure ionization time-of-flight LC/MS. Anal Chem 72:800–806, 2000.

72. F Dubois, R Knochenmuss, R Zenobi. Optimization of an ion-to-photon detector for large molecules in mass spectrometry. Rapid Commun Mass Spectrom 13:1958–1967, 1999.

73. JA Loo, R Pesch. Sensitive and selective determination of proteins with electrospray ionization magnetic sector mass spectrometry and array detection. Anal Chem 66:3659–3663, 1994.

74. H Tong, D Bell, K Tabei, MM Siegel. Automated data massaging, interpretation, and e-mailing modules for high throughput open access mass spectrometry. J Am Soc Mass Spectrom 10:1174–1187, 1999.

75. G Perkins, F Pullen, C Thompson. Automated high-resolution mass spectrometry for the synthetic chemist. J Am Soc Mass Spectrom 10:546–551, 1999.

76. WA Korfmacher, CA Palmer, C Nardo, K Dunn-Meynell, D Grotz, K Cox, C-C

Lin, C Elicone, C Liu, E Duchaslav. Development of an automated mass spectrometry system for the quantitative analysis of liver microsomal incubation samples: a tool for rapid screening of new compounds for metabolic stability. Rapid Commun Mass Spectrom 13:901–907, 1999.

77. JR Yates, SF Morgan, CL Gatlin, PR Griffin, JK Eng. Method to compare collision-induced dissociation spectra of peptides: potential for library searching and subtractive analysis. Anal Chem 70:3557–3565, 1998.

78. W Weinmann, A Wiedemann, B Eppinger, M Renz, M Svoboda. Screening of drugs in serum by electrospray ionization/collision-induced dissociation and library searching. J Am Soc Mass Spectrom 10:1028–1037, 1999.

79. TA Baillie, PG Pearson. The impact of drug metabolism in contemporary drug discovery: new opportunities and challenges for mass spectrometry. In: AL Burlingame, SA Carr, MA Baldwin, eds. Mass Spectrometry in Biology and Medicine. Totowa, NJ: Humana Press, 1999.

80. GK Poon, G Kwei, R Wang, K Lyons, Q Chen, V Didolkar, CECA Hop. Integrating qualitative and quantitative liquid chromatography/tandem mass spectrometric analysis to support drug discovery. Rapid Commun Mass Spectrom 13:1943–1950, 1999.

# 4
# Ion Chemistry and Fragmentation

**Richard A. J. O'Hair**
*University of Melbourne, Victoria, Australia*

> Progress in mass spectrometry has been so rapid since 1939 that one can
> hardly predict even the immediate future. . . . Many present difficulties will
> be overcome when apparatus with better design and more simple operation
> is generally available, but even before the perfect spectrometer is built, the
> chemist will find that this tool of the physicist has already earned a place in
> the chemical laboratory.
>
> —*David W. Stewart*, Physical Methods of Organic Chemistry,
> Volume II, *Chapter 26, 1946.*

## I. INTRODUCTION

Stewart's quote, from 1946, was quite prophetic: Mass spectrometry has certainly
become one of the important structural tools for chemists and is increasingly
becoming an important tool for those in the biological sciencies, including those
whose aim is drug discovery.

Regardless of the application, it is important to remember that mass spectrometers are nothing more than gas-phase reactors in which the unimolecular
fragmentation reactions and/or bimolecular reactions of ionic species can be observed. Thus, a given mass spectrum is a unique record of all the events (ionization, "in source" reactions, and postsource reactions) prior to mass analysis and
detection. It stands to reason that if we fully understand the gas-phase chemistry
of ions, then we should be able to understand (and ultimately predict) mass spectra. Indeed, ever since the advent of "organic mass spectrometry," a long-term
goal of analytical mass spectrometrists has been the ability to effectively and
efficiently predict the fragmentation pathways for any compound with a given

**85**

molecular structure. Despite impressive recent attempts such as the Mass Frontier™ 1.0 program by HighChem [1], no expert computer software program has yet been designed to predict the complete mass spectrum (positive or negative ion mode, including the molecular ion and all fragment ions together with their relative abundance) for all compounds under any ionization condition. This goal may in fact never be completely achievable for a number of reasons, including the following:

1. The cornucopia of ionization methods [electron impact ionization (EI), chemical ionization (CI), field desorption (FD), fast atom bombardment (FAB), electrospray ionization (ESI), and matrix-assisted laser desorption/ionization (MALDI)] yield gas-phase ions for mass analysis via quite different physical and chemical processes. Unfortunately, many of these processes are still not well understood.
2. Even for a given ionization mode, the ion source is "manufacturer specific," which can lead to different "in source" reactions. For example, desolvation in ESI is achieved in different ways by different manufacturers.
3. Key physical organic data, such as the thermochemistry and kinetics of gas-phase unimolecular and bimolecular reactions, are often unknown.
4. Postsource reactions are governed by a number of factors such as the internal energy of the ions, the time between exit of ions from the source to mass analysis, and the pressure of the mass analyzer. Since there is no ideal mass analyzer [2], a range of analyzers are commercially available which have been interfaced with most of the common ionization methods. Each mass analyzer has unique properties which can influence the actual mass spectrum observed.

Regardless of this seemingly gloomy view, future programs for qualitative or quantitative prediction of mass spectra must be based on sound principles of "gas-phase" physical organic chemistry. With this in mind, the aim of this chapter is to provide analytical mass spectrometrists with a brief "tutorial" in ion chemistry and fragmentation reactions.

## II. SCOPE OF THE CHAPTER

A comprehensive review of the gas-phase ion–molecule reactions and fragmentation reactions of compounds of direct interest to those working in the area of drug discovery is beyond the scope of this chapter. Rather, our main aim is to use selective examples of simple biomolecules which help illustrate how certain tools can be used to unravel the mechanisms of gas-phase ion reactions and to illustrate some key concepts. Both positive and negative ions are considered because these often yield complementary information. Finally, those interested in the latest analytical developments in drug discovery can turn to the biannual *Analytical Chemistry Reviews in Mass Spectrometry* [3].

## III. THERMOCHEMISTRY AND KINETICS IN THE GAS PHASE

The unimolecular and bimolecular reactions of gas-phase ions that occur both during the ionization process and while the ions are inside the mass spectrometer are controlled by both thermochemistry [4] and kinetics [5]. Some key thermochemical quantities associated with cations and anions are presented below.

### A. Thermochemistry

#### 1. Thermochemistry of Cations

In mass spectrometry, the main types of cations arising from ionization of a neutral molecule M are radical cations $M^{+\bullet}$ (typically formed via EI) and protonated ions $[M + H]^+$ (formed in several ionization processes, including CI, FAB, ESI, and MALDI). The energies associated with the formation of these gas-phase ions are the ionization potential and the proton affinity, respectively. These quantities are defined in more detail below.

*Ionization Potential.* Ionization potential (IP) is defined as the energy required to remove an electron from a molecule or atom [Eq. (1)]. A distinction between adiabatic and vertical IPs is required. The *adiabatic IP* is the lowest energy required to effect the removal of an electron from a molecule or atom and corresponds to the transition from the lowest electronic, vibrational, and rotational level of the isolated molecule to the lowest electronic, vibrational, and rotational level of the isolated ion. Thus an adiabatic IP can be used in conjunction with the enthalpy of formation of the neutral (M) to derive the enthalpy of formation of the ion, $M^{+\bullet}$. In contrast, the *vertical IP* is the energy change corresponding to an ionization reaction leading to formation of the ion in a configuration which is the same as that of the equilibrium geometry of the ground-state neutral molecule. Note that from the Franck–Condon principle, when a molecule is ionized by photoionization or by interaction with energetic electrons, the highest probability configuration of the resulting ion will be the configuration of the precursor neutral molecule. Thus vertical IP > adiabatic IP.

$$M \rightarrow M^{+\bullet} + e^- \quad \Delta H_{rxn} = IP\,(M) \tag{1}$$

*Proton Affinity.* The formal relationship between the enthalpy of formation of $[M + H]^+$ and its neutral counterpart, M, is defined in terms of a quantity
called the proton affinity, PA [Eq. (2)]. Note that unlike the adiabatic IP, which
represents the 0 K enthalpy change, the PA is a quantity defined at a finite temperature (typically 298 K) as follows:

$$M + H \rightarrow [M+H]^+ \quad \Delta H_{rxn} = \text{-PA (M)} \tag{2}$$

$$\Delta H_f^\circ([M+H]^+) = \Delta H_f^\circ(M) + \Delta H_f^\circ(H^+) - PA$$

## 2.  Thermochemistry of Anions

In mass spectrometry, the main types of anions arising from ionization of a neutral
molecule M are deprotonated ions, $[M - H]^-$. Key energies associated with the
formation and stability of these gas-phase anions are the anion proton affinity
(APA) and the electron affinity (EA).

*Anion Proton Affinity.* This enthalpy change measures the ease of losing
a proton from a compound [Eq. (3)] and is related to the inherent acidity of a
gas-phase molecule. It is usually defined at 298 K as follows:

$$M \rightarrow [M - H]^- + H^+ \quad \Delta H_{rxn} = \Delta H_{acid} (M) \tag{3}$$

*Electron Affinity.* Whether the resultant $[M - H]^-$ anion formed from
loss of a proton [Eq. (3)] is actually stable in the gas phase with respect to loss
of an electron depends on the the negative ion equivalent of IP, which is termed
the electron affinity and is defined by Eq. (4). As with the IP, EAs represent the
0 K enthalpy change and it is possible to have either vertical or adiabatic EAs.
A major difference from IPs is that for stable ("bound") negative ions, the ion
is lower in energy than the corresponding neutral. If the negative ion is higher
in energy than the neutral, the neutral is said to have a negative electron affinity,
and the ion will undergo spontaneous loss of the electron. An important consequence from an analytical perspective is that any species which has a negative
EA will not be observed in a negative ion mass spectrum.

$$[M - H]^- \rightarrow [M - H]^\bullet + e^- \quad \Delta H_{rxn} = EA (M - H) \tag{4}$$

## B.  Kinetics

Even though a reaction may be favored by thermochemistry, this does not mean
that it will occur inside the mass spectrometer. The rates of gas-phase reactions
are governed by the energy barriers on their potential energy surfaces. Furthermore, whether a product ion is observed in the mass spectrometer will also be
dependent on the "time frame" of the mass spectrometry experiment relative to
the rate of the reaction.

## 1. Unimolecular Rates (k in s$^{-1}$)

The ionization of a neutral sample can result in the formation of molecular ions with a wide range of internal energies (especially in EI). As a result, some of the molecular ions can have sufficient energy to undergo fragmentation via one or more processes. The extent of the competition between the available dissociation process is determined by the value of the rate constant ($k$) for each process at a given internal energy ($E$). A simplified expression for $k$ is given by Eq. (5) [11a]. Considerable effort has gone into modeling the rates of these unimolecular reactions using theories such as Rice, Ramsperger, Kassel, Marcus (RRKM), and recent efforts have focused on large polyatomic ions [6].

For EI on a magnetic sector, $k > 10^{-6}$ generally occurs in the source while $10^{-6} > k > 10^{-4}$ generally occurs outside the source region, giving rise to "metastable" ions. There are two main types of fragmentation reactions, as illustrated in Fig. 1: (a) simple bond cleavage (Fig. 1a), which has a small or no reverse activation barrier; and (b) rearrangement, which has a reverse activation barrier

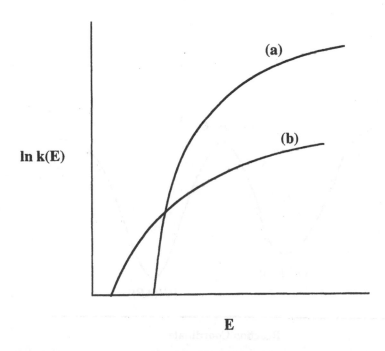

**Figure 1** Comparison of two hypothetical fragmentation reactions. These are plotted as the ln of the rate constant [Eq. (5)] as a function of the internal energy of the ion for (a) a simple cleavage reaction and (b) a rearrangement reaction.

(Fig. 1b). For rearrangement reactions to compete with simple bond cleavages they must have low $E_0$.

$$k = \upsilon \left( \frac{E - E_0}{E} \right)^{s-1}, \tag{5}$$

where $k$ is the rate constant, $\upsilon$ is the frequency factor, $E$ is the internal energy, $E_0$ is the activation energy, and $s$ is the effective number of oscillators.

## 2. Bimolecular Rates ($k$ in $cm^3$ molecule$^{-1}$ s$^{-1}$)

Gas-phase ion–molecule reactions often proceed via potential energy surfaces with multiple wells [7a], involving the intermediacy of various species including ion–neutral complexes [8]. A hypothetical example showing a "double well" potential is shown in Fig. 2. Note that the initial collision complex between an ion and a neutral (exemplified by $[A^+(BC)]$ in Fig. 2) is more stable than the separated ion and neutral and this excess energy of association can be used to overcome reaction barriers, thus driving the formation of products. This is one of the main reasons

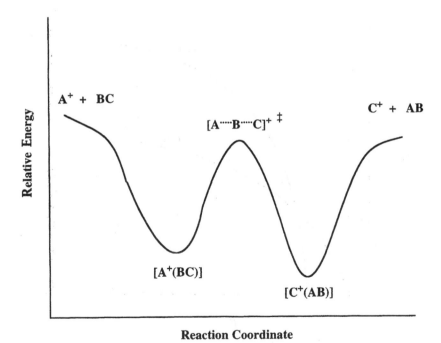

**Reaction Coordinate**

**Figure 2**  Double-well potential model for a hypothetical ion–molecule reaction: $A^+$ + BC → $C^+$ + AB. Note the reaction proceeds via a transition state and involves the intermediacy of ion–neutral complexes in both the entrance $[A^+ (BC)]$ and exit $[C^+ (AB)]$ channels.

why the rates of gas-phase ion–molecule reactions can be very fast compared to solution (where ions are solvated and where the solvent is used to thermalize intermediates). The overall reaction exothermicity of a gas-phase reaction can result in fragmentation of the product ions, which is common in proton transfer reactions in CI which involve highly acidic CI reagents such as $H_3^+$ [9].

Once again, considerable effort has gone into the theoretical modeling of ion–molecule reactions [7b] in order to establish the reaction efficiency (expressed as the experimental rate divided by the theoretical rate—thus an efficiency of 1 means that every collision between an ion and a neutral will result in the formation of products) of these reactions.

## IV. FRAGMENTATION REACTIONS OF IONS: USE OF TANDEM MASS SPECTROMETRY TO PROBE STRUCTURE

Apart from molecular weight information, the organic mass spectrometrist has always been interested in the use of fragment ions as indicators of structure. The mode of fragmentation is generally governed by the nature of the charge of the precursor ion (positive, negative, radical, or even electron ion). In terms of qualitative predictions of fragmentation modes, the best understood class of ions are radical cations of organic compounds formed under 70 eV EI conditions. These are not reviewed here because McLafferty's seminal book has been updated [10] and the Mass Frontier™ 1.0 program [1] provides a useful tool for examining these processes for any given compound. It is, nonetheless, worth reminding readers that the two main pathways are α *cleavages* directed by the radical site (termed α cleavages) and *charge site directed cleavages* (termed *inductive or i cleavages*).

### A. Tandem Mass Spectrometry Instrumentation and Activation Techniques

With the advent of "soft ionization methods" (CI, FAB, ESI, and MALDI), which produce pseudomolecular ions (e.g., protonated $[M + H]^+$ and deprotonated $[M - H]^-$ ions) with little or no fragmentation, came an interest in the use of mass spectrometric methods to examine fragmentation reactions [11]. There are two common ways this can be achieved:

1.  The ion is mass selected (isolated) using one stage of mass analysis and subjected to fragmentation, with the product ions being analysed via a second stage of mass analysis. This is called a tandem mass spectrometry (MS/MS) experiment and is discussed in more detail later.

2. The ion is not mass selected but rather subjected to fragmentation before mass analysis (a widespread current example is "in source" fragmentation under ESI conditions). A potential pitfall is discussed in
   further detail below.

The types of tandem mass spectrometers capable of performing MS/MS
experiments fall into two basic categories: tandem in space and tandem in time.
Tandem-in-space instruments have discrete mass analyzers for each stage of mass
spectrometry; examples include multisector, triple-quadrupole, and hybrid instruments (instruments having mixed types of analyzers such as a magnetic sector
and a quadrupole). Tandem-in-time instruments have only one mass analyzer
where each stage of mass spectrometry takes place in the same analyzer but is
separated in time via a sequence of events. Examples of this type of instrument
include Fourier transform ion cyclotron resonance (FT-ICR) mass spectrometers
and quadrupole ion traps, described in Chapter 3.

There are several ways of depositing energy into the mass selected ion to
cause it to fragment. These include (1) studying those ions which are long lived
and have sufficient internal energy from the ionization process to fragment after
the first stage of mass selection [these ions are called metastable and the resultant
MS/MS spectrum is called a metastable ion (MI) spectrum [12]]; (2) collisions
with a collision gas such as helium [this is known as collision-induced dissociation (CID), collisionally activated dissociation, or collisional activation [13]]; (3)
collisions with a surface (this is known as surface-induced dissociation [14]); (4)
interaction with photons (e.g., via a laser), resulting in photofragmentation [15];
(5) interaction with an electron beam, resulting in electron-induced dissociation
for singly charged ions [16] or electron capture dissociation when multiply
charged cations are allowed to interact with low-energy electrons [17]; and (7)
thermal/black body infrared radiation dissociation (BIRD) [18]. By far the most
common MS/MS technique used for organic and biomolecule ions has been CID.

## B.  Charge Remote versus Charge Directed
## Fragmentation Reactions

The term *charge remote* reaction was introduced by Gross to describe those fragmentation reactions which are essentially uninfluenced by the charge (i.e., the
charge acts as a "spectator"). Such reactions generally occur in the high-energy
CID (keV) regime of sector instruments. Under the "slow heating" conditions
[11b] of low-energy CID experiments (ion traps, triple quadrupoles, and FT-
ICRs), *charge-directed fragmentation* reactions often dominate. As these types
of experiments are increasingly becoming the main type of MS/MS experiments
routinely carried out, we focus our efforts on understanding key concepts associated with these processes. Those interested in charge remote fragmentations are
directed to several useful reviews [19].

## C. Physical Organic Concepts and Tools Used to Unravel the Mechanisms of Charge-Directed Fragmentation Processes

### 1. The "Bowie Rules"

No review has thoroughly addressed the key concepts and tools required to unravel the mechanisms of charged-directed fragmentation reactions for all charge types. Perhaps the best guidelines are for organic negative ions, as expounded by Bowie's empirical fragmentation rules [20]. The genesis of the rules are numerous high-energy CID studies with careful isotopic labeling on $[M - H]^-$ ions formed via CI using the strong bases $HO^-$ and $NH_2^-$. Briefly stated they are as follows:

1.  Simple homolytic cleavage reactions where loss of a radical forms a stable radical anion.
2.  Reactions that occur by initial formation of an anion-neutral complex that could then undergo a variety of reactions, including direct displacement of the anion as well as reactions induced by the anion (e.g., deprotonation and elimination).
3.  Reactions that do not occur via the first formed deprotonated species, but where intramolecular proton transfer (usually endothermic) leads to a new deprotonated species which fragments via an anion-neutral complex.
4.  Rearrangement reactions, including internal nucleophilic substitution/displacement and skeletal rearrangement reactions.

These rules contain several key concepts which are readily transferable to other charged systems such as protonated species (i.e., $[M + H]^+$ ions). Before discussing these and other key concepts, several important tools used to help unravel gas-phase ionic fragmentation mechanisms are described.

### 2. Tools

In order to probe the mechanisms of the fragmentation reactions of gas-phase ions, it is necessary to use several of the tools listed in Table 1. Each of these tools are further discussed below.

*MS/MS Experiments Coupled with Labeling.*   These experiments are the mainstays for the determination of gas-phase fragmentation mechanisms. Often there are similar functional groups in molecules and thus it is important to differentiate between them using either isotopic or structural labeling.

ISOTOPIC LABELING.   Although isotopes are of value for the measurement of isotope effects in the gas phase [21], mostly they have been used in labeling studies to distinguish between various mechanisms. Such studies have proven to be extremely valuable in determining fragmentation mechanisms of organic ions and ions derived from simple amino acids and peptides.

**Table 1**   Tools Available to Examine Gas Phase Fragmentation Mechanisms

MS/MS experiments coupled with labeling
   Isotopic labeling
   Structural labeling (e.g., convert carboxylic acid group into ester)
MS$^n$ experiments
   Comparison of MS$^3$ spectra of product ions with MS/MS spectra of proposed
      product ion structure formed via independent synthesis
   Gas-phase ion molecule reaction of product using an MS$^3$ experiment
   Probes of the structures of neutrals: NRMS and NFR using an MS$^3$ experiment
Comparison of activation energies for competing fragmentation processes
   Energy resolved CID
   Measurement of dissociation kinetics
Theoretical modeling techniques
   For example, use of *ab initio* calculations to gain insights into the structures and
      energies of reactants, transition states, intermediates, and products

STRUCTURAL LABELING.   A common structural label is to replace a hydrogen atom attached to a heteroatom such as N, O, or S with a methyl group (e.g., convert carboxylic acid group into ester). This can not only help elucidate mechanisms in a similar fashion to isotope labeling, but in some instances can also be used to "switch off" reactions or induce new reactions.

*MS$^n$ Experiments: Comparison of MS$^3$ Spectra of Product Ions with MS/MS Spectra of Proposed Product Ion Structures Formed via Independent Sythesis.*   With the development of tandem-in-time instruments such as FT-ICRs and ion traps, multistage MS experiments are becoming routine. Thus, the structures of CID product ions can be interrogated via CID using a further stage of MS. In some instances, if a suitable independent synthesis can be achieved, the resultant MS$^3$ spectrum can be compared with the MS/MS spectra of ions of known structure, thereby facilitating ion structure assignment.

*Probes of the Structures of Neutrals: NRMS and NFR Using an MS$^3$ Experiment.*   Most tandem mass spectrometry experiments that cause fragmentation of a mass selected ion result in the formation of both ionic and neutral fragments. Generally, only the ionic fragments are detected in conventional MS/MS experiments and potentially relevant structural information on the coproduced neutral(s) is lost. This problem can be overcome through the use of neutral fragment reionization (NFR) mass spectrometry, which utilizes a tandem mass spectrometer with two collision cells between each stage of mass spectrometry. Applications of these techniques in structure and mechanism determination have recently been reviewed [22].

*Thermochemical Considerations: Comparison of Activation Energies for Competing Fragmentation Processes.* Inducing gas-phase fragmentation reactions of low-energy protonated ions requires these ions to be "heated up." The resultant energy is used to overcome activation barriers associated with bond cleavage. While a considerable body of thermochemical data exists for ions and neutrals that can often be used to predict the overall reaction thermochemistry, activation energies must be obtained via experiment, using either of the two techniques described below.

ENERGY RESOLVED CID VIA GUIDED ION BEAM EXPERIMENTS. Guided ion beam experiments have been used to measure a wealth of gas-phase thermochemistry [23]. Because these experiments measure thresholds for fragmentation, they can be probes of the activation energy [24].

MEASUREMENT OF DISSOCIATION KINETICS VIA BIRD EXPERIMENTS. In trapping instruments, fragmentation reactions can be studied as a function of both time as well as temperature of the cell. An Arrhenius analysis of the resultant unimolecular rate constants as a function of temperature allows activation energies to be determined for each fragmentation process [18].

*Theoretical Modeling Techniques.* Another important tool for understanding the structure and reactivity of gas-phase ions is through the judicious use of theoretical modeling techniques. An array of theoretical modeling techniques are available, including molecular modeling, semiempirical techniques, as well as ab initio molecular orbital theory. It is beyond the scope of this chapter to describe the merits and weaknesses of each of these techniques in detail. Suffice it to say that with advances in computational hardware and software, it is now possible to carry out theoretical modeling on biomolecules. The ultimate aim is to use ab initio calculations to gain insights into the structures and energies of reactants, transition states, intermediates, and products.

## 3.  Key Gas-Phase Physical Organic Concepts

Given that we are interested in fragmentations that are initiated by or directly involve the charge site, we need to discuss the key concepts in Table 2 in detail. Note that gas-phase fragmentation reactions tend to be under kinetic control rather than thermodynamic control and thus for competing processes, the relative transition state barrier heights are important. Also note that even electron ions generally fragment via heterolytic bond cleavage (inductive cleavages) rather than homolytic bond cleavage [10].

*Effects of Protonation or Deprotonation.*
CONCEPT OF "LOCAL PROTON AFFINITIES" AND "LOCAL ACIDITIES." In complex organic molecules or biomolecules, there are multiple functional groups and thus multiple sites of protonation or deprotonation. As is shown below, pro-

**Table 2**  Important Gas-Phase Physical Organic Concepts to
Help Rationalize Charge-Directed Fragmentation Reactions

Effects of protonation
  Site of protonation
  Proton mobility via intramolecular proton transfer
  Bond weakening versus strengthening upon protonation
Charge-directed fragmentation
  Leaving groups
  Neighboring-group participation
  Ion—molecule complexes

tonation or deprotonation at one site can weaken bonds, making them more susceptible to fragmentation. To a first approximation, we can model "local proton affinities" or "local acidities" by considering small model systems that contain the functional group of interest. The PA and APAs of simple systems should be readily available from the National Institute of Standards and Technology (NIST) database [4]. If we consider cysteine (Scheme 1), the order of local PA is $NH_2$ > $CO_2H$ > SH, while the order of local acidity is $CO_2H$ > SH > $NH_2$.

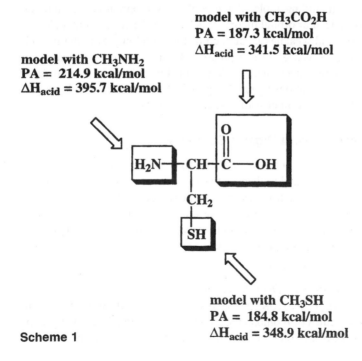

**model with $CH_3CO_2H$**
**PA = 187.3 kcal/mol**
**$\Delta H_{acid}$ = 341.5 kcal/mol**

**model with $CH_3NH_2$**
**PA =  214.9 kcal/mol**
**$\Delta H_{acid}$ = 395.7 kcal/mol**

**model with $CH_3SH$**
**PA =  184.8 kcal/mol**
**$\Delta H_{acid}$ = 348.9 kcal/mol**

**Scheme 1**

CHARGE MIGRATION AND THE MOBILE PROTON MODEL. Although experiments provide important data on the magnitude of the PA, they often cannot directly probe a key issue, viz. determining at which site protonation [25] or deprotonation occurs in molecules with several possible basic or acidic sites. This remains a challenging question to address experimentally because a way must be found to probe the site of protonation or deprotonation while avoiding the possibility of intramolecular proton transfer from one site to another. Several clever experiments have been developed to probe this issue. For negative ions, trimethylsilyl derivatives can be subjected to fluoride ion induced desilylation to "synthesize" specific anions in the gas phase. An example is the formation of the acetic acid enolate anion [26]. For positive ions, a combination of CI/MS and CI/MS/MS techniques with rigid [27] or conformationally restricted [28] species can shed light on protonation under condition of kinetic control. To avoid the possibility of intramolecular proton transfer, the protonation of para disubstituted benzenes **A** under CI conditions to produce $[M + H]^+$ ions **B** and **C** was examined (Scheme 2) [27]. Some of these $[M + H]^+$ ions decompose to form benzyl cations **D** and **E**, which reflect the site of protonation (**B** and **C**). The site of protonation in the remaining ions can be probed via MS/MS experiments.

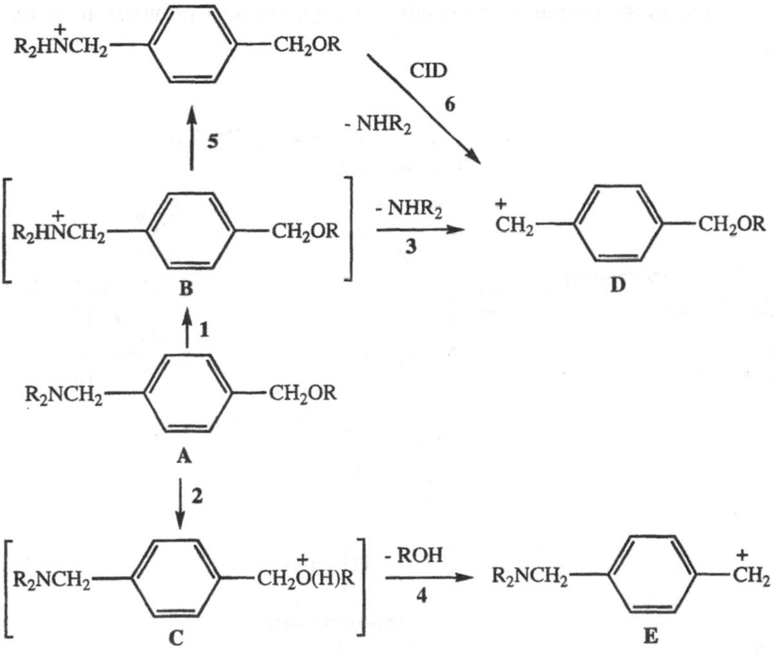

**Scheme 2**

    A similar methodology has been used to examine site of protonation and
proton mobility in *cis*- and *trans*-1-butyl-3- and -4-dimethylaminocyclohexanols
and their methyl ethers (Schemes 3 and 4) [28]. Thus, while intramolecular proton
transfer cannot occur between **G** and **H** in Scheme 3, such a process is possible
for **L** and **M**, procceding via **N** (Scheme 4).

    An alternative way of addressing this issue is to carry out theoretical model-
ing of the protonation process. If the theoretical modeling satisfactorily repro-
duces the experimentally observed PA (or APA) for the thermodynamically favored site of protonation (or deprotonation), then optimization of all the key
intermediates and transition states can provide valuable data on the energetics
for these processes. Even for a "simple" system such as glycine, a full and
detailed examination of the complete $[M + H]^+$ potential energy surface has not
yet been published. A significant complication is that protonation of glycine can
occur at three possible sites [the amino ($NH_2$) group, the carbonyl oxygen ($C=O$)
atom of the carboxyl group, or the hydroxyl oxygen (OH) atom of the carboxyl
group] and protonation at each site can yield a multitude of conformations. With
ever-increasing computational power and sophiscated molecular modeling programs, such studies are likely to appear.

    Perhaps one of the biggest successes in the use of MS/MS for structure
elucidation has been in the sequencing of peptides via CID of their $[M + H]^+$
ions. To gain the most sequence information for peptides, a population of iso-

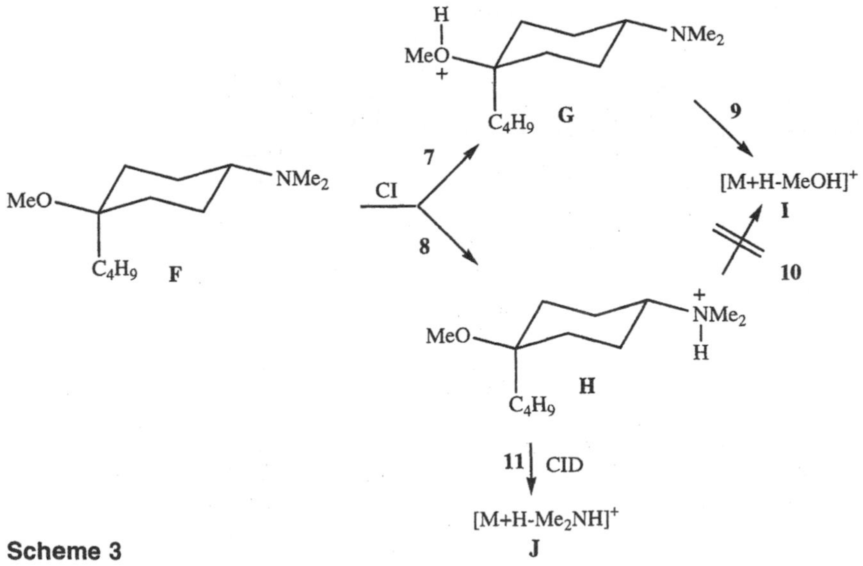

**Scheme 3**

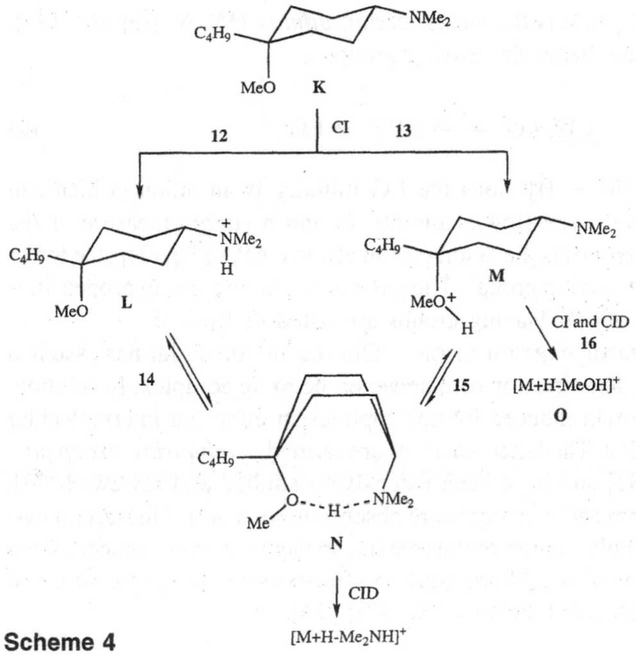

**Scheme 4**

meric $[M + H]^+$ ions in which protonation occurs at each of the sites required to trigger cleavage of each peptide bond. It is unlikely that the ionization process produces the required population of isomeric $[M + H]^+$ ions, but that such a population arises via CA-induced intramolecular proton transfer prior to fragmentation. This concept has been called the mobile proton model of fragmentation [29] (note that this is essentially a restatement of Bowie rule no. 3).

BOND WEAKENING VERSUS STRENGTHENING UPON PROTONATION OR DE-PROTONATION. Certain bonds are strengthened in a molecule upon protonation or deprotonation, while other bonds are weakened. A classic example involves protonation of the amide bond. When the carbonyl oxygen atom is protonated, it leads to a shortening and concomitant strengthening of the amide C-N bond. In contrast, N protonation of the amide bond results in lengthening and weakening of the amide C-N bond.

*Charge-Directed Fragmentation.* A key questions is this: Once we have the charge located at a particular site, how does the charge help induce fragmentation? A number of factors play a role.

LEAVING GROUPS. The likelihood of a specific gas phase bond cleavage is also dependent on another factor, viz. the nature of the leaving group (LG). For $[M + H]^+$ ions the LG is a neutral and thus a simple measure of the leaving-group order in the gas phase is the *methyl cation affinity* (MCA) [Eq. (6)] [31]. The lower the MCA, the better the leaving group.

$$CH_3\text{-}LG^+ \xrightarrow{\text{MCA}} CH_3^+ + LG \qquad (6)$$

In contrast, for [M − H]⁻ ions the LG initially is an anion (which can undergo further reaction—see Bowie rule no. 2) and a simple measure of the order of these leaving groups is the anion proton affinity APA [Eq. (3)] (the lower the APA, the better the leaving group). The gas-phase leaving-group propensities for a few neutral and anionic leaving groups are listed in Table 3.

NEIGHBORING GROUP PARTICIPATION. Our discussion so far has assumed simple bond cleavage. The situation can, however, be more complex. In solution, cleavage of bonds is often induced by nucleophiles in either an intermolecular or intramolecular fashion. The latter reactions are termed *neighboring-group participation reactions* [32] and have been extensively studied and reviewed [33]. In the gas phase, when reactive neutrals are absent, these neighboring-group participation reactions can play an important role in cleavage reactions. Indeed, some of the first observations of neighboring-group effects in the gas phase involved loss of water from protonated alcohols [Eq. (7)] [34].

$$\text{(7)}$$

$$X = Cl, Br, OR, SH, SR, NH_2$$

**Table 3**   Measure of Leaving Group Ability in the Gas Phase

| Neutral leaving group | MCA (kcal/mol)[a] | Anionic leaving group | APA (kcal/mol)[b] |
|---|---|---|---|
| $H_3N$ | 104.3 | $H_2N^-$ | 403.6 |
| $CH_3OH$ | 83.1 | $CH_3O^-$ | 380.6 |
| $H_2O$ | 67.5 | $HO^-$ | 390.7 |
| $H_2S$ | 83.4 | $HS^-$ | 350.7 |
| $HCl$ | 56.2 | $Cl^-$ | 328 |

[a] From Ref. 31.
[b] From Ref. 4.

Other examples of neighboring-group reactions include the competition between water [Eq. (8)] and ammonia [Eq. (9)] loss in amino alcohols and amino acids, which has been probed as a function of chain length [35].

$$\tag{8}$$

$$\tag{9}$$

*Ion–Molecule Complexes.* Ion–molecule complexes (IMCs, also called ion–neutral complexes and sometimes collision complexes) are important intermediates in gas-phase ion chemistry (of both cations and anions). These species consist of an incipient ion "coordinated" to a putative neutral. The two components are not connected by a normal covalent bond, but instead are "bound" together via ionic forces such as ion-dipole attraction. As we noted above, IMCs play an important role in bimolecular ion-molecule reactions; indeed the entrance and exit channels of many of these reactions are IMCs. IMCs also are significant in unimolecular dissociation reactions, especially those involving rearrangement. This is because the components are free to rotate about each other, thereby allowing reactions that would be geometrically impossible if they were joined by a covalent bond. Although it can be difficult to unequivocally prove the existence of a specific IMC (criterion for their proof are reviewed by Bowen [8c]), it is convenient to rationalize the gas-phase reactions of many organic and biomolecule ions as occuring via the intermediacy of IMCs. We note that Bowie's rule no. 2 highlights the importance of using IMCs to help rationalize fragmentation reactions

## 4. Toward a Predictive Model for MS/MS Fragmentation Behavior of Protonated Organic Molecules

Using similar key physical organic concepts, Cheng and coworkers have developed a method for predicting the collision energy requirements and fragmentation patterns for organic molecules [36]. The method assumes that there is facile proton movement among all possible protonation sites during CID and that there is no reverse activation barrier for simple ionic bond dissociation so the dissociation propensity can be predicted based on the analysis of the energetics of the products and precursor. Their procedure involves the following steps:

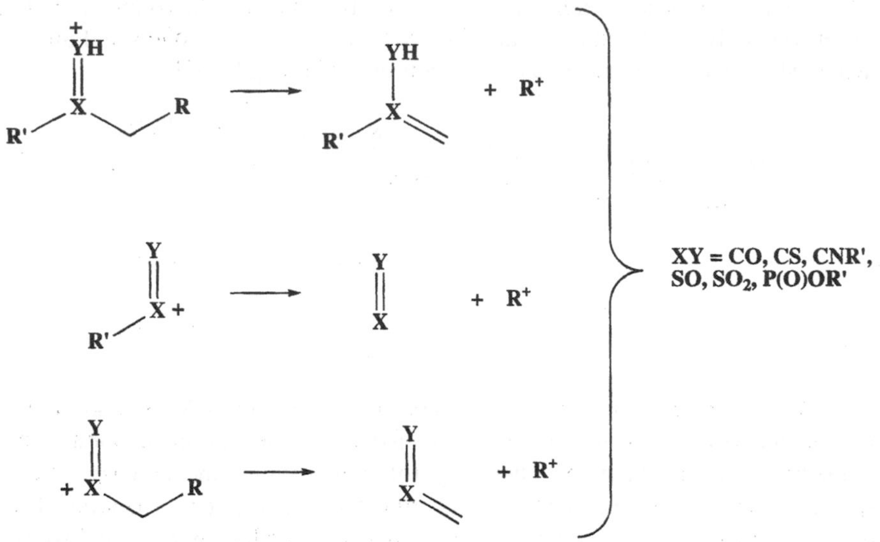

**Scheme 5**

1. The most likely bond cleavage locations are identified by examining the functional groups of the organic molecule (some common examples were identified and these are shown in Scheme 5).
2. Model structures are selected for the precursor and products (for each functional group).
3. the relative bond dissociation energies (RBDE) are estimated (for each functional group) for the model structure using known or estimated heats of formation and proton affinities (selected examples were given, which are listed in Scheme 6).
4. RBDEs are corrected for possible intramolecular interactions.

The collision energy requirement is then determined by the lowest energy dissociation channel while the fragmentation propensities are determined by the ranking of the RBDE value for each of the dissociating bonds. Using microelectrospray on a triple quadrupole with Ar as collision gas (2–3 mTorr), they applied this predictive model to more than 100 combinatorial library compounds and found that this procedure correctly predicts 80% of the major dissociation channels.

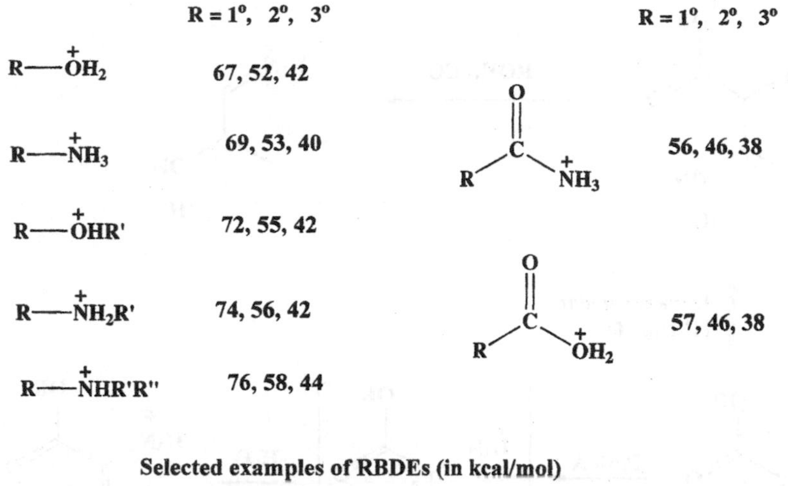

$$R = 1°, \ 2°, \ 3° \qquad\qquad R = 1°, \ 2°, \ 3°$$

$$R{-\overset{+}{O}H_2} \qquad 67, 52, 42$$

$$R{-\overset{+}{N}H_3} \qquad 69, 53, 40$$

$$R{-\overset{+}{O}HR'} \qquad 72, 55, 42$$

$$R{-\overset{+}{N}H_2R'} \qquad 74, 56, 42$$

$$R{-\overset{+}{N}HR'R''} \qquad 76, 58, 44$$

Structure with $R-C(=O)-\overset{+}{N}H_3$: 56, 46, 38

Structure with $R-C(=O)-\overset{+}{O}H_2$: 57, 46, 38

**Selected examples of RBDEs (in kcal/mol)**

**Scheme 6**

## 5. Selected Examples of CID of Protonated and Deprotonated Amino Acids

As noted above, a first step is to identify potential sites of protonation and deprotonation within the molecule of interest in order to establish local proton affinities and acidities. Although the fragmentation reactions of the $[M + H]^+$ [37] and $[M - H]^-$ [38] ions of all of the amino acids have been examined, we focus on only threonine and cysteine as illustrative examples.

*[M + H]⁺ Ion of Threonine.* There are three sites of protonation in threonine: the amino group, the hydroxyl group, and the carboxylic acid group. Labeling studies [using, for example, the methyl ester ($R = CH_3$ in Scheme 7)] have identified two main fragmentation pathways: loss of the combined elements of ROH and CO to give the immonium ion **H** in Scheme 7 and loss of $H_2O$ from the side chain. The latter loss can potentially proceed via two mechanisms: a charge remote reaction to give **C** (Scheme 7), or a neighboring-group mechanism to yield **F** (Scheme 7). Using a precursor ion in which all the exchangable hydrogens have been replaced by deuterium atoms, MS/MS studies reveal that $D_2O$ rather than HOD is lost, thereby favoring the neighboring-group mechanism (Path B) [39a].

*[M − H]⁻ Ion of Threonine.* The $[M - H]^-$ ion of threonine undergoes several reactions under MS/MS conditions. Perhaps the most structurally diagnostic reaction of threonine and threonine-containing peptides is the loss of acetaldehyde. A mechanism involving an elimination induced by intramolecular proton transfer has been proposed [Eq. (10)] [38].

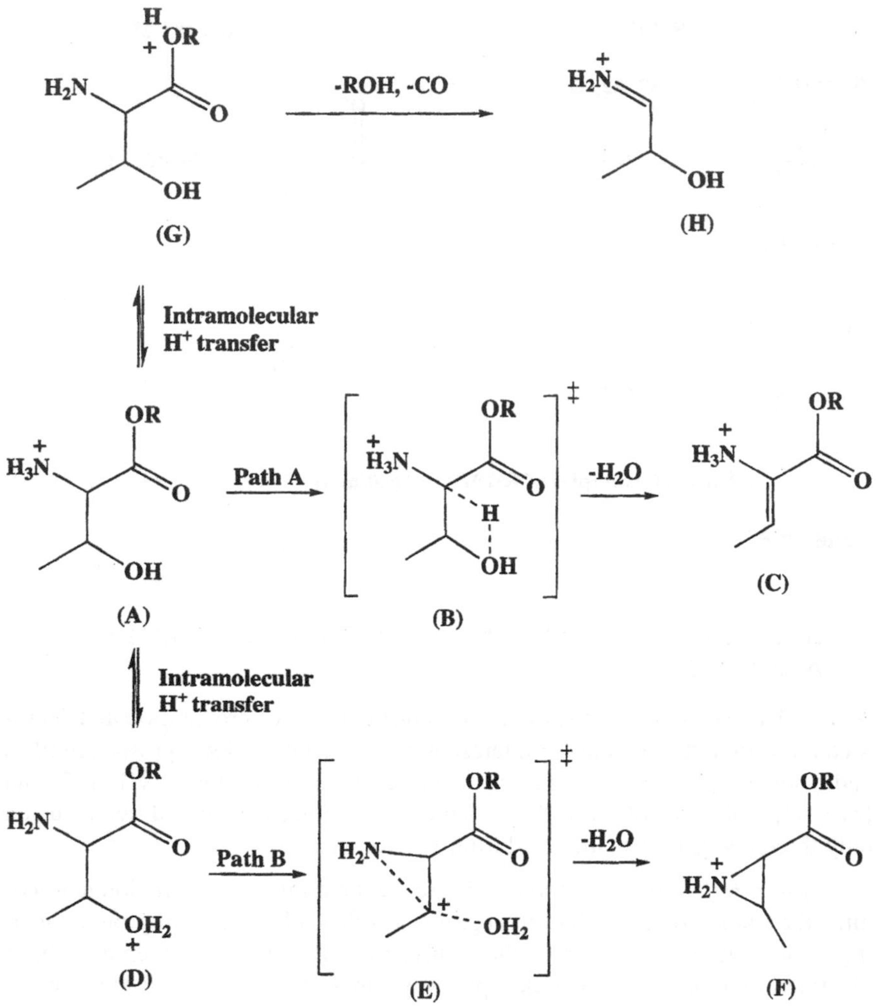
H
OR
+
H2N
O
OH
(G)
-ROH, -CO
+
H2N
OH
(H)
Intramolecular
H+ transfer
OR
+
H3N
O
OH
(A)
Path A
OR
+
H3N
O
H
OH
(B)
‡
-H2O
OR
+
H3N
O
(C)
Intramolecular
H+ transfer
OR
H2N
O
OH2
+
(D)
Path B
OR
H2N
O
+
OH2
(E)
‡
-H2O
OR
+
H2N
O
(F)

Scheme 7

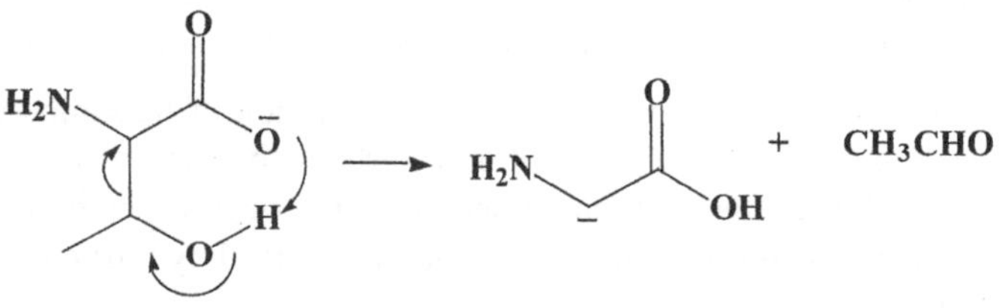
O
H2N
O
H
O
O
H2N
OH
+
CH3CHO

(10)

*[M + H]$^+$ Ion of Cysteine.* There are three sites of protonation in cysteine: the amino group, the sulfhdryl group, and the carboxylic acid group. Tandem mass spectrometry (MS/MS) on the [M + H]$^+$ ion indicates two main fragmentation pathways: loss of the combined elements of $H_2O$ and CO to give the immonium ion (cf. **H** in Scheme 7) and loss of $NH_3$. Studies utilizing MS$^3$ (that is, MS/MS/MS) reveal that the product ion from the latter is an episulfonium ion and that this reaction proceeds via another neighbouring-group mechanism [Eq. (11)] [39b].

$$\text{(11)}$$

*[M − H]$^-$ Ion of Cysteine.* A structurally diagnostic reaction for cysteine and cysteine containing peptides is the loss of $H_2S$ from their [M − H]$^-$ ions. A possible mechanism, induced by deprotonation at the $\alpha$ carbon of cysteine, is shown in Eq. (11) [38].

$$\text{(12)}$$

## 6. Sequencing of Peptides

Tandem mass spectrometry is replacing the Edman method as a sequencing tool. Due to the fact that peptides possess both basic and acid functional groups, they readily form [M + H]$^+$ and [M − H]$^-$ ions under FAB, ESI, and MALDI conditions, which can be subjected to CID to yield the partial or complete sequence of the peptide.

*Positive Ions.* A nomenclature for positive sequence fragment ions is well accepted and is shown in Scheme 8 for a tetrapeptide [40a]. The main types of fragment ions are b ions (peptide cleavage with charge retention on the N-terminal side) and y ions (peptide cleavage with charge retention on the C-terminal side). Several studies have examined the mechanisms for forming sequence ions [40]. Harrison has shown that b$_n$ ions (where $n \geq 2$) possess oxazolone structures that arise from neighboring-group attack by an amide carbonyl to facilitate cleavage of the adjacent peptide bond (Scheme 9, X = O) [40b–40d]. One reason b$_1$

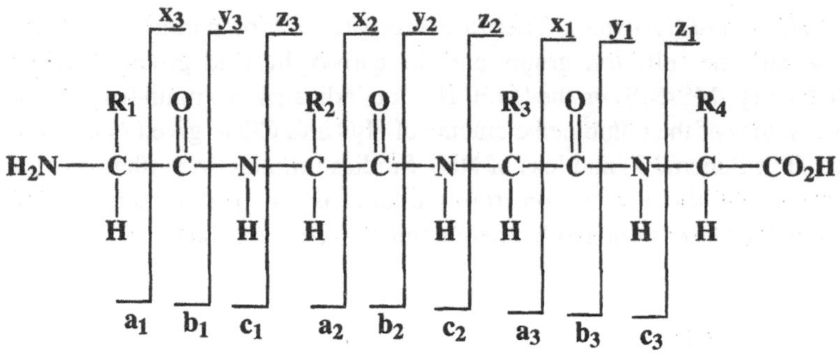

**Scheme 8**

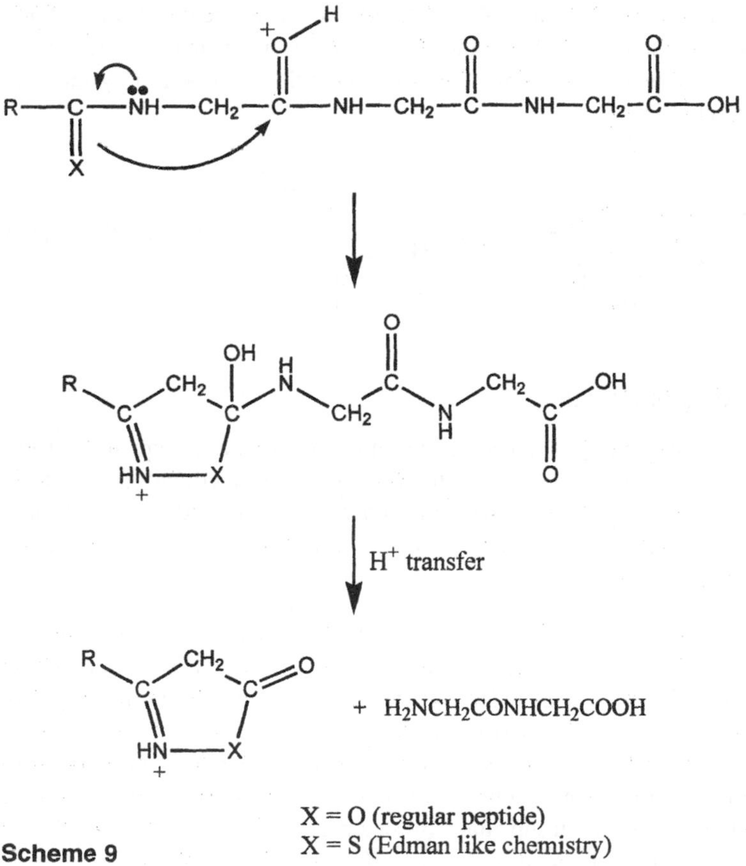

X = O (regular peptide)
X = S (Edman like chemistry)

**Scheme 9**

ions are often not observed in MS/MS spectra is that these acylium ions can be unstable with respect to CO loss with concomitant $a_1$ ion formation [40e]. Apart from sequence ions, there are some structurally useful fragmentations due to side-chain cleavages, particularly for posttranslationally modified residues such as methionine sulfoxide (which loses $CH_3SOH$ [40f]) and serine-$O$-phosphate (which loses $H_3PO_4$ [40g]). Whether these latter reactions proceed by neighboring-group mechanisms or other processes needs to be established.

*Negative Ions.* Sequencing of peptides via MS/MS of their $[M - H]^-$ ions has received much less attention, although Bowie has done extensive study on the fragmentation mechanisms [41]. For tetrapeptides, the $[M - H]^-$ ions generally show two different collision-induced backbone cleavages, which allow the determination of the amino acid sequence of the peptide. The first of these involves the formation of the carboxylate anions of either constituent amino acids or fragment peptides (Bowie has termed these $\alpha$ cleavages [31c]). In the second, amino acids or fragment peptides are eliminated as neutrals (Bowie has termed these $\beta$ cleavages [31c]). These classes of sequence ions are related because they can be rationalized as arising from ion–neutral complexes, as shown in Scheme 10. In addition, there are a number of residues that undergo characteristic side-chain fragmentations irrespective of their position in the tetrapeptide, including Ser (which loses $CH_2O$), Thr (which loses $CH_3CHO$), Cys (which loses $H_2S$), Met (which loses $CH_3SH$), Phe (which forms the benzyl anion), Tyr (which loses $OC_6H_4CH_2$), and Trp (loss of $C_9H_7N$). However, there are also some residues that, when situated at the C-terminal end of the peptide, promote pronounced fragmentation at the C-terminal position which occurs to the exclusion of the normal backbone cleavages. In a comparison study between CID of $[M + H]^+$ and $[M - H]^-$ ions, it was concluded that the data obtained from the negative-ion cleavages are analytically useful and are complementary those provided by the positive-ion mode [41c].

## 7. Directing Fragmentation of Peptides via Derivatization

Gaskell has cleverly exploited neighboring-group reactions to form modified $b_1$ ions from N-terminal phenylthiocarbamoyl (PTC) derivatives of peptides [42]. Low-energy collisional activation of peptide PTC derivative $[M + 2H]^{2+}$ ions during electrospray tandem mass spectrometry results in highly favored cleavage of the N-terminal peptide bond, yielding complementary $b_1$ ions and $y_{n-1}$ fragments. The apparently close mechanistic similarity between the gas-phase fragmentation reactions and the related condensed-phase Edman degradation processes can be readily understood by a neighboring-group attack by the nucleophilic thiocarbonyl sulfur atom onto the adjacent peptide bond (Scheme 9, X = S), a process directly related to that discussed for $b_n$ (where $n \geq 2$) ion formation in regular peptides (Scheme 9, X = O). Replacement of the PTC deriv-

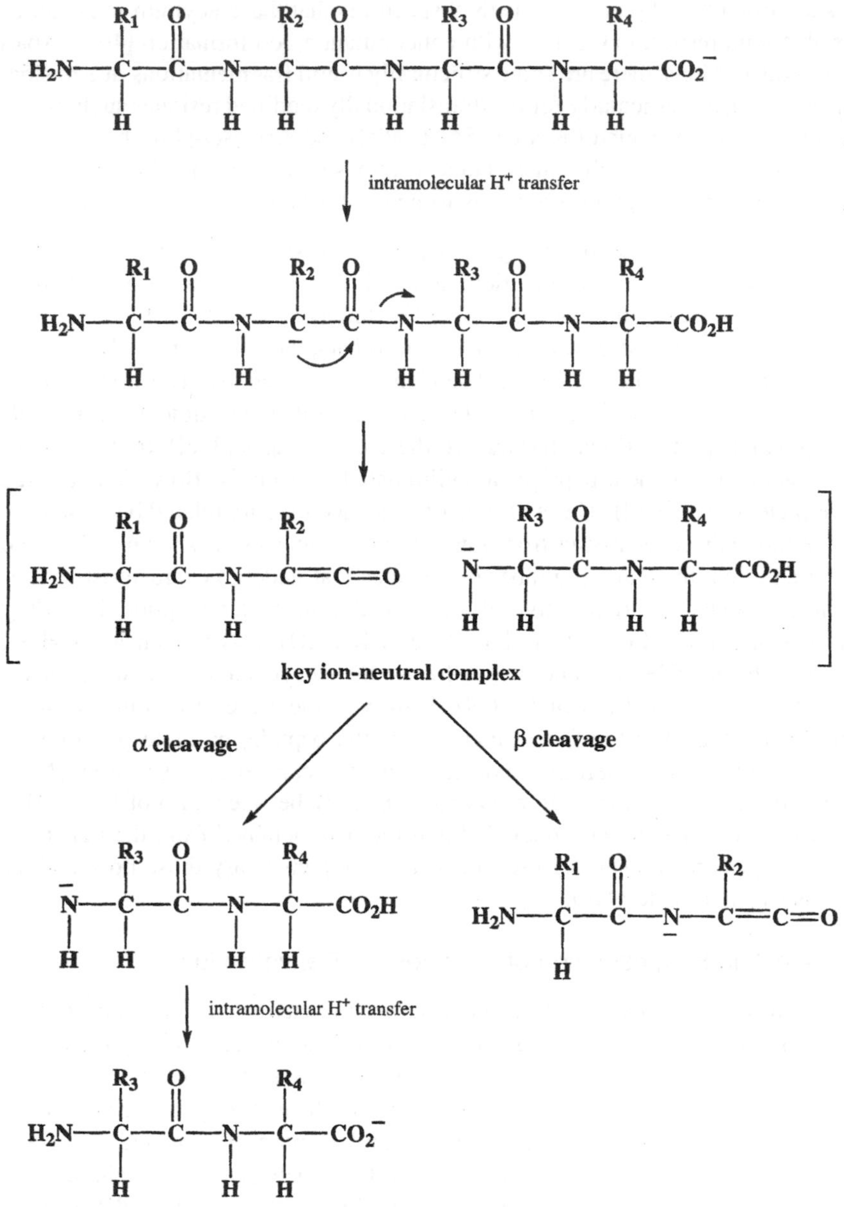

**Scheme 10**

ative by the pentafluoro-PTC analog results in similar fragmentation chemistry but with preferential loss of the derivatized N-terminal residue as a neutral fragment [42b]. This demonstrates that a judicious choice of derivatization procedure enables not only the direction of fragmentation but also of charge retention.

## 8. Sequencing of Oligonucleotides

There has also been considerable interest in the use of MS/MS to sequence oligonucleotides. In a similar fashion to peptides, oligonucleotides possess both basic (i.e., the nucleobase sites) and acidic (i.e., the phosphate sites) functional groups. They readily form $[M + H]^+$ and $[M - H]^-$ ions under ESI conditions, which can be subjected to CID to yield the partial to complete sequence information. The sequence ion nomenclature of McLuckey [43] has gained widespread acceptance (Scheme 11).

*Comparisons between MS/MS of the $[M - H]^-$ and $[M + H]^+$ Ions of Oligodeoxynucleotides.* Under low-energy CID conditions in the ion-trap, similar classes of ions, such as the base loss "nonsequence" and the $w_n$ "sequence" ions, are observed in the MS/MS spectra of both the $[M - H]^-$ and $[M + H]^+$ ions [44]. The key physical organic concepts discussed above can readily be used to provide mechanistic rationals for these fragmentation reactions. For the $[M + H]^+$ ions the most likely site of protonation is at the nucleobase sites, which have the higher proton affinity. Thus protonation at these sites acts as a "trigger" for neutral base loss via E1 or intramolecular E2 mechanisms [44]. The order of nucleobase loss for the $[M + H]^+$ ions is $C > G > A > T$. In contrast, for the

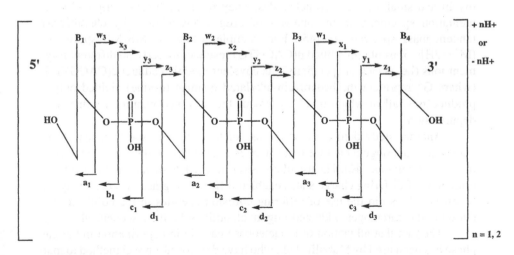

**Scheme 11**

[M − H]⁻ ions, the most likely site of deprotonation is at the most acidic sites, namely the phosphate sites. Deprotonation at these sites exposes an anionic oxygen base that facilitates charged anionic base loss via an intramolecular E2 mechanisms. Initial CID of the [M − H]⁻ ions showed neutral base loss predominantly from the 5' terminus, followed by the 3' and internal positions in the order A > T > G > C. Notwithstanding the differences in mechanisms between the [M − H]⁻ and [M + H]⁺ ions, the net effect is the same with regard to the subsequent formation of sequence ions. Thus loss of the 5' base triggers formation of the $w_2$-type ion, loss of the internal base yields $w_1$- and the ($a_2$-$B_2$)-type ions, while 3' base loss results in the formation of the ($a_3$-$B_3$)-type ion.

Using MS/MS and MS$^3$ experiments the sequences of 64 trimer and 16 tetramer oligodeoxynucleotide were determined via CID of their [M − H]⁻ ions. In those cases where their sequences cannot be completely assigned, CID of the [M + H]⁺ ions provides the required sequence information. This illustrates the complementarity of MS/MS studies on both the [M − H]⁻ and [M + H]⁺ ions and suggests that sequencing strategies for oligodeoxynucleotides should consider collecting MS$^n$ data on both charge types. Larger oligodeoxynucleotides offer the potential for the formation of [M − $n$H]$^{n-}$ and [M + $n$H]$^{n+}$ ions and further work needs to be carried out on the effect that the number of charges has on the formation of sequence and nonsequence ions.

## 9.  Applications in Determining Modified Oligodeoxynucleotides

Using a triple-quadrupole instrument, Iannitti and coworkers examined the ESI/MS/MS spectra of covalent adducts formed between the alkylating agent hedamycin and small oligodeoxynucleotides. They found high sensitivity and fragmentation specificity in the analysis of drug–DNA adducts by electrospray tandem mass spectrometry [45]. For example, while the MS/MS spectra of [M − 2H]$^{2-}$ ions of unmodified d(CACGTG) resulted in nearly 30 different fragment ions (i.e., a lack of specificity), its covalent modified adduct d(CACG*TG) (where G* represent the hedamycin-alkylated guanine residue) yielded only 5 product ions, all of which involved loss of the alkylated guanine residue (i.e., depurination).

Another study demonstrated the use of tandem mass spectrometry to sequence a 7-mer oligonucleotide that had been adducted with the benzene metabolite p-benzoquinone [46]. The use of collision-induced dissociation of oligonucleotide ions of differing charge states enabled different fragmentation pathways to be observed, as did the use of differing collision energies in the collision cell, allowing the carcinogen-adducted oligonucleotide to be fully sequenced.

The fact that alkylation of a nucleobase can aid in depurination in the gas phase has been used by Marzilli et al., who have developed a novel method to map guanine bases in short oligonucleotides via in situ guanine-specific methylation followed by gas-phase fragmentation reactions [47]. Thus CID of the monomethylated oligonucleotide strand promotes rapid depurination and further collision (MS3) of the apurinic oligonucleotide leads to preferential cleavage of the backbone at the site of depurination.

## V. ION–MOLECULE REACTIONS AS STRUCTURAL PROBES

Gas phase ion–molecule reactions are complementary to CID reactions and are gentler structural probes because if these reactions are carried out at room temperature, the only energy available is the thermal energy plus the ion–neutral complexation energy. Most ion–molecule reactions are directed by the charge site (exceptions include the reactions of radical sites in distonic ions [67] and McLuckey's HI adduction to $[M + H]^+$ ions of peptides [52]). As we noted when discussing sites of protonation and how these effect fragmentation reactions, simple biomolecules such as amino acids and peptides have multiple sites of reactivity and these may also play a role in ion–molecule reactions [48]. Unlike CID reactions, where only charged precursors are involved, for ion–molecule reactions, the biomolecule can be charged (either an anion or a cation) or neutral. If the bimolecule is charged then its companion in an ion–molecule reaction will be a neutral chemical reagent. Conversely, if the bimolecule is neutral, then a charged chemical reagent will be required to induce an ion–molecule reaction. Many chemical reagents that have been successfully used to probe structures of biomolecules in the condensed phase are used because they employ site-specific reactions. How might this be achieved in a gas-phase ion–molecule reaction? If the biomolecule is a small neutral system, then regioselectivity for the reaction will depend on both the nature of the chemical reagent and the relative nucleophilicities of the various sites within the biomolecule (the reactivity of less reactive sites can be probed by moderating the reactivity of the most reactive sites). In charged biomolecule systems, the mode of reaction will depend on both the type of chemical reagent (is it a base or acid? does it have electrophilic sites?) as well as the nature of the charge site (is it protonated or cationized by a metal ion or is it deprotonated?). Note that a key factor for charged biomolecules is the location of the charge (for some reactions the charge could be in the wrong place!).

Some of the reactions that have been studied are shown in Table 4. While the area of gas phase ion–molecule reactions of biomolecules is still in its infancy, it is interesting to note that some regioselective reactions have been developed; thus reactions that target either the N-termius [61] or C-termius [62] of peptides have been identified. Ion–molecule reactions that offer the potential to sequence peptides have also been examined [59,60].

**Table 4**   Gas-Phase Ion–Molecule Reactions of Biomolecules

| Chemical reagent | Type of biomolecule | Form of biomolecule | Type of reaction | Reference |
|---|---|---|---|---|
| Base (e.g., Et₃N) | Nucleobases<br>Amino acids, peptides, and proteins | $[M + nH]^{n+}$ | Deprotonation | 49 |
| Acid (e.g., CF₃CO₂H) | Nucleic acids<br>Peptides and proteins | $[M - nH]^{n-}$ | Protonation | 50 |
| XD<br>(X=DO, D₂N, etc.) | Nucleobases, nucleic acids<br>Amino acids, peptides, and proteins | $[M + nH]^{n+}$<br>$[M - nH]^{n-}$ | H/D exchange | 49,51 |
| HI | Peptides and proteins | $[M + nH]^{n+}$ | HI adduction | 52 |
| $[A + H]^+$ (A = CH₄, etc.) | Amino acids, peptides | $M$ | Proton transfer induced dissociation (CI) | 53 |
| H₃N | Peptides | $[M + H]^+$ | Endothermic proton transfer induced dissociation | 54 |
| (CH₃)₂X⁺ (X=Cl, Br) | Nucleobases<br>Amino acids | $M$ | S$_N$2 methylation | 55 |
| (CH₃)₃SiCl | Nucleic acids<br>Peptides | $[M - nH]^{n-}$ | S$_N$2(Si) | 56,57 |
| CH₃OCH₂⁺ | Amino acids, peptides | $M$ | Formation of $[M + CH]^+$ and $[M + CH_3]^+$ | 58 |
| CH₃OCH₂⁺ | N-acetylated peptides | $M$ | Peptide bond cleavage | 59 |
| CF₃COSEt | Peptides models | $[M - H]^-$ | Peptide bond cleavage [eq. (1)] | 60 |
| (CH₃)₂CO | Simple peptides | $[M + H]^+$ | Identification of N-terminus via Schiff's base formation | 61 |
| CF₃COSMe | Peptides | $[M - H]^-$ | Identification of C-terminal carboxylate | 62 |
| (CX₃CO)₂O (X=H, F) | Peptides | $[M - H]^-$ | Identification of C-terminal carboxylate | 62 |
| RC(O)⁺ (R=CH₃, Ph) | Nucleobases, amino acids, and peptides | $M$ | Formation of $[M + RCO]^+$ and cleavage reactions | 63 |
| (CH₂)₂X⁺ (X=Cl, Br) | Amino acids | $M$ | Alkylation | 64 |
| NO⁺ | Amino acids | $M$ | Hydride transfer | 65 |
| H₂O | Nucleic acids<br>Amino acids, peptides | $[M - nH]^{n-}$<br>$[M + nH]^{n+}$ | Hydration | 66 |

A very interesting reaction in terms of its broad applicability involves the reaction of $[M + nH]^{n+}$ ions of small peptides or even proteins with hydroiodic acid (HI) [52]. In small peptides this reaction can be used to "count" the number of basic residues [52a,52b], but in larger systems the three dimensional structure of the $[M + nH]^{n+}$ ion can limit the availability (or exposure) of basic sites to reactivity with HI. Indeed, the kinetics of attachment of hydroiodic acid (HI) to the $(M + 6H)(6+)$ ions of the native form of bovine pancreatic trypsin inhibitor (BPTI) exhibit distinctly nonlinear (pseudo-first-order) reaction kinetics, indicating two or more noninterconverting structures in the parent ion population. In contrast, the reduced form shows very nearly linear reaction kinetics. Both forms of the parent ion attach a maximum of five molecules of hydroiodic acid, which is expected based on the amino acid composition of the protein (there is a total of 11 strongly basic sites in the protein: 6 arginines, 4 lysines, and 1 N-terminus; thus an ion with protons occupying 6 of the basic sites has another 5 available for hydroiodic acid attachment). A key difference is that the kinetics of successive attachment of HI to the native and reduced forms of BPTI differ, particularly for the addition of the fourth and fifth HI molecules. A very simple kinetic model was developed, which describes the behavior of the reduced form reasonably well, suggesting that all of the neutral basic sites in the reduced BPTI ions have roughly equal reactivity. In contrast, the behavior of the native ion is not well described by this simple model. Thus the HI reaction kinetics appear to have potential as a chemical probe of protein ion three-dimensional structure in the gas phase. Hydroiodic acid attachment chemistry is significantly different from other gas-phase chemistries used to probe three-dimensional structure (e.g., H/D exchange) and thus yields complementary information.

## VI. POTENTIAL PITFALLS IN ESI/MS

Electrospray ionization mass spectrometry (ESI/MS) has proven to be an extremely versatile and useful tool in examining precharged and readily ionizable species [68]. It is easy to overlook that during the desolvation process, ions are transferred from the solution phase to the gas phase and at some intermediate gas phase regime, the previously innocuous solvent molecules become reactive entities that can change the ion structure. This can become a significant problem when fragile charged species undergo "in source" CID and/or ion–molecule reactions. A recent example from our laboratories (Ivan Vrionis and Richard A. J. O'Hair, unpublished observations) highlights this problem. Figure 3 shows the ion abundances for ESI-generated ions from a solution of the diazonium salt $p$-BrC$_6$H$_4$N$_2^+$BF$_4^-$ in methanol as a function of the tube lens offset potential in an ion-trap mass spectrometer. The tube lens region is the key region where ions continue to be desolvated prior to transferal to the octapole ion guide for subse-

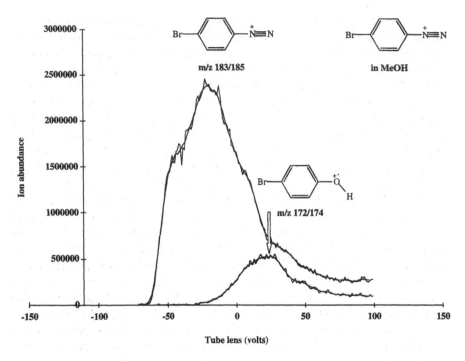

**Figure 3**  Ion abundances as a function of tube lens voltage in the ESI/MS.

quent mass analysis via the quadrupole ion trap. By setting the mass analyzer at a specific $m/z$ value, the effect of the tube lens voltage on the ion abundance of that ion can be ascertained. An examination of Fig. 3 reveals that the abundance of $p$-BrC$_6$H$_4$N$_2^+$ ion (**A**) peaks at about $-20$ volts and then decreases (note that at low voltages essentially only $p$-BrC$_6$H$_4$N$_2^+$ ion is observed). As the abundance of $p$-BrC$_6$H$_4$N$_2^+$ ion decreases, other ions are observed, as shown in the mass spectrum in Fig. 4. The ion abundance of one of these ions, $p$-BrC$_6$H$_4$OH$^+$ (**G**), has been examined as a function of the tube lens voltage (Fig. 3) and it is clear that as the ion abundance of (**A**) decreases, that of (**G**) increases. A similar situation occurs when acetonitrile is used as the solvent and the complex gas-phase chemistry that leads to the mass spectrum shown in Fig. 4 has been unraveled using detailed MS/MS experiments and ion–molecule reactions and is summarized in Scheme 12. Briefly, the diazonium ion readily loses N$_2$ to form a phenyl cation (**B**), which is highly reactive toward nucleophiles, including the ESI solvent acetonitrile as well as traces of adventitious water. The high exoethermicity of bond formation via addition of nucleophiles is sufficient to break other bonds homolytically to yield radical cations, including the distonic ion (**D**) and (**I**).

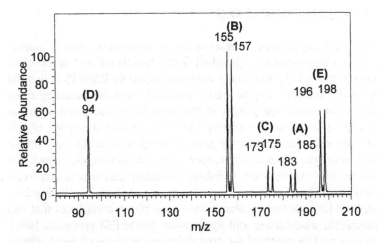

**Figure 4** In-source CID of $p$-BrC$_6$H$_4$N$_2^+$ in tube lens region of the LCQ followed by ion–molecule reactions with reactive gases derived from the ESI solvent (CH$_3$CN). Ions are labeled according to Scheme 12.

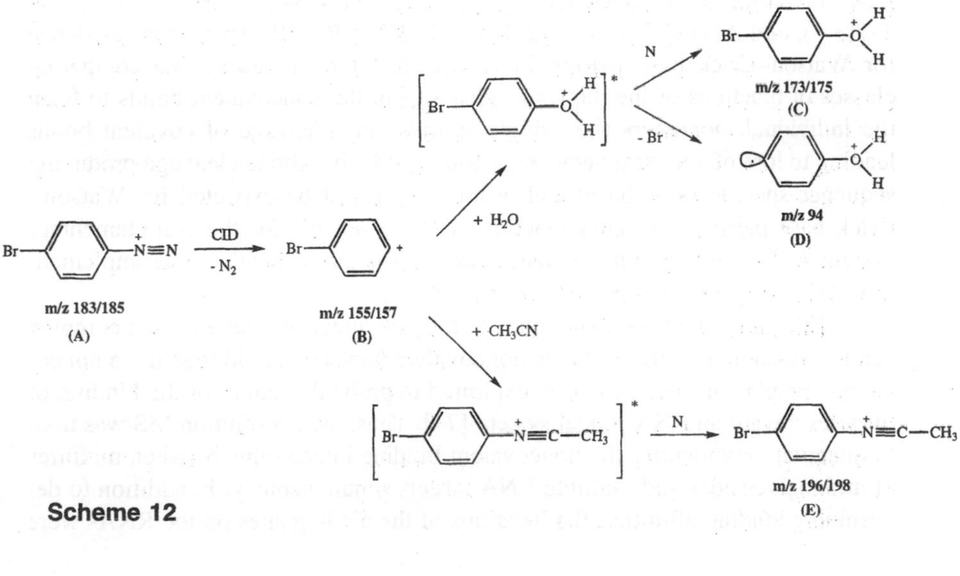

**Scheme 12**

## VII.  NONCOVALENT COMPLEXES

The fact that ESI/MS can be used to observe noncovalent complexes of bioma-
cromolecules from solution has been ascribed to the ''gentle nature'' of the electrospray process [69b,69c]. An extremely attractive aspect of ESI/MS is that the
stoichiometry of the complex can be easily obtained from the resulting mass
spectrum because the molecular weight of the complex is measured with the
additional benefits of speed and sensitivity [69]. It is beyond the scope of this
chapter to critically examine whether the ions observed in ESI/MS truly reflect
solution phase binding phenomena. The reader is directed to several reviews that
have examined the key experimental variables (including ESI source and interface conditions) [69]. Thus Smith has paid attention to methods for distinguishing, and conditions for avoiding, artifacts due to gas-phase complexes that can
arise from nonspecific associations and aggregation due to ESI processes [69c].
Clearly this is an attractive potential for drug discovery in terms of rapid affinity
screening of combinatorial libraries and conducting competitive binding studies.
Once again, the ultimate analytical usefulness will depend on a thorough understanding of fragmentation reactions and gas-phase ion–molecule reactions. Some
recent highlights on these two fronts are described below.

### A.  Unusual Fragmentation Processes

A study of the activation energies for dissociation of double-strand oligonucleotide anions has examined the dissociation kinetics of a series of complementary and noncomplementary DNA duplexes, (TGCA)(2)(3=), (CCGG)(2)(3=),
(AATTAAT)(2)(3=), (CCGGCCG)(2)(3=), A(7)•T-7(3=), A(7)•A(7)(3=), T-7•
T-7(3=), and A(7)•C-7(3=) using BIRD in FT-ICR [70]. Apart from evidence
for Watson–Crick base pairing in vacuo, this study revealed two competing
classes of reactions of the duplexes: cleavage of the noncovalent bonds to form
the individual monomers (i.e., single strands) and cleavage of covalent bonds
leading to loss of a neutral nucleobase followed by backbone cleavage producing
sequence-specific (a = base) and w ions. As might be expected for Watson–
Crick base pairing, the latter reaction is ''shut down'' for the complementary
system A(7)•T-7(3=), but was particularly pronounced for the noncomplementary A(7)•A(7)(3=) and A(7)•C-7(3=) dimers.

This notion that covalent bonds of the individual components of a complex
can be broken in preference to the noncovalent bonds that hold together a noncovalent complex has been cleverly exploited to probe the nature of the binding of
libraries toward an RNA model system [71]. Thus, high-resolution MS was used
to quantitatively identify the noncovalent binding interactions between mixtures
of aminoglycosides and multiple RNA targets simultaneously. In addition to determining binding affinities, the locations of the binding sites on the RNAs were

identified from a protection pattern generated by fragmenting the aminoglycoside–RNA complex. Specific complexes were observed for the prokaryotic rRNA A-site subdomain with ribostamycin, paromomycin, and lividomycin, whereas apramycin preferentially formed a complex with the eukaryotic subdomain. Differences in binding between paromomycin and ribostamycin were probed using an MS/MS protection assay. The impact of specific base substitutions in the RNA models on binding affinity and selectivity was examined. It was found that binding of apramycin to the prokaryotic subdomain strongly depends on the identity of position 1408, as evidenced by the selective increase in affinity for an A1408G mutant. An A1409–G1491 mismatch pair in the prokaryotic subdomain enhanced the binding of tobramycin and bekanamycin. These observations demonstrate the potential of MS/MS based methods to provide molecular insights into small molecule–RNA interactions useful in the design of selective new antimicrobial drugs.

## B. Ion–Molecule Reactions

Chemical reagents have been used in solution to "footprint" sites of interaction in noncovalent complexes. Only a limited set of studies have been performed in the gas phase using ion–molecule reactions between a charged noncovalent complex and various neutral reagents [72,73]. They can be classified into two areas: H/D exchange reactions [72] and ligand switching reactions [73]. Recent results in these two areas are discussed below.

### 1. H/D Exchange

The handful of studies in this area show contrasting mechanisms of exchange in the complexes. Thus Lebrilla was the first to examine the hydrogen/deuterium exchange reactions of a complex between a protonated amino acid and a monosaccharide. Rate constants were determined and compared for complexed and uncomplexed amino acids and the overall rate constant, which corresponds to exchange of a specific number of hydrogens, was deconvoluted to yield site-specific rate constants. Comparisons of these site-specific rate constants revealed that complexation of amino acids with saccharides significantly decreases the rate constants of the exchange [72a]. Beauchamp observed a similar inhibition of H/D exchange for crown ether attachment to protonated ethylenediamine and 1,4-diaminobutane [72b]. These results suggest that if one partner in the complex "shields" the charge site from the other, then H/D exchange is inhibited.

In contrast, simple proton bound dimers $[A + H + B]^+$ show an enhancement (i.e., catalysis) for H/D exchange over their corresponding monomers. This has been shown for simple symmetrical dimers of amines [72c], homo- and heterodimers between sarcosine and glycylglycine [73d], and for the antibiotic–peptide complexes [72e].

## 2. Ligand Switching Reactions

Lebrilla has developed a "chiral mass spectrometry" method that has potential to determine enantiomeric excess of mixtures. The method involves an ion–molecule reaction between a neutral amine and a noncovalent charged complex (formed via ESI) consisting of a selector (cyclodextrin) and the desired selectand (chiral substrate, typically an amino acid). The protonated selector–selectand complex undergoes a guest–ligand exchange reaction with the amine. The rate of this reaction is sensitive to the chirality of the bound selectand and thus can be used to the quantify enantiomeric excess of other mixtures using suitable calibration curves.

## VIII. CONCLUSIONS

Taking advantage of the ever-increasing opportunities that mass spectrometry affords in drug discovery requires an appreciation of the fundamental unimolecular and bimolecular gas phase chemistry of ions. This chapter has attempted to draw the reader's attention to a number of important physical organic concepts that can be applied to understanding the fragmentation reactions of organic ions and ions derived from small biomolecules under MS/MS conditions. The potential for ion–molecule reactions as structural probes has been described and recent developments in the analysis of noncovalent complexes using MS/MS techniques and ion–molecule reactions have also been highlighted.

## ACKNOWLEDGMENTS

I thank the Australian Research Council for financial support (Grant A29930202) and the University of Melbourne for funds to purchase the ion-trap mass spectrometer.

## REFERENCES

1. The Mass Frontier™ 1.0 program by HighChem, and marketed by Finnegan has a Fragments & Mechanisms module which features automated generation of possible fragments from either M+. or [M + H]+ ions starting from a user supplied chemical structure. See: http://www.highchem.com.
2. C. Brunnee. The ideal mass analyzer: Fact or Fiction? Int J Mass Spectrom Ion Proc 76:121–237, 1987.
3. (a) AL Burlingame, RK Boyd, SJ Gaskell. Mass spectrometry Anal Chem 70:647R–

716R (1998); (b) AL Burlingame, RK Boyd, SJ Gaskell. Mass spectrometry. Anal Chem 68:599R–651R, 1996.

4. For detailed thermochemical definitions as well as experimental data for gas phase ions and neutrals see: http://webbook.nist.gov/chemistry.

5. Kinetic data involving reactions of gas phase ions are scattered through the literature. For the only "recent" comprehensive compilation of bimolecular rate constants for both anions and cations see Y Ikezoe, S Matsuoka, M Takebe, A Viggiano. Gas Phase Ion–Molecule Reaction Rate Constants Through 1986. Ion Reaction Research Group, Tokyo, 1987.

6. V Bernshtein, I Oref. Unimolecular dissociation of very large polyatomic molecules. J Phys Chem 98:136–140, 1994 (and references cited therein).

7. (a) JI Brauman. Some historical background on the double-well potential model. J Mass Spectrom 30:1649–1651, 1995; (b) WJ Chesnavich, MT Bowers. Theory of ion–neutral interactions: applications of transition state theory concepts to both collisional and reactive properties of simple systems. Prog React Kinet 11:137–267, 1982.

8. For reviews on the role of ion–molecule complexes in gas-phase reactions see (a) T Morton. Gas phase analogues of solvolysis reactions. Tetrahedron 38:3195–3243, 1982; (b) DJ McAdoo. Mass Spectrom Rev 7:363, 1988; (c) RD Bowen. Ion–neutral complexes. Acc Chem Res 24:364–371, 1991, (d) P Longevialle. Ion-neutral complexes in the unimolecular reactivity of organic cations in the gasphase. Mass Spectrom Rev 11:157, 1992.

9. AG Harrison. Chemical Ionization Mass Spectrometry. 2nd ed. Boca Raton, FL: CRC Press, 1992.

10. FW McLafferty. F Turecek. Interpretation of Mass Spectra. 4th ed. Mill Valley, NY: University Science Books, 1993.

11. (a) KL Busch, GL Glish, SA McLuckey. Mass Spectrometry/Mass Spectrometry: Techniques & Applications of Tandem Mass Spectrometry. New York: VCH. (1988); (b) SA McLuckey, DE Goeringer. Slow heating methods in tandem mass spectrometry. J Mass Spectrom 32:461–474, 1997; (c) E de Hoffman. Tandem mass spectrometry: a primer. J Mass Spectrom 31:129–137, 1996.

12. RG Cooks, JH Beynon, RM Caprioli, GR Lester. Metastable Ions. Amsterdam/New York: Elsevier, 1973.

13. (a) J Bordas-Nagy, KR Jennings. Collision-induced dissociation of ions. Int J Mass Spectrom Ion Proc 100:105–131, 1990; (b) RN Hayes, ML Gross. Collision-induced dissociation. Meth Enzymol 193:237–263, 1990.

14. AR Dongre, A Somogyi, VH Wysocki. Surface-induced dissociation: an effective tool to probe structure, energetics and fragmentation mechanisms of protonated peptides. J Mass Spectrom 31:339–350, 1996.

15. (a) GS Gorman, IJ Amster. Photodissociation studies of small peptide ions by Fourier mass spectrometry. Org Mass Spectrom 28:437–444, 1993; (b) DP Little, JP Spier, MW Senko, PB O'Connor, FW McLafferty. Infrared multiphoton dissociation of large multiply charged ions for biomolecule sequencing. Anal Chem 66:2809–2815, 1994 (and references cited therein).

16. BH Wang, FW McLafferty. Electron impact excitation of ions from larger organic molecules. Org Mass Spectrom 25:554–556, 1990.

17. RA Zubarev, NA Kruger, FW McLafferty. Electron capture dissociation of gaseous

multiply-charged proteins is favored at disulfide bonds and other sites of high hydrogen atom affinity. J Am Chem Soc 121:2857, 1999.

18.  RC Dunbar, TB McMahon. Activation of unimolecular reactions by ambient blackbody radiation. Science 279:194, 1998.

19.  (a) ML Gross. Charge-remote fragmentations: method, mechanism, and applications. Int J Mass Spectrom Ion Proc 118/119:137, 1992; (b) J Adams. Charge-remote fragmentations: analytical applications and fundamental studies. Mass Spectrom Rev 9: 141, 1990.

20.  (a)RAJ O'Hair, JH Bowie. Organic negative ions: structure, reactivity and mechanism. In: Mass Spectrometry (Specialist Periodical Report), London: The Royal Society of Chemistry, 1989, pp 145–180; (b) JH Bowie. The fragmentations of [M − H]⁻ ions derived from organic compounds. In: DH Russell, ed. Experimental Mass Spectrometry. New York: Plenum, 1994, pp 1–38; (c) JH Bowie. The fragmentations of even electron organic negative ions. Mass Spectrom Rev 9:349–379, 1990.

21.  PJ Derrick. Isotope effects in fragmentation. Mass Spectrom Rev 2:285, 1983.

22.  For reviews of ionization of neutral fragments as a structural probe see: (a) MJ Polce, S Beranova, MJ Nold, C Wesdemiotis. Characterization of neutral fragments in tandem mass spectrometry: a unique route to mechanistic and structural information. J Mass Spectrom 31:1073–1085, 1996; (b) MM Cordero, C Wesdemiotis, Reionization and characterization of neutral losses from biomolecular ions. In: Biological Mass Spectrometry: Present and Future. New York: Wiley, 1994, pp 119–126.

23.  PB Armentrout. Thermochemical measurements by guided ion beam mass spectrometry. In: N Adams, LM Babcock, eds. Advances in Gas Phase Ion Chemistry. Greenwich, CT: JAI Press, 1992, Vol. 1, p 83.

24.  JS Klassen, P Kerbarle. Collison-induced dissociation threshold energetics of protonated glycine, glycinamide, and some related small peptides and peptide amino amides. J Am Chem Soc 119:6552–6563, 1997.

25.  AG Harrison. Chemical Ionization Mass Spectrometry. 2nd ed. Boca Raton, FL: CRC Press, 1992, pp 76–81.

26.  RAJ O'Hair, S Gronert, CH DePuy, JH Bowie. Gas-phase ion chemistry of the acetic acid enolate anion [CH₂CO₂H]⁻. J Am Chem Soc 111:3105–3106, 1989.

27.  (a) H Nakata, N Arakawa, R Mizuno. Significant differences in site of protonation and extent of fragmentations in chemical ionization and fast atom bombardment mass spectrometry of simple bifunctional compounds: A mechanistic implication for formation of protonated molecules. Org Mass Spectrom 29:192–196, 1994; (b) H Nakata, Y Suzuki, M Shibata, K Takahashi, H Konishi, N Takeda, A Tatematsu. Chemical ionization mass spectrometry of bifunctional compounds: The behaviour of bifunctional compounds on protonation. Org Mass Spectrom 25:649–654, 1990.

28.  V Vais, A Etinger, A Mandelbaum. Intramolecular proton transfers in stereoisomeric gas-phase ions and the kinetic nature of the protonation process upon chemical ionization. J Mass Spectrom 34:755–760, 1999.

29.  AR Dongre, JL Jones, A Somogyi, VH Wysocki. Influence of peptide composition, gas-phase basicity, and chemical modification on fragmentation efficiency: Evidence for the mobile proton model. J Am Chem Soc 118:8365–8374, 1996.

30.  A Somogyi, VH Wysocki, I Mayer. The effect of protonation site on bond strngths in simple peptides: Application of ab initio and modified neglect of differential over-

lap bond orders and and modified neglect of differential overlap energy partitioning. J Am Soc Mass Spectrom 5:704–717, 1994.

31. For a discussion and compilation of MCAs see JE Bartmess. Gas phase equilibrium affinity scales and chemical ionization mass spectrometry. Mass Spectrom Rev 8: 297–343, 1989.

32. For a rigorous definition of neighbouring groups see IUPAC, Commission on Physical Organic Chemistry. P Muller, ed. Glossary of terms used in physical organic chemistry. Pure App Chem 66:45, 1994.

33. (a) B Capon. Neighboring group participation. Quart Rev 18:45, 1964; (b) J March. Advanced Organic Chemistry. 4th ed. New York: Wiley, 1992, pp 308–312.

34. JK Kim, MC Findlay, WG Henderson, MC Caserio. Ion cyclotron resonance spectroscopy: Neighbouring group effects in the gas-phase ionization of beta-substituted alcohols. J Am Chem Soc 95:2184–2193, 1973.

35. VH Wysocki, DJ Burinsky, RG Cooks. Competitive dehydration and deamination of $\alpha$, $\omega$-amino alcohols and $\alpha$, $\omega$ amino acids in the gas phase. J Org Chem 50: 1287–1291, 1985.

36. X Cheng, L Gao, A Bukp, L Miesbauer. Prediction of CID-MS/MS behaviour of protonated organic molecules. Proceedings of the 46th ASMS Conference, Orlando, Florida, 1998, p 85.

37. (a) W Kulik, W Heerma. Biomed Mass Spectrom 15:419, 1988; (b) NN Dookeran, T Yalcin, AG Harrison. Fragmentation reactions of protonated alpha-amino acids. J Mass Spectrom 31:500–508, 1996; (c) AG Harrison, T Yalcin. Proton mobility in protonated amino acids and peptides. Int J Mass Spectrom 165:339–347, 1997.

38. M Eckersley, JH Bowie, RN Hayes. Collision-induced dissociations of deprotonated alpha amino acids. The occurrence of specific proton transfers preceding fragmentation. Int J Mass Spectrom Ion Proc 93:199, 1989.

39. (a) RAJ O'Hair, GE Reid. Does side chain water loss from protonated threonine yield N-protonated dehydroamino-2-butyric acid? Rapid Commun Mass Spectrom 12:999–1002, 1998; (b) RAJ O'Hair, ML Styles, GE Reid. Role of the sulfhydryl group on the gas phase fragmentation reactions of protonated cysteine and cysteine containing peptides. J Am Soc Mass Spectrom 9:1275–1284, 1998.

40. (a) RS Johnson, SA Martin, K Biemann. Collision induced fragmentation of [M + H]+ ions of peptides: Side chain specific sequence ions. Int J Mass Spectrom Ion Proc 86:137–154, 1988; (b) T Yalcin, C Khouw, IG Csizmadia, MR Peterson, AG Harrison. Why are B ions stable species in peptide spectra? J Am Soc Mass Spectrom 6:1165–1174, 1995; (c) T Yalcin, IG Csizmadia, MR Peterson, AG Harrison. The structure and fragmentation of Bn (n equal to or greater than 3) ions in peptide spectra. J Am Soc Mass Spectrom 7:233–242, 1996; (d) B Paizs, G Lendvay, K Vekey, S Suhai. Formation of b2+ ions from protonated peptides: an ab initio study. Rapid Commun Mass Spectrom 13:525–533, 1999; (e) WD van Dongen, W Heerma, J Haverkamp, CG de Koster. The b1 fragment ion from protonated glycine is an electrostatically-bound ion/molecule complex of $CH_2 = NH_2+$ and CO. Rapid Commun Mass Spectrom 10:1237–1239, 1996; (f) X Jiang, JB Smith, EC Abraham. J Mass Spectrom 31:1309, 1996; (g) J Qin, BT Chait, Anal Chem 69:4002, 1997.

41. (a) RJ Waugh, JH Bowie. A review of the collision-induced dissociations of deprotonated dipeptides and tripeptides: an aid to structure determination. Rapid Commun

Mass Spectrom 8:169, 1994; (b) AM Bradford, RJ Waugh, JH Bowie. Characterization of underivatized tetrapeptides by negative-ion fast-atom bombardment mass spectrometry. Rapid Commun Mass Spectrom 9:677–685, 1995; (c) ST Steinborner, JH Bowie. A comparison of the positive- and negative-ion mass spectra of bio-active peptides from the dorsal secretion of the Australian Red Tree frog Litoria rubella. Rapid Commun Mass Spectrom 10:1243–1247, 1996.

42.  (a) SG Summerfield, MS Bolgar, SJ Gaskell. Promotion and stabilization of b1 ions in peptide phenylthiocarbamoyl derivatives: analogies with condensed-phase chemistry. J Mass Spectrom 32:225–231, 1997; (b) SG Summerfield, H Steen, M O'Malley, SJ Gaskell. Phenyl thiocarbamoyl and related derivatives of peptides: Edman chemistry in the gas phase. Int J Mass Spectrom 188:95–103, 1999.

43.  SA McLuckey, GJ van Berkel, GL Glish. Tandem mass spectrometry of small, multiply charged Oligonucleotides J Am Soc Mass Spectrom 3:60–70, 1992.

44.  AK Vrkic, RAJ O'Hair, S Foote, GE Reid. Fragmentation reactions of all 64 protonated trimer oligonucleotides and 16 mixed base tetramer oligonucleotides via tandem mass spectrometry in an ion trap. Int J Mass Spectrom 194:145–164, 2000.

45.  P Iannitti, MM Sheil, G Wickham. High sensitivity and fragmentation specificity in the analysis of drug-DNA adducts by electrospray tandem mass spectrometry. J Am Chem Soc 119:1490–1491, 1997.

46.  RP Glover, JH Lamb, PB Farmer. Tandem mass spectrometry studies of a carcinogen modified oligodeoxynucleotide. Rapid Commum Mass Spectrom 12:368–372, 1998.

47.  LA Marzilli, JP Barry, T Sells, SJ Law, P Vouros, A Harsch. Oligonucleotide sequencing using guanine-specific methylation and electrospray ionization ion trap mass spectrometry. J Mass Spectrom 34:276–280, 1999.

48.  RAJ O'Hair. Designer reactions: biomolecules in the gas phase. Chem Australia September:50–53, 1998.

49.  MK Green, CB Lebrilla. Ion–molecule reactions as probes of gas-phase structures of peptides and proteins. Mass Spectrom Rev 16:53–71, 1997.

50.  (a) SA McLuckey, DE Goeringer. Ion–molecule reactions for improved effective-mass resolution in electrospray mass spectrometry. Anal Chem 67:2493–2497, 1995; (b) XH Cheng, DG Camp, QY Wu, R Bakhtiar, DL Springer, BJ Morris, JE Bruce, GA Anderson, CG Edmonds, RD Smith. Molecular weight determination of plasmid DNA using electrospray ionization mass spectrometry. Nucl Acid Res 24:2183–2189, 1996.

51.  MA Freitas, SDH Shi, CL Hendrickson, AG Marshall. Gas-phase RNA and DNA ions. 1. H/D exchange of the $[M - H](-)$ anions of nucleoside 5′-monophosphates (GMP, dGMP, AMP, dAMP, CMP, dCMP, UMP, dTMP), ribose 5-monophosphate, and 2-deoxyribose 5-monophosphate with $D_2O$ and $D_2S$. J Am Chem Soc 120:10187–10193, 1998.

52.  (a) JL Stephenson, SA McLuckey. Gaseous protein cations are amphoteric. J Am Chem Soc 119:1688–1696, 1997; (b) TG Schaaff, JL Stephenson, SA McLuckey. The reactivity of gaseous ions of bradykinin and its analogues with hydro- and deuteroiodic acid. J Am Chem Soc 121:8907–8919, 1999; (c) JL Stephenson, TG Schaaff, SA McLuckey. Hydroiodic acid attachment kinetics as a chemical probe of gaseous protein ion structure: bovine pancreatic trypsin inhibitor. J Am Soc Mass

Spectrum 10:552–556, 1999; (d) JL Stephenson, SA McLuckey. Counting basic sites in oligopeptides via gas-phase chemistry. Anal Chem 69:281–285, 1997.

53. RJ Beuhler, LJ Greene, L Friedman. Solvated proton mass spectra of tripeptide derivative. J Am Chem Soc 93:4307, 1971; JP Speir, IJ Amster. An investigation of the energetics of peptide ion dissociation by laser desorption chemical ionization Fourier transform mass spectrometry. J Am Soc Mass Spectrom 6:1069, 1995.

54. (a) R Orlando, C Fenselau, RJ Cotter. Endothermic ion molecule reactions. J Am Soc Mass Spectrom 2:189, 1991; (b) JC Tabet, et al. Reactivity of molecular species of leucine-enkephalin towards nucleophilic reagents in EBQQ/MS and FT/MS. Spectrosc Int J 5:253, 1987.

55. RAJ O'Hair, MA Freitas, S Gronert, JAR Schmidt, TD Williams. Concerning the regioselectivity of gas phase reactions of glycine with electrophiles: the cases of the dimethylchlorinium ion and the methoxymethyl cation. J Org Chem 60:1990–1998, 1995.

56. RAJ O'Hair, SA McLuckey. Trimethylsilyl derivatization of nucleic acid anions in the gas phase. Int J Mass Spectrom Ion Proc 162:183–202, 1997.

57. F Chew, S Gronert, CB Lebrilla. Preliminary studies of the reactions of peptide dianions in a mass spectrometer: applications for sequencing. Proceedings of the 47th ASMS Conference on Mass Spectrometry and Allied Topics, Dallas, TX, 1999.

58. MA Freitas, RAJ O'Hair, TD Williams. Gas phase reactions of cysteine with charged electrophiles: regioselectivities of the dimethylchlorinium ion and the methoxymethyl cation. J Org Chem 62:6112–6120, 1997.

59. MA Freitas, RAJ O'Hair, S Dua, JH Bowie. The methoxymethyl cation cleaves peptide bonds in the gas phase. Chem Commun 1409–1410, 1997.

60. X Cheng, JJ Grabowski. Designing a reagent to cleave carbonyl-nitrogen bonds in peptides in an anion-molecule reaction. Proceedings of the 39th ASMS Conference on Mass Spectrometry and Allied Topics, Nashville, TN 1991, pp 477–478.

61. RAJ O'Hair, GE Reid. Derivatization of protonated peptides via gas phase ion-molecule reactions with acetone. J Am Soc Mass Spectrom 11:244–256, 2000.

62. EM Marzluff, S Campbell, MT Rodgers, JL Beauchamp. Low-energy dissociation pathways of small deprotonated peptides in the gas phase. J Am Chem Soc 116: 7787–7796, 1994.

63. (a) S Prabhakar, P Krishna, M Vairamani. Acetone chemical ionization studies. Acetone chemical ionization studies-IX: amino acids and nucleobases. Rapid Commun Mass Spectrom 11:1945, 1997; (b) GE Reid, SE Tichy, J Pérez, RAJ O'Hair, RJ Simpson, HI Kenttämaa. N-terminal derivatization and fragmentation of neutral peptides via ion-molecule reactions with acylium ions: towards gas-phase edman degradation? J Am Chem Soc 123:1184–1192, 2001.

64. RAJ O'Hair, MA Freitas, TD Williams. Gas phase reactions of the cyclic ethylene halonium ions $(CH_2)_2X^+$ (X = Cl, Br) with glycine. J Org Chem 61:2374–2382, 1996.

65. MA Freitas, RAJ O'Hair, JAR Schmidt, SE Tichy, BE Plashko, TD Williams. Gas phase reactions of glycine, alanine, valine, and their N-methyl derivatives with the nitrosonium ion, $NO^+$ J Mass Spectrom 31:1086, 1996.

66. (a) J Woenckhaus, RR Hudgins, M Jarrold. Hydration of gas phase proteins: a special

hydration site on gas phase BPTI. J Am Chem Soc 119:9586, 1997; A Blades, Y Ho, P Kebarl. J Phys Chem 100:2443, 1996. (b) Free energies of hydration in the gas phase of some phosphate singly and double charged anions.

67.  KM Stirk, LKM Kiminkinen, HI Kenttamaa. Ion–molecule reactions of distonic radical cations. Chem Rev 92:1649, 1992.

68.  RB Cole, ed. Electrospray Ionization Mass Spectrometry: Fundamentals, Instrumentation, and Applications. New York: Wiley, 1997.

69.  (a) M Przybylski, MO Glocker. Electrospray mass spectrometry of biomacromolecular complexes with noncovalent interactions—new analytical perspectives for supramolecular chemistry and molecular recognition processes. Angew Chem Int Ed Engl 35:807–826, 1996; (b) JA Loo. Studying noncovalent protein complexes by electrospray ionization mass spectrometry. Mass Spectrom Rev 16:1–23, 1997; (c) RD Smith, X Cheng, BL Schwartz, RD Chen, SA Hofstadler. Noncovalent complexes of nucleic acids and proteins studied by electrospray ionization mass spectrometry. In: Biochemical and Biotechnological Applications of Electrospray Ionization Mass Spectrometry. ACS Symp 619:294–314, 1996; (d) DL Smith, ZQ Zhang. Probing noncovalent structural features of proteins by mass spectrometry. Mass Spectrom Rev 13:411–429, 1994; (e) RD Smith, JE Bruce, Q Wu, QP Lei. New mass spectrometric methods for the study of noncovalent associations of biopolymers (review). Chem Soc Rev 26: 191–202, 1997.

70.  PD Schnier, JS Klassen, EE Strittmatter, ER Williams. Activation energies for dissociation of double strand oligonucleotide anions: evidence for Watson–Crick base pairing in vacuo. J Am Chem Soc 120:9605–9613, 1998.

71.  (a) RH Griffey, MJ Greig, HY An, H Sasmor, S Manalili. Targeted site-specific gas-phase cleavage of oligoribonucleotides: application in mass spectrometry-based identification of ligand binding sites. J Am Chem Soc 121:474–475, 1999; (b) RH Griffey, SA Hofstadler, KA Sannes-Lowery, DJ Ecker, ST Crooke. Determinants of aminoglycoside-binding specificity for rRNA by using mass spectrometry. Proc Natl Acad Sci USA 96:10129–10133, 1999.

72.  (a) MK Green, SG Penn, CB Lebrilla. The complexation of protonated peptides with saccharides in the gas phase decreases the rates of hydrogen/deuterium exchange reactions. J Am Soc Mass Spectrom 6:1247–1251, 1995; (b) SW Lee, HN Lee, HS Kim, JL Beauchamp. Selective binding of crown ethers to protonated peptides can be used to probe mechanisms of H/D exchange and collision-induced dissociation reactions in the gas phase. J Am Chem Soc 120:5800–5805, 1998; (c) MK Green, CB Lebrilla. The role of proton-bridged intermediates in promoting hydrogen-deuterium exchange in gas-phase protonated diamines, peptides and proteins. Int J Mass Spectrom 175:15–26, 1998; (d) GE Reid, RAJ O'Hair, ML Styles, WD McFadyen, RJ Simpson. Gas phase ion–molecule reactions in a modified ion trap: H/D exchange of noncovalent complexes and coordinatively unsaturated platinum complexes. Rapid Commum Mass Spectrom 12:1701–1708, 1998; (e) AJR Heck, TJD Jorgensen, M O'Sullivan, M von Raumer, PJ Derrick. Gas-phase noncovalent interactions between vancomycin-group antibiotics and bacterial cell-wall precursor peptides probed by hydrogen/deuterium exchange. J Am Soc Mass Spectrom 9:1255–1266, 1998.

73.  J Ramirez, F He, CB Lebrilla. Gas-phase chiral differentiation of amino acid guests in cyclodextrin hosts. J Am Chem Soc 120:7387–7388, 1998.

# 5

# The LC/MS Experiment

**Lucinda R. H. Cohen and David T. Rossi**
*Pfizer Global Research and Development, Ann Arbor, Michigan*

> When the only tool you have is a hammer, every problem looks like a nail.
> —*Home improvement proverb*

## I. WHY SHOULD SEPARATIONS BE DONE PRIOR TO MASS SPECTROMETRY?

The simplest justification for conducting separations prior to mass spectrometry (MS) is because we must, like it or not. Even the most avid proponents of mass spectrometry will concede that separations are often necessary prior to mass spectrometric analysis: The mass spectrometer is incapable of directly determining every analyte in all possible types of samples. An efficient chromatographic separation prior to atmospheric pressure ionization (API) mass spectrometry can add many valuable aspects, including improved linearity and accuracy; better sensitivity; and, in the case of coeluting metabolites or related substances, better assay selectivity. In fact, the importance of liquid chromatography (LC) compared to mass spectrometry in the LC/MS experiment is a matter of perspective. For the mass spectrometrist, the LC step may be regarded as part of a preparative procedure required for sample cleanup. For the chromatographer, the mass spectrometer functions only as an expensive and high-maintenance detector for complex separations of metabolites and impurities from the peak of interest. Regardless of perspective, the main goal of this chapter is to describe how to effectively use LC/MS when addressing analytical problems in pharmaceutical drug discovery.

Several excellent reviews and books on the development of LC/MS have already been written [1–13] and are recommended as resources. The applications of liquid chromatography to mass spectrometry, using techniques such as particle beam, thermospray ionization, or the monodisperse aerosol generating interface, are not discussed because the majority of most experiments today utilize either electrospray or atmospheric pressure chemical ionization (APCI) techniques. There is a vast array of reference materials available regarding liquid chromatography and method development [14–18]. These should be read with the understanding that many mainstream LC techniques are not directly applicable to atmospheric pressure ionization LC/MS.

## A.  Mass Spectrometric Selectivity

As described in Chapter 3, one of the primary advantages of mass spectrometry is the selectivity provided by the selected reaction monitoring or multiple reaction monitoring (SRM/MRM) experiment. By monitoring not only the specific precursor mass of the ion of interest but also a characteristic product ion, interference from other components within the sample can be eliminated in most cases. This selectivity also provides increased sensitivity by reducing the chemical noise produced by ions arising from the sample matrix, mobile phase, and impurities. The development of multicomponent procedures for such analytical procedures as cassette dosing (Chapter 11) or simultaneous identification and quantitation of metabolites is possible primarily because of this selectivity. Traditional HPLC/ UV and fluorescence detectors suffer primarily from limited selectivity and susceptibility to sample interferences. An example of this is shown in Fig. 1, which contains chromatograms of a low-molecular-weight compound detected by either mass spectrometry or UV absorbance. An intense analyte peak eluting at approximately 1 min is observed with MRM/MS detection (Fig. 1a), whereas extensive interferences and broad sloping backgrounds are observed within that same retention time window at 250, 220 or 210 nm (Figs. 1b–1d). In some ways, LC/MS allows for greater flexibility in chromatographic development because endogenous matrix components can coelute with the analyte, yet not interfere, as long as these components possess unique precursor-product masses.

## B.  Ion Suppression and Effects on Ionization

One of the most undesirable processes that can occur during atmospheric pressure mass spectrometric analysis is a nonlinear decrease of ionization by sample or mobile phase. This *ion suppression*, or *ionization suppression*, is an effect whereby the extent of ionization for an analyte is decreased due to competition between analyte and sample matrix components within the atmospheric pressure ion source. Studies have shown ion suppression to be a somewhat proportional effect [19]. That is, a quasilinear relationship is observed between the amount of salt present in a sample and the loss of analyte molecular ion signal until a limiting amount of salt is reached, whereby the response is constant with increas-

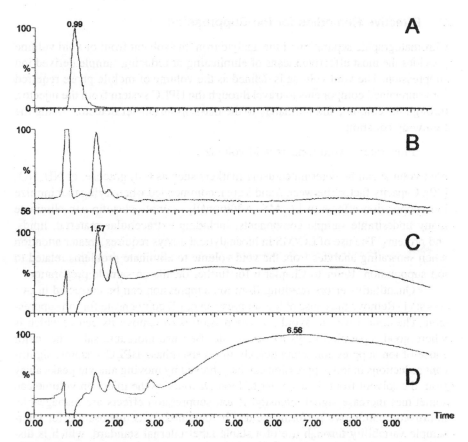

**Figure 1** Comparison of UV versus MS detection. Compound prepared using liquid–liquid extraction with MTBE, 1000 ng/mL in rat plasma. Symmetry $C_{18}$ 3.5 μm, 2.0 × 50 mm, 0.2 mL/min, 50/50 ACN/0.1% formic acid, Micromass Quattro II, electrospray ionization. (A) MRM/MS detection, (B) 250 nm, (C) 220 nm, (D) 210 nm.

ing amounts of salt added. If the analyte is not sufficiently separated from these endogenous materials, significant decreases in signal intensity can be observed. Additionally, ion suppression will occur when an excessive concentration of salts, buffers, acids, detergents or ion pair reagents are present. Ion suppression is caused by different mechanisms in APCI and electrospray. During electrospray ion formation, excessive charge buildup on the surface of the droplet may prevent desorption of analyte molecular ions. In contrast, during APCI ion formation, gas-phase cationization mechanisms can compete with $(M + H)^+$ or $(M - H)^-$ pathways.

## C.  Effective Remedies for Ion Suppression

Chromatographic separation of the analyte from the solvent front or void volume provides the most effective means of eliminating or reducing sample derived ion suppression. The void volume is defined as the volume of mobile phase required for nonretained components to travel through the HPLC system from the injector, through the column, and ultimately to the mass spectrometer detector. Using the following equation:

Flow rate $\times$ void time = void volume

void volume can be determined using markers such as CsI, glucose, or $NH_4NO_3$ [20]. Capacity factors between 2 and 5 are recommended when trying to minimize ion suppression effects in LC/MS. Additionally, sample cleanup can eliminate many undesirable sample components, including extracellular material, lipids, and proteins. The use of LC/MS in bioanalytical assays requires greater attention when separating analytes from the void volume to eliminate problems related to ion suppression. Refer to Chapter 6 for further details on sample preparation.

Quantitative errors resulting from ion suppression can be corrected in two ways: (1) Remove the cause of ion suppression or (2) compensate for ion suppression. The easiest way to avoid ion suppression is to remove its cause, which is where good separation plays a critical role. Because inorganic salts, the major cause of ion suppression, elute quickly in reverse-phase HPLC systems, significant reductions in ion suppression can be achieved by moving analyte peaks away from the solvent front. In a similar fashion, *decreasing* the injection volume can sometimes increase signal intensity. If ion suppression effects are manageable, improved quantitative performance can be obtained by compensating for inter-sample variability through use of a stable label internal standard, which is discussed later.

## II.  REQUIREMENTS FOR DEVELOPING EFFECTIVE LC/MS METHODS

### A.  Mobile-Phase Requirements

Mobile-phase degassing is an important step in the LC/MS experiment and can be accomplished via on-line membrane or vacuum devices, sonication, helium sparging or as part of the mobile-phase filtration step. Degassing will eliminate pump cavitation, ensure reproducible retention times and minimize possible sputtering from the ion source.

### 1.  Organic Components

Acetonitrile and methanol are almost exclusively chosen in LC/MS methods as organic mobile-phase components. Methanol has a greater gas-phase acidity, polarity, and volatility than acetonitrile and may be preferred for some types of separations. In positive ion mode, methanol has been shown to deliver 10 to 50% better sensitivity than acetonitrile, while in negative ion mode there is little difference in sensitivities for most analytes [21]. A typical mobile phase to start the experiments can be 90/10 mixtures of MeOH/$H_2O$ or ACN/$H_2O$ and can be varied until the desired capacity factor is achieved. Although there is a small chance for methanol:analyte adduct formation, methanol is often the solvent of choice in LC/MS because higher percentages of this solvent in the mobile phase will give equivalent chromatographic capacity relative to the same percentage of acetonitrile. For example, 50% acetonitrile gives the same elutropic strength as 60% methanol. Higher organic composition is desirable in LC/MS due to improved effluent evaporation at a given temperature, thereby decreasing the background.

Ethanol, 2-propanol, and tetrahydrofuran are less frequently used as mobile-phase components. Toluene, hexane, benzene, cyclohexane, dichloromethane, and carbon tetrachloride have been used for normal-phase separations [22–24] although many of these organic solvents require addition of more polar solvents for good ionization. Normal-phase HPLC solvents are compatible with LC/MS and generally pose no threat of fire or explosion in the nitrogen atmosphere that blankets the ion source. These solvents are often more effective when used with APCI rather than electrospray because as a solvent-mediated ionization process, electrospray will only produce an ion beam if preformed ions are present. Also, if air is allowed to enter the ion source as a result of loss of nitrogen pressure, it is possible to damage the components of the instrument vacuum region due to mobile-phase oxidation.

## 2. Aqueous Components

Careful choice of an appropriate acid or buffer will help ensure success in the LC/MS experiment. Nonvolatile aqueous components, whether salts, acids, bases, or buffers, will greatly decrease and even prevent the detection of analyte ions. These nonvolatile buffers can also foul ion sources and vacuum regions of mass spectrometers. Nonvolatile phosphate or citrate buffers are strongly discouraged for both ionization and practical reasons. Although many instrument manufacturers have developed ion sources rugged enough to tolerate deposition of nonvolatile components, frequent cleaning is often necessary. Ion suppression and decreased sensitivity will also be observed when nonvolatile buffers are used. In most cases, volatile acids or buffers can be substituted for more traditional HPLC choices.

## 3. Buffers

Ammonium acetate or formate buffers can be used with concentrations ranging from 2 to 50 mM, although a maximum concentration of 10–20 mM is recommended to avoid ion suppression. The acceptable buffer concentration that does not adversely affect ionization has been observed to be highly compound dependent. A useful rule of thumb is to use as low a concentration of buffers as possible to give reasonable chromatographic performance [21].

Buffers should be chosen such that the pH of the mobile phase falls within the buffer's natural $pK_a$ range for maximum buffering capacity. For example, the $pK_a$ of acetate buffer is 4.8, with a range 3.8–5.8; the useful pH range for formate buffers is 2.8–4.6. Buffer capacity outside of these ranges is limited, particularly at lower concentrations. $N$-methyl morpholine may be used at higher pH ranges and has shown utility in improving sensitivity and chromatographic peak shape for negative ion APCI analysis [25]. Ammonium adducts can be observed in positive ion mode and formate or acetate adducts in negative mode at higher buffer concentrations. Adduct formation is detrimental only in that it can cause greater variability and loss of detection sensitivity for the analyte. These adducts can sometimes be destroyed by increasing the source temperature, voltage, or both.

## 4.  Acids and Bases

Formic or acetic acid concentrations of 0.1–1% (v/v) are recommended when preparing low pH mobile phases to enhance ionization in electrospray. Trifluoro-acetic acid is preferred for protein and peptide separations but should be avoided when negative ion mode is utilized. Ammonium hydroxide, or, in rare cases, triethylamine, are recommended for high pH mobile phases.

For basic compounds, 0.1% acid should be mixed with the organic component, whereas water or neutral buffer should be used for neutral or acidic species. The previously discussed buffers can be used as the aqueous component to improve peak shape and resolutions by providing greater control of pH.

## B.  Mobile-Phase Components That Are Not
##    Recommended

Certain types of traditional LC mobile phase additives should be avoided due to nonvolatility and ion suppression effects. Mobile-phase related ion suppression will not depend on the analyte proximity to the solvent front, or capacity factor. These additives include detergents; surfactants; ion pairing agents; inorganic acids such as sulfuric, phosphoric, hydrochloric, and sulfonic acids; nonvolatile salts such as phosphates, citrates, and carbonates; strong bases; and quaternary amines. Complete suppression of ionization as well as interferences in both positive and negative ion mode will occur when these agents are utilized.

Tetrahydrofuran usage is discouraged as a major mobile-phase component in LC/MS. Safety concerns due to flammability and peroxide formation restrict usage of this solvent in heated nebulizer ion sources. Additionally, mobile-phase solvents containing high percentages of tetrahydrofuram will cause PEEK tubing, commonly used in LC/MS plumbing, to swell with extended usage and often degrade polymeric components in HPLC seals and valves. However, tetrahydrofuran is commonly used as a mobile phase for HPLC [26,27] and has been coupled with mass spectrometric detection [28]. In some experiments, a dilute solution of NaI in methanol was added postcolumn to allow analyte cationization. Common sense should prevail when tetrahydrofuran is considered as a mobile-phase component; up to 10% can be used, with care, to effectively alter chromatographic selectivity.

Table 1 summarizes the commonly used and recommended LC mobile-phase components as well as those that should not be used.

## C.  Useful HPLC Column Dimensions and Parameters

## 1.  Flow Rates

The initial impulse of an analyst conducting an LC separation with mass spectrometric detection can be to use the analytical column, mobile-phase, and flow-

**Table 1**  Recommendations for LC/MS Mobile-Phase Preparation

|  | Recommended | Not recommended/highly discouraged |
|---|---|---|
| Organic solvents | Acetonitrile, methanol, ethanol, isopropanol, dichloromethane | THF |
|  | Normal phase: toluene, hexane, benzene, cyclohexane carbon tetrachloride |  |
| Buffers | Ammonium acetate, ammonium formate, triethylammonium acetate (10 mM) | Phosphate, citrate, carbonate |
| Acids | Acetic acid, formic acid, trifluoroacetic acid (positive ion mode only) | Sulfuric acid, perchloric acid, phosphoric acid, hydrochloric acid, sulfonic acids |
| Bases | Ammonium hydroxide | Quaternary amines, strong bases, generally triethylamine |
| Other components |  | Detergents, surfactants, ion pairing agents, nonvolatile salts |

rate conditions from an LC/UV method. As discussed above, mobile-phase components should be carefully considered in light of their mass spectrometric compatibility. Additionally, the flow-rate requirements of electrospray versus APCI should be considered. Atmospheric pressure chemical ionization can be operated at flow rates from 0.1 to 2 mL/min and provide optimum sensitivity at ~1 mL/min, whereas electrospray is better suited to flow rates in the range of 10–500 µL/min and is optimal at ~200 µL/min. Flow-rate limits for electrospray are defined by source design; some instruments operate well up to 1 mL/min, provided nebulizer gas temperatures and gas flow rates are proportionally increased. Optimized flow rates can be achieved by splitting the LC effluent prior to introduction into the mass spectrometer or by decreasing overall flows concurrent with decreased column dimensions. Splitting can be accomplished by diverting part of the LC flow to an additional detector or waste.

Column manufacturers have developed a wide range of column bore size, length, and column packing particle size to accommodate the increased demand for smaller and more efficient columns. The column bore size will dictate the appropriate flow rate and can be tailored to meet the analyst's needs. Table 2 lists the general column bore size (inner diameter, or i.d.) and appropriate flow rates.

## D.  Column Dimension Trade-Offs

Decreasing the size of the column inner diameter will increase sensitivity compared to analytical columns when all other parameters are kept constant and the system is not limited by extracolumn band broadening. The injection volume should also be proportionally decreased to minimize any band-broadening effects. Because overall sensitivity is increased, interfering as well as analyte peaks will increase in intensity. The extracolumn volume within an HPLC system will have greater effect on band broadening for smaller bore columns, which will decrease resolution between peaks. One disadvantage to using smaller columns is loss of

**Table 2**  Column and Flow Rate
Parameters

| Column i.d. (mm) | Flow rate |
| --- | --- |
| Capillary | <10 µL/min |
| 1.0 | 40–50 µL/min |
| 2.1 | 0.1–0.5 mL/min |
| 3.0 | 0.5 mL/min |
| 4.6 | 1.0 mL/min |

ruggedness because smaller columns will clog more rapidly from "dirty" samples. Use of a guard column prior to the analytical column will help reduce column replacement costs and prolong column lifetimes. For greater chromatographic capacity and resolution, the column length can be increased at the price of longer run times.

One of the more impressive advantages of LC/MS is that less resolution and selectivity are required from the LC separation. For assay methods involving complex matrices such as plasma, the separation of analyte from matrix components is not as critical, so the separation can be much shorter. Assay methods requiring 15 to 20 min with LC/UV or fluorescence can be shortened to only a few minutes. Methods that previously required large numbers of theoretical plates and long (25-cm) analytical columns can now be implemented with less resolution and shorter (1- to 15-cm) columns.

## E.  Simple Strategies for Choosing an Appropriate HPLC Column

Once again, the wide depth and breadth of knowledge regarding HPLC method development and assay optimization available in the literature cannot be covered here. The first suggestion is to check the literature for well-documented assays of known or related compounds. In many cases, the method used for HPLC/UV provides a helpful starting point for LC/MS methods with the same compound. In the interest of brevity, method development discussions are limited to reverse-phase separations. The unique requirements of normal phase and chiral chromatography are best discussed in other forums.

The first step in selecting a column for LC/MS analysis, once the column size has been determined, is to consider the analyte molecule's characteristics, such as hydrophobicity/hydrophilicity, $pK_a(s)$, or salt form. For hydrophobic compounds, the traditional $C_{18}$ column allows selectivity for polar versus nonpolar compounds. Column manufacturers have made significant progress and refinements of manufacturing techniques for uniform high-surface-area supports. Shorter chain packings such as $C_8$ or $C_4$ present fewer hydrophobic surfaces for interaction with analyte molecules. Aromatic compounds may be well suited for analysis by phenyl columns, which allow separation via $\pi$-$\pi^*$ interactions. Cyanopropyl columns allow the greatest degree of separation for polar compounds, although they generally are less rugged and reproducible from column to column than other reverse-phase columns.

Assay sensitivity can often be an issue, even with LC/MS. Some ways to improve sensitivity in LC/MS center on the composition of the mobile phase. A mobile phase that is high in organic content can be readily volatilized and pumped away in the atmospheric pressure region, which lessens the mass spectral background and improves the signal-to-noise ratio. Also, chromatographic peaks that are rapidly eluted from the column ($k'$ of 1 to 3) are much narrower than those with longer retention. From this perspective, it is therefore useful to run mobile phases high in organic content.

To gain adequate retention with high organic mobile phases, highly retentive columns should be selected initially. These columns are characterized by high carbon loads and are usually $C_{18}$-sorbent material. Some widely available, high-retention column packing media include Jsphere $C_{18}$, Adsorsphere $C_{18}$, Alltima $C_{18}$, Symmetry $C_{18}$, and Zorbax XDB $C_{18}$, although other useful varieties are also available.

In many cases, the quality of the column is dictated by a combination of manufacturing factors such as silica particle size, batch reproducibility, and surface pore size. Chromatographers agree that the additional price of purchasing columns with rigorously controlled characteristics is well worth the cost in terms of time savings and decreased frustration.

## F.  Silanol Effects and Solutions

Although bonded silica-based chromatographic media are the most efficient and versatile packing materials currently available, they are not without problems. One important feature of these materials is the presence of silanol functional groups in the silica backbone. Because these silanol groups are present in all silica (monomeric as well as polymeric) and because it is not possible to deactivate all of them through endcapping, silanol activity can dramatically affect separations. In reverse-phase separations this so-called secondary interaction primarily occurs in the form of mixed mechanism retention, a combination of reverse-phase partitioning and ion exchange [29–31]. If controlled and utilized, silanol-derived cation exchange can allow greater chromatographic retention and selectivity for amine-containing analytes. If not correctly controlled, silanol effects will lead to peak tailing, variable retention, and loss of chromatographic efficiency.

The number and nature of unreacted surface silanols affects the character of a stationary phase. Initially free, geminol or associated silanols are minimized through a process known as endcapping, which bonds various species to the residual silanols. Hydrophilic endcaps or bulky ''steric'' endcaps that separate the hydrocarbon chains and prevent analyte interaction with the silica surface can be used. If residual silanols are left unreacted (and some always are), the analyte will be separated based on a combination of interactions with both the reverse-phase support and the highly polar silanol groups. Increased retention, changes in elution order, and tailing will result for basic compounds.

For strongly cationic compounds, such as aliphatic amines, where silanol effects are most dramatic, the separation can be developed by several good approaches, the first of which is column selection. Choosing a column packed with polymerically bound silica and exhaustive endcapping will deactivate most of the silanol groups. Modern chromatographic media such as Zorbax Rx™ and Waters Symmetry™ have been effective in this regard. A classic way to manage and decrease silanol interactions is to add a small percentage (0.01 to 0.1%) of triethylamine (TEA) or a related amine to the mobile phase. This approach is generally not useful in LC/MS separations because the presence of TEA will suppress ionization of positively charged analytes and cause great loss of sensitivity. TEA also tends to bind to reagent lines and internal pump surfaces, leaching out slowly over several days. This is problematic in that it will add greater response variability (50% or more) to the LC/MS experiment and will affect determinations with other analytical methods on the same instrument in unpredictable ways.

A third approach to modulate silanol effects is through mobile-phase pH control. At pHs below 4, most silanol groups will be protonated and less available as cation exchange sites. Addition of acetic or formic acid to the mobile phase is useful for this purpose. Buffer concentrations of about 25 mM will promote retention of buffer cations rather than analyte molecules, but can promote ion suppression.

Silanol effects can also manifest as columns "age." At low pHs (<2), the bonded phase and endcapping can be hydrolyzed from the silica support, leaving behind free silanols. At higher pHs (>8), the silica itself can slowly dissolve. This hydrolysis limits the useful mobile-phase pH range to 2–8. Greater column stability is observed as a function of chain length, with $C_{18}$ columns exhibiting longer lifetimes than $C_8$ or phenyl columns [32]. Some new silica-based columns such as the Zorbax Bonus™ or the Waters Xterra™ are reported to operate at higher pH ranges than conventional silica columns. This could be useful when greater retention of amines is required. Polymeric stationary phases that do not use silica can be a useful choice if silanol effects are problematic, but are generally not as chromatographically efficient or technically well developed.

## G. Retaining Ionic Compounds

A significant challenge for users of API LC/MS is achieving adequate retention for ionic compounds when using the limited mobile-phase additives that are compatible with this technique. This is much less of an issue with other detection techniques such as UV absorbance, fluorescence, or amperometry because ionic compounds can be readily retained in reverse-phase mode using ion-pair reagents or through the use of ion-exchange chromatography. Because each of these approaches typically requires nonvolatile mobile phase additives, they are not generally useful with API LC/MS. A few approaches have, however, been adapted for use with API LC/MS and these are described below.

*Chromatographic ionization suppression* is useful for increasing the retention of organic acids in reverse-phase systems. With this approach, the apparent pH of the mobile phase is decreased by the addition of acetic or formic acid.

Acetic acid (0.1% by volume) will give an apparent mobile-phase pH of slightly less than 3 and this is capable of protonating most carboxylic analytes. In this way, the organic acid analytes become neutral molecules and are better retained under reverse-phase conditions. It has been reported that there is little or no loss in sensitivity for carboxylic acids when using electrospray ionization under acidic conditions [24]. In a similar way, it is possible to suppress the ionization of basic drug molecules by elevating the mobile phase to a pH above its $pK_a$. For amines, this would be pH 10 or higher. To do this effectively, a special HPLC sorbent such as polymer or a surface-modified zirconium could be used. Chromatographic ionization suppression is not practical for compounds that are unstable under acidic and/or basic conditions.

The majority of ion-pair reagents are either nonvolatile and will quickly lead to fouling of the ion source and the vacuum region of the mass spectrometer or will lead to ion suppression in the mass spectrometer source and are thereby unsuitable. A few papers [33,34] have reported the effective use of ion-pair reagents as additives to the injection solvent. Although this approach is not as rugged, in terms of reproducibility of chromatographic retention, as addition of the ion-pair reagent to the mobile phase, it does achieve adequate analyte retention and can also improve assay sensitivity.

Figure 2 demonstrates the effects of adding octane sulfonic acid to the injection solvent for a reverse-phase separation of pyridinium and deoxypyridinium, components of collagen, in rat urine. These polyamine containing compounds are protonated and poorly retained under the usual acidic or neutral mobile-phase conditions. The sample preparation method is simple dilution and does not afford the removal of salts from the sample. Therefore if the analytes were inadequately retained the sensitivity, as well as the method accuracy, would suffer. As the concentration of octane sulfonic acid is increased to 50 mM (Fig. 2a), both the retention time and the peak response for the analytes improve significantly. Little improvement is obtained at higher concentrations of octane sulfonic acid and retention is not strongly dependent on the injection volume (Fig. 2b). To protect the ion source from the fouling effects of octane sulfonic acid in the injection solvent, a timed divert valve was inserted before the ion source to shunt the excess ion pair reagents to waste during the first few minutes of each injection.

Although weak cation-exchange separations, such as those obtained on crosslinked benzoic acid/divinyl benzene or bare silica, have been used for many years with more conventional HPLC detectors, they are starting to gain some attention as alternative separations for API LC/MS [35,36]. These ion-exchange separations have advantages for small zwitter ionic compounds such as amino acids that cannot be retained by chromatographic ionization suppression or by addition of an ion-pair reagent to the injection solvent. In an ion-exchange mode, a cationic analyte is exchanged on the active silanol sites of silica or on the cation-exchange sites of other materials. The separation parameters are limited

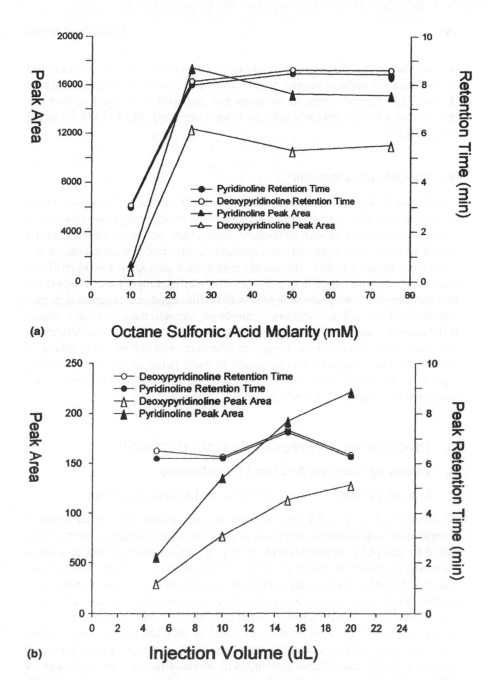

**Figure 2** Effect of (a) injection solvent concentration of octane sulfonic acid and (b) injection volume containing 50 mM octane sulfonic acid on retention time and electrospray peak area response for pyridinoline and deoxypyridinoline in a reverse-phase LC/MS/MS separation. (From Ref. 34.)

because the highest useable concentrations of counter ions are 25 to 50 mM and the practical selection of counter ions is limited to volatile components such as $H^+$ or $NH_4^+$. Improvements in the types and strengths of available cation exchange resins would definitely help the researcher using API LC/MS for small ionic species.

## H.  Sample Requirements

Sample purity plays an important role in speed and accuracy of analysis as well as method ruggedness. In general, the endogenous matrix components present in most biological samples such as proteins, lipids, salts, and extracellular material should be removed via precipitation, extraction, filtration, centrifugation, or any convenient means possible. The sample preparation procedure should purify as well as release drug bound to proteins or other matrix components. After extraction and sample cleanup, the sample should be dissolved or reconstituted in mobile phase or in a solvent with lower eluotropic strength than the mobile phase. If the sample is dissolved in solvents stronger than the mobile phase, differences in solvent strength can cause changes in retention behavior and peak tailing or fronting. Sample preparation is discussed in more detail in Chapter 6; specific types of samples from studies in microdialysis, permeability, and metabolism are discussed in Chapters 8, 9, and 12.

## III.  DEVELOPING AN EFFECTIVE LC/MS METHOD

## A.  Choosing Between APCI and Electrospray

### 1.  Analyte Volatility, Thermal Stability, and Molecular Weight

As shown in Fig. 3, APCI and electrospray are suitable for a wide range of compounds, with complementary advantages for each technique. APCI is best suited for less polar compounds with molecular weights below 1000 Da, whereas electrospray should be chosen for mid- to high-molecular-weight polar or ionic analytes. Volatility and thermal stability are more important factors in APCI than electrospray because APCI is a gas-phase ionization technique. In fact, the molecular weight limit of approximately 1000 is defined by the compound volatility. At molecular weights greater than approximately 1000 Da, the molecule will not volatilize well and thus cannot be readily ionized in the gas phase. Increased analyte volatility and thermal stability will increase the ease of ionization for either technique.

In general, APCI is better suited for heteroatom-containing molecules, and electrospray for ionic compounds. Atmospheric pressure chemical ionization is

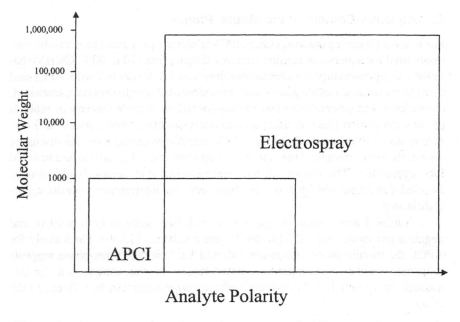

**Figure 3** Comparison of APCI and electrospray applicability to analyte molecular weight as a function of polarity.

also less prone to matrix effects and performs better at higher flow rates (>0.5 mL/min), which has led to its reputation as a more robust technique than electrospray.

## B. Ionizability

The ability of a given analyte to form a gas-phase ion plays a critical role in mass spectrometry. The presence of highly polar functional groups such as carboxylic acid groups or primary amines indicates high ionizability. Molecular gas-phase acidity or basicity can generally be predicted from general considerations of acidic or basic groups within the molecule. Other functional groups such as hydroxyls, thiols, ethers, sulfones, or amides are also easily ionized. More ionic/polar species are amenable to electrospray due to their ability to accept or donate a proton from/to the mobile phase during the ionization process. Atmospheric pressure chemical ionization requires donation or acceptance of a proton during gas-phase reaction with a mobile phase ion but less polar compounds with no readily ionizable functional group have also been determined by this technique.

## C. Aqueous Content of the Mobile Phase

For reverse-phase separations, either APCI or electrospray ionization can be routinely used for aqueous or organic contents ranging from 10 to 90%. Chiral separations using normal-phase chromatography can be conducted with APCI and completely organic mobile phases such as acetonitrile, tetrahydrofuran, methanol, chloroform, and others. Some general considerations include the ease of mobile-phase vaporization (thus, limiting a completely aqueous mobile phase) and surface tension effects that will interfere with electrospray ionization. As discussed above, the ionic strength of the aqueous component should be minimized to avoid ion suppression. The use of ion-pair reagents and surfactants should also be avoided if at all possible due to ion suppression and interference with the vaporization step.

Tables 3 and 4 show the gas-phase acid–base scale in both positive and negative ion modes for LC/MS mobile-phase solvents [13,37]. Particularly for APCI, the mobile phase components should be considered gas-phase reagents to promote either protonation or deprotonation of the analyte ion. In this manner, appropriately volatile mobile-phase components can be tailored to the assay.

**Table 3**  Gas Phase Acid–Base Scale for Positive Ions

|             | Reagent ions       |             | Neutral molecules |
|-------------|--------------------|-------------|-------------------|
| Strong acid | $CH_5^+$           | Weak base   | Methane           |
|             | $N_2OH^+$          |             | $N_2O$            |
|             | $C_2H_5^+$         |             |                   |
|             |                    |             | Formaldehyde      |
|             | $H_3O^+$           |             | Water             |
|             |                    |             | Formic acid       |
|             | $CH_3OH_2^+$       |             | Methanol          |
|             |                    |             | Benzene           |
|             | $CH_3CNH^+$        |             | Acetonitrile      |
|             |                    |             | Acetic acid       |
|             |                    |             | Acetone           |
|             |                    |             | Phenol            |
|             | $t\text{-}C_4H_9^+$ |            | Ethyl acetate     |
|             |                    |             | Diethyl ether     |
|             | $NH_4^+$           |             | Ammonia           |
|             |                    |             | Methylamine       |
|             | $C_5H_5NH^+$       |             | Pyridine          |
| Weak acid   |                    | Strong base | Trimethylamine    |

**Table 4** Gas Phase Acid–Base Scale for Negative Ions

|            | Reagent ions   |             | Neutral molecules |
|------------|----------------|-------------|-------------------|
| Strong base | $NH_2^-$       | Weak acid   | Ammonia           |
|            | $OH^-$         |             | Water             |
|            | $C_6H_5CH_2^-$ |             | Toluene           |
|            | $CH_3O^-$      |             | Methanol          |
|            |                |             | Ethanol           |
|            | $NCH_2C^-$     |             | Acetonitrile      |
|            | $CH_3COH_2C^-$ |             | Acetone           |
|            | $CH_3S^-$      |             | $CH_3SH$          |
|            | $O_2NH_2C^-$   |             | $CH_3NO_2$        |
|            | $CN^-$         |             | HCN               |
|            | $C_6H_5O^-$    |             | Phenol            |
|            | $CH_3COO^-$    |             | Acetic acid       |
| Weak base  | $Cl^-$         | Strong acid | HCl               |

## D. Dynamic Range Requirements

The dynamic range required for a given assay will depend on extra-assay requirements. For example, trace determinations of drugs in brain microdialysates can require great sensitivity but not a particularly wide dynamic range. A combination of background noise level as well as sample extraction efficiency will typically determine the limit of quantitation. Correspondingly, mass effects during ionization will influence the upper limit of the dynamic range if electrospray ionization is used. This effect is caused by the limit on the number of analytes produced per droplet during ionization if the concentration is too high. Typical drug compound linear ranges can vary anywhere from 1 to 100,000 ng/mL, with lower or higher limits of quantitation (LOQs) reported, depending on the molecule. The dynamic range observed is typically wider for APCI, with 3 to 4 orders of magnitude routinely observed compared to 3 orders of magnitude for electrospray. For many polar, nonionic compounds, APCI can provide more universal ionization, better sensitivity, and lower limits of quantitation.

## E. Sample Cleanliness

APCI has been shown to be more readily influenced by the presence of negative or positive species from salts or buffers which will prevent protonation or deprotonation by competition with the analyte molecules [25,38–39]. Salts are sometimes deliberately added to samples to be assayed by electrospray in order to increase analyte ionization through cation adduct formation (($M + Na)^+$ or $(M +$

K)$^+$). Buffers are not necessary for APCI, although they can be required for adequate chromatographic separation of sample components.

## F.  Gradient Use

Gradients can be used with equal ease for either ionization technique. In most cases, cycle time for system reequilibration (determined by the overall system dead volume) provides the practical limitation to their usage. If, for example, a particular HPLC pump/autosampler combination has 1.0 mL of dead volume (or dwell volume, the volume of all plumbing between where the solvents are mixed and the column head) and is operating at a flow rate of 1.0 mL/min (typical for APCI), then the lag time between when the gradient is initiated and when the correct solvent composition reaches the pump head is 1 min (1.0 mL/(1.0 mL/ min)). If the flow rate is only 0.2 mL/min (typical for electrospray), then the lag time will be 5 min. This means that a typical gradient run would require 5 min to initiate reequilibration plus whatever time is required for elution and final reequilibration (usually 10 to 20 column volumes). This is clearly an unacceptable time delay.

Two solutions to this problem are feasible; one is to purchase a low-dead-volume pump. Pumps with dead volumes as low as 10 µL are available for low-flow-gradient work. A second solution is to construct and introduce a splitter before the column or the ion source. In this way, the pump can be operated at 1.0 mL/min and the lag time shortened to 1 min. By using a 5:1 split, the ion source will be exposed to 0.2 mL/min, a typical flow rate for an electrospray source.

Gradient separations in LC/MS offer three distinct advantages. First, poorly shaped chromatographic peaks can be sharpened. Tailing and fronting can be eliminated to a considerable extent. Second, the time required for separation of dissimilar analytes can be shortened extensively. Early eluting compounds can exhibit adequate retention while late eluting compounds can be more quickly eluted. Third, gradients can serve as a rudimentary sample cleanup technique, especially for samples with high salt content but low amounts of proteins, cellular components, or lipids.

Step or "ballistic" gradients have been successfully utilized for many separations, including in vitro buffer samples, and are discussed in greater detail in Chapter 6. The step square wave or curve allows more rapid changes in organic/ aqueous composition compared to a simple linear gradient, thereby providing an opportunity for salts and polar matrix components to elute at or near the chromatographic void volume, while analytes are retained.

### 1.  Generic Gradient Methods for Compound Classes

Depending on sample mixture complexity, a gradient method is easily tailored to affect separation of closely and widely spaced chromatographic peaks. For

**Table 5**  Comparison of APCI Versus Electrospray

|  | APCI | Electrospray |
|---|---|---|
| Advantages | Neutral molecules ionize | Wide range of compounds |
|  | Less ion suppression | No problems with thermal |
|  | Wider dynamic range $10^3$ | stability/volatility |
|  | $-10^4$ | Higher MW |
|  | Mass sensitive detection | Concentration sensitive detection |
|  | Flow 0.5–2.0 mL/min | Better sensitivity |
|  | Higher buffer conc. (up to 20 mM) OK |  |
| Disadvantages | Thermal stability necessary (130–150°C) | Mobile-phase ion suppression |
|  | Volatility necessary | Lower dynamic range |
|  | More sensitive for nonvolatile components | LC flow <200–300 μL/min |
|  |  | Multimer formation |
|  |  | Corona discharge problems |

basic compounds, a gradient consisting of 10–90% of 0.1% formic or acetic acid with acetonitrile can be used over any run time desired. Basic or neutral compounds can be separated using water/organic gradients in the same manner.

A summary of the similarities and differences between electrospray and APCI is shown in Table 5.

## IV.  ESTABLISHING THE MASS SPECTRAL PROPERTIES

### A.  Choosing Ionization Polarity

Ionization polarity should be chosen in conjunction with the mobile phase because pH can influence ease of ionization in positive or negative ion mode. Basic or neutral compounds are readily ionized in positive ion mode at a pH below 7. Because many potential drug molecules contain amine moieties, the majority of LC/MS methods are conducted in positive ion mode. Acidic compounds are most amenable to negative ion mode unless the molecule exists as a zwitterion. Negative ion mode provides an additional advantage of selectivity because chemical background is lower compared to positive ion mode. In some cases, the LC/MS experiment can be conducted in both positive and negative ion mode. If, for example, the analyte is best suited for positive ion mode analysis but its corresponding hydroxylated metabolite is best analyzed in negative ion mode, both polarities can be monitored in tandem. The LC conditions should be developed to sufficiently separate the positive and negative components to allow polarity switching. The first portion of the chromatogram may be measured in positive ion mode, with the second portion in negative, or vice versa. Rapidly cycling between the two modes during the entire chromatographic acquisition is possible with some instruments. However, stabilization of the electronics after switching requires significant delay times. Insufficient delay times between polarity switching leads to extensive electronic noise and reduced sensitivity. An application of polarity switching is shown in Fig. 4 for a mixture of gabapentin and naproxen. For the first few minutes of the chromatographic run, gabapentin

**Rat 4 PO II 15m PLM-1**

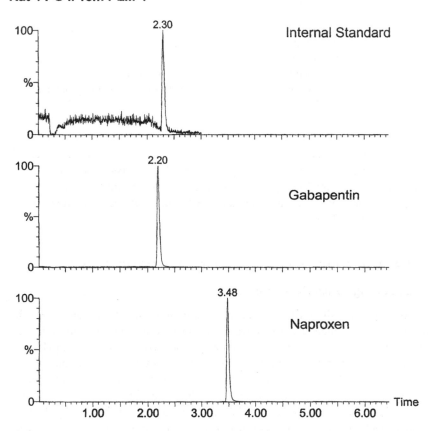

**Figure 4**    LC/MS/MS of a mixture of gabapentin and naproxen using polarity switching. Micromass Quattro II, electrospray ionization rat plasma sample prepared by protein precipitation/extraction using a MetaChem 4.6 × 30-mm Polaris column with 3-μm packing with a MetaChem 2000-MG2 guard column. Flow rate = 2.0 mL/min split 1:10 to MS. Injection volume 10 μL; step gradient: 100% 0.1% acetic acid, 0–1.0 min; 70/30 0.1% acetic acid/acetonitrile, 1.0–1.5 min; 100% acetonitrile, 1.5–2.0 min. (Courtesy of Laura F. Polchinski, Jon Ebright, and Scott T. Fountain, personal communication.)

and its internal standard are monitored in MRM positive ion mode. At 3 min, the polarity of the instrument is switched to negative ion mode, and naproxen is detected.

## B. Precursor and Product Ions

For quantitative determinations, the precursor and product ions should be chosen to minimize the possibility of interference and maximize specificity and reproducibility. Interference generated by selection of nonspecific precursor–product ions will increase the background noise and thus decrease sensitivity. Generally the molecular ion (either $(M + H)^+$ or $(M - H)^-$) is preferred as a precursor although $(M + Na)^+$ or $(M + K)^+$ can be more effective under certain circumstances.

Selection of a molecular fragment as a precursor should only occur in cases in which fragmentation is unavoidable, even at low cone/orifice voltages. Frequently, higher molecular weight conjugates such as metabolites or degradation products will produce fragments identical to those observed for the parent compound. The product ion chosen should represent a significant loss (i.e., greater than 50 Da) from the precursor molecule. Loss of water or carbon dioxide is generally less compound specific than losses of phenyl groups or rings. Very low molecular weight product ions such as $NH_4^+$, $COO^-$, or $Br^-$ should also be avoided based on increased potential for interferences. Despite careful attention to the selection of precursor and product ions, a real potential exists for interference from higher molecular weight metabolites, such as glucuronides.

## C. Compound Profiling Techniques

Prior to the sample preparation and chromatographic separation, a clean solution of the compound to be determined will be "profiled" by the analytical chemist to determine the appropriate precursor–product masses, polarity, and ionization mode (electrospray versus APCI). The appropriate mass spectrometric parameters such as orifice or cone voltage, collision energy, lens voltages, and nebulizer gas flows will also be optimized. Solution concentrations of 500–1000 ng/mL should be used to avoid contaminating the source with the analyte (thus increasing background in subsequent experiments). Ideally the sample should be dissolved in mobile phase to mimic experimental HPLC conditions, but neat organic solvents often perform equally well.

Profiling should be conducted in "high resolution" mode in order to precisely determine mass and verify precursor isotopic distributions of $M + 1$ and $M + 2$ isotopic patterns where appropriate. Excessively high standard concentrations will interfere with isotopic distribution verification and accurate mass determination if the signal for the molecular ion is saturated. Once the precursor and

product ion masses have been determined, the resolution can be lowered for
SRM/MRM experiments to increase sensitivity.

## D. Infusion versus Loop Injection

To determine the correct mass spectrometer operational conditions for a given
compound, a clean standard solution can be introduced into the ion source by
either syringe pump infusion at a low flow rate (10–30 µL/min) or by multiple
loop injections. An actual loop (50–100 µL) can be partially filled with solution
and a Rheodyne valve manually switched by the analyst to inject the sample.
Alternatively, small-volume ( ~1 µL) loop injections can be introduced by an
autosampler while mobile phase is pumped into the mass spectrometer ion source.
The choice of continuous infusion versus loop injection depends on both the type
of instrument and analyst preference. Infusion techniques require larger volumes
of solution but provide additional time for parameter optimization. The maximum
practical flow rates provided by the syringe pump typically limit this type of
introduction to electrospray sources because APCI requires flow rates of 500 µL/
min or greater. If APCI conditions are to be obtained in this manner, the analyte
can be added to the mobile phase and delivered by the HPLC pump or a simple
mixing tee. While profiling with loop injections is not as precise as continuous
infusion, it can provide great time saving and higher throughput.

   In the following example, a compound of molecular weight 454 (structure
shown in Fig. 5a) is profiled using the loop injection approach and without a
chromatography column. As shown in Fig. 5b, the first step in the profiling experiment is to obtain the precursor ion spectrum. An intense peak corresponding to
the pseudomolecular $(M + H)^+$ ion is observed at 455 Da. Next, the product ion
spectrum is obtained as shown in Fig. 5c. The most abundant product ion is
detected at 100 Da, corresponding to a portion of the molecule containing a morpholine ring substituted with an ethylene group. Both the precursor and product
ion spectra are obtained using a general ion source voltage (cone or orifice) setting
that should be low enough to minimize in-source fragmentation yet still allow
detection of some precursor ions. For the product ion spectrum, the collision
energy is also held at the minimum value to avoid observation of less specific
low-molecular-weight fragments.

   Once the precursor and product ion masses are determined, the cone or
orifice voltage can be optimized by an MRM experiment that will simultaneously
monitor the precursor–product transitions for a range of voltages. The results are
shown in Fig. 5d for cone voltages ranging from 10 to 40 V, with the collision
energy held constant at 20 V. The MRM channel with the greatest intensity (25
V) is then chosen as the optimal cone voltage. A similar experiment is then conducted for collision energies ranging from 10 to 40 V, the results of which are
shown in Fig. 5e, and the optimal voltage was determined to be 20 V.

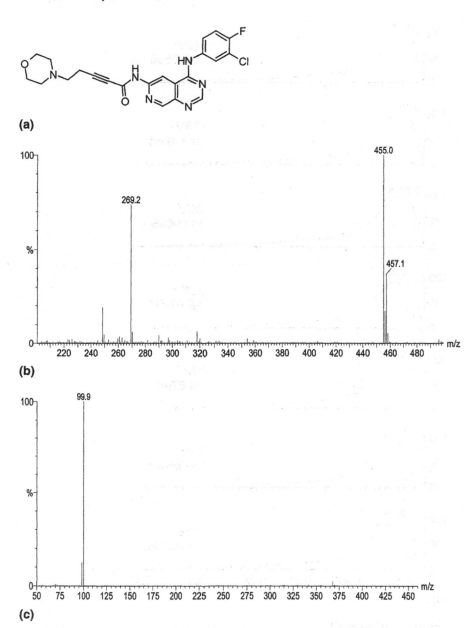

**Figure 5** Step-by-step process of profiling a typical compound of molecular weight 454 Da using electrospray positive ionization, Micromass Quattro II, loop injections with 50/50 ACN/0.1% formic acid, flow rate = 0.25 mL/min, no column. (a) Structure of analyte, (b) precursor ion spectrum, (c) product ion spectrum, (d) cone voltage optimization, (e) collision energy optimization.

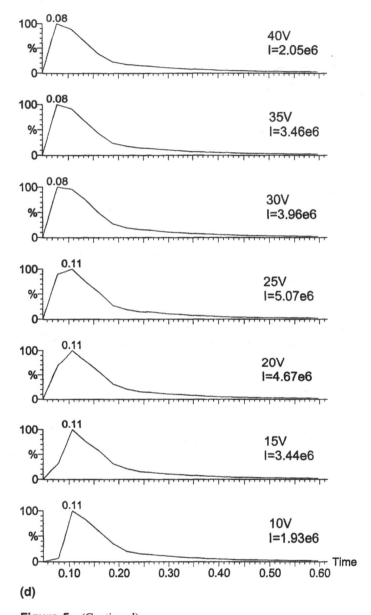

**(d)**

**Figure 5**   (Continued)

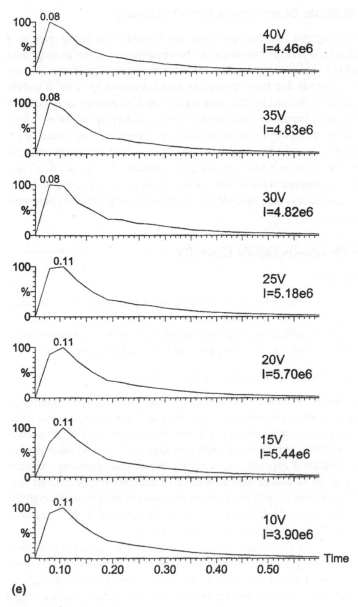

**(e)**

**Figure 5** (Continued)

### E.  Profiling Multiple Components Simultaneously

To gain higher throughput in signal profiling, as it pertains to development of quantitative methods, several compounds can be combined in a multicomponent mixture and profiled simultaneously. The multicomponent mixture can contain eight or more compounds that have molecular ions separated by several atomic mass units and, ideally, ionized by the same mode (such as electrospray positive ion). As this solution is introduced into the ion source, either by infusion or loop injection, each of the components to be profiled is sequentially optimized by a process similar to that described in the preceding section. At least two commercial LC/MS vendors (PE Sciex and Micromass) have introduced software that automates this process. Although some of the results obtained by this type of multiplexed procedure can be less than optimal, the resulting throughput for compound profiling is improved.

### F.  Adequate Chromatographic Capacity

#### 1.  Ion Suppression

As discussed in Sec. I, ion suppression can be a great hindrance to a sensitive, linear and rugged LC/MS method. Optimization of the chromatographic and sample preparation steps to sufficiently separate the analyte from endogenous interferences in the sample and the void volume will help guarantee success.

#### 2.  Cross Talk—What It Is and How to Deal with It

Despite the enormous selectivity provided by multiple-reaction monitoring through targeting individual precursor and product ion mass combinations, interference from *cross talk* can still confound mixture analysis in LC/MS. Cross talk is defined as contributions from ions of similar or identical mass to other precursor–product ion channels. Cross talk can occur via two distinct pathways. Chemical interference due to in-source fragmentation of metabolites or related substances will yield ions identical to the analyte precursor which pass through the first quadrupole. This type of interference can only be detected if the metabolite is chromatographically resolved from the analyte peak.

    A second source of cross talk occurs for mixtures of precursor–product ions with similar masses. Figure 6b shows the interference occurring when product and precursor ions within 1 Da of each other are monitored simultaneously for an analyte with the structure shown in Fig. 6a. The actual analyte signal should be observed only for a channel monitoring the 416.9-to-244.9 transition. Similar, although less intense, peaks are observed, however, for a precursor mass of 415.9 and/or a product ion mass of 243.9. This cross talk is caused by insufficient mass selectivity in the first quadrupole that allows the $M - 1$ to be transmitted.

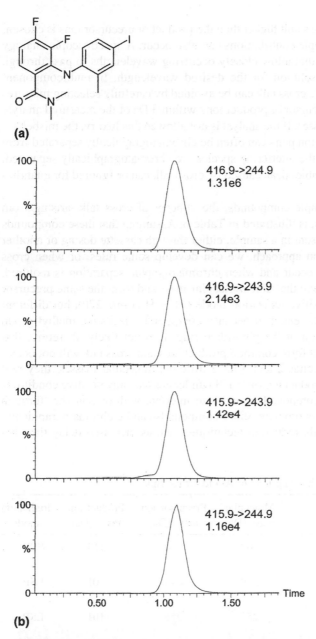

**(a)**

**(b)**

**Figure 6** Cross talk due to closely spaced precursor and product ions. Compound analyzed on Micromass Quattro II, 3-μL injection with 70/30 ACN/0.1% acetic acid, flow rate = 0.25 mL/min, MetaChem Phenyl 3,5 μm, 2.0 × 50 mm. (a) Compound structure, (b) intensity of interference from precursor or product ions within one Da of the correct MRM conditions.

If a channel 1 mass unit higher than the product or precursor ions is chosen, interference from isotopic contributions can also occur. A spectroscopic analogy is slit-width problems that allow closely occurring wavelengths to pass through the slit due to low resolution for the desired wavelength. In multicomponent determinations, isotopic cross talk can be avoided by carefully selecting mixtures that do not contain precursor or product ions within 3 Da of the measured analyte precursor–product masses. If the analyst is not allowed this luxury, the misbehaving precursor–product ion pairs can often be chromatographically separated from nonoffending pairs. If the interfering species are chromatographically separated, the additional (undesirable) peak caused by cross talk can be ignored for quantitative purposes.

Using four example compounds, the concept of cross talk, arising from common daughter ions, is illustrated in Table 6. Assuming that these compounds are simultaneously present in a sample, either through cassette dosing or another concomitant medication approach, we can develop some rules for when cross talk interferences will occur and when chromatographic separation is required. Compounds 1 and 2 have the same molecular mass and form the same precursor ion in electrospray positive ionization mode (M + H at $m/z$ 270), but different product ion masses are generated because compound 2 lacks the methyl group present at the 3 position of the piperidine ring. It is not likely, therefore, that compounds 1 and 2 will form common product ions and cross talk will not occur.

Although compounds 2 and 3 have different molecular masses, they will give rise to the same product ion ($m/z$ 101) under electrospray positive conditions and it is possible for compounds 2 and 3 to interfere with one another through cross talk. To avoid this problem, these compounds can be chromatographically separated or a cross talk reduction technique, such as increased delay time be-

**Table 6**  Example of Choosing Ions to Avoid Cross Talk

| Compound | Molecular mass (Da) | Precursor ion mass (Da) | Product ion mass (Da) | Ionization mode |
|---|---|---|---|---|
| | 269 | 270 | 114 | ESP + |
| | 269 | 270 | 101 | ESP + |
| | 255 | 256 | 101 | ESP + |
| | 292 | 291 | 136 | ESP − |

tween channels, could be used. To avoid this type of cross talk on some older instruments, product ion masses should differ by at least 3 Da. Compound 4, although structurally related to compounds 1–3, will not produce a useful ion in electrospray positive mode because it lacks the piperidine moiety and it will not interfere with the determination of the three other compounds in this series.

Interchannel delay or the insertion of dummy channels can be used to solve cross talk between compounds 2 and 3. Interchannel delay is the insertion of additional time for electronics to settle when mass channels are switched. This allows the collision cell to empty of gaseous ions. Fourth-generation LC/MS instrumentation (PE Sciex API 3000 or Micromass Ultima) has improved ion optics to facilitate emptying of the collision cell and make large interchannel delays unnecessary [40]. Although increased time for interchannel delay (from 3 to 50 μsec, for example) translates to less points sampled across a chromatographic peak, this increase in sampling time is generally manageable. An important consideration to bear in mind is the possible chromatographic sampling error introduced into data acquisition by increasing the chromatographic sampling interval, as shown in Fig. 7. As the sampling interval increases, the percentage error in the peak area also increases [41]. For a 10-sec-wide exponentially modified Gaussian (signal-to-noise ~500) chromatographic peak considered in this example, sampling error remains small and manageable if the sampling interval

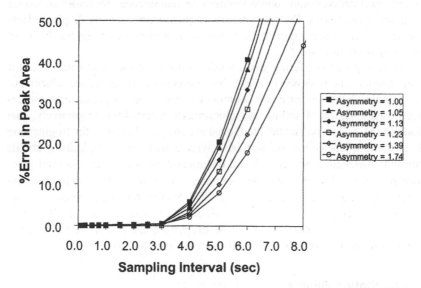

**Figure 7** Percentage error in peak area as a function of sampling rate for a 10-sec-wide chromatographic peak (*S/N* ~500) with chromatographic asymmetry values from 1.00 to 1.74.

remains under a couple of seconds. This effect is less pronounced for asymmetric peaks caused by tailing or fronting, which will have greater peak widths and have more points sampled per peak.

## 3. Problems with Metabolites

Metabolism mechanisms can wreak havoc on drug discovery programs via multiple routes; the most benign of which can be interferences in the quantitative analytical method. From the analytical chemist's point of view, metabolites cause problems primarily through coelution and subsequent interference with analyte peak detection. Because phase II metabolites are higher molecular weight conjugates of the original analyte, these molecules are prone to *in-source fragmentation* thereby generating additional parent drug. Because the compound is now parent drug in every way, it cannot be differentiated by mass spectrometry and will interfere with accurate quantitation of the analyte unless the metabolite is sufficiently separated chromatographically. A similar argument can be made for oxidative degradation products such as $N$-oxides. Metabolites are generally more polar than the parent molecule and usually elute first. Sufficient separation of the analyte from these peaks by increasing the chromatographic capacity is a practical solution to this problem.

It is sometimes possible to decrease or eliminate in-source fragmentation cross talk by decreasing ion source voltages or temperature, therefore softening the ionization conditions. Different brands of instruments usually display different degrees of in-source fragmentation cross talk, depending on the harshness of the ionization environment.

An example of interfering glucuronide, sulfate, and quinone metabolites for the drug troglitazone is shown in Fig. 8. The parent drug (troglitazone) chromatogram is shown in the top trace and contains two early eluting peaks from source-induced fragmentation of sulfate and glucuronide metabolites, respectively, that produce an $(M - H)^-$ ion identical to the parent drug in addition to the troglitazone peak. Deuterium-labeled internal standards were used for both troglitazone and its glucuronide metabolite. Source-induced fragmentation of the stable-labeled glucuronide produces a low-intensity early-eluting peak in the stable-labeled troglitazone internal standard chromatogram (Fig. 8C). No interference from the quinone metabolite is observed. Due to the cross talk caused by source-induced fragmentation, the troglitazone analyte could not be properly quantified if the parent drug were not chromatographically separated from the metabolites.

## G.  Quantitation Basics

## 1.  Definitions

The nomenclature and basic definitions describing a quantitative method are discussed to lay the groundwork for readers unfamiliar with these concepts. The

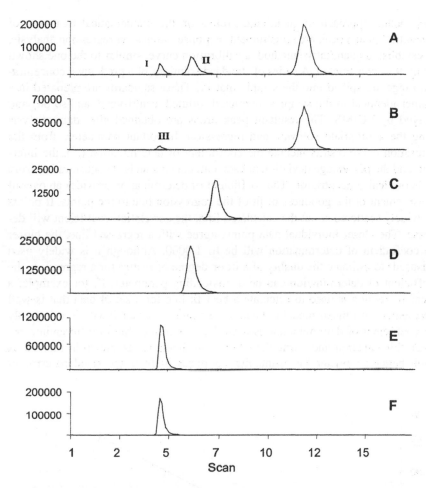

**Figure 8** Metabolite interferences for analysis of troglitazone and metabolites in human plasma. (A) troglitazone, (B) $d_4$-troglitazone (internal standard), (C) troglitazone glucuronide metabolite, (D) troglitazone sulfate metabolite, (E) troglitazone glucuronide metabolite, (F) $d_4$-troglitazone glucuronide metabolite (internal standard). Source-induced fragmentation from I: troglitazone glucuronide metabolite; II: troglitazone sulfate metabolite, and III: $d_4$-troglitazone glucuronide metabolite. Analysis performed with a Supelco RP Amide $C_{16}$ 2.0 × 50-mm column with 5-$\mu$m packing 60/40 ACN/10 mM NH$_4$OAc, flow rate = 0.3 mL/min, electrospray negative ion, PE Sciex API 3000. (Courtesy of Gerry Pace, Steven Michael, and Roger Hayes, personal communication.)

fundamental approach to quantitation relies on the mathematical treatment of concentration and peak area (instrument response) values via regression analysis. To establish a quantitative method, a calibration curve, similar to the one shown in Fig. 9, is constructed. A series of standard solutions of a predefined concentration range are spiked into the sample matrix. These standards are prepared in a manner identical to the samples (extracted, diluted, centrifuged, and so on) and assayed by LC/MS. The resultant peak areas are obtained after data analysis using the appropriate software and regression. Important parameters from the calibration curve results include the coefficient of determination ($r^2$), the intercept, and the percentage deviation of each calculated standard concentration from its theoretical concentration. The coefficient of determination provides an overall measurement of the goodness of fit of the regression line to the points. If points are widely scattered about the calculated line, the correlation coefficient will decrease. The closer individual data points agree with a regressed line, the closer the coefficient of determination will be to 1.0000. Although it is widely used (misused) to estimate the quality of a fit or degree of scatter for a regression, the coefficient of determination can be a misleading parameter. If, for example, a linear regression is used to calculate a best fit line for a set of data that is well represented by a linear model but has some scatter associated with it, using only the coefficient of determination as a guide, it is impossible to differentiate between this situation and a true deviation from linearity, as shown in Fig. 10. A much better parameter for estimating goodness of fit is the standard error of

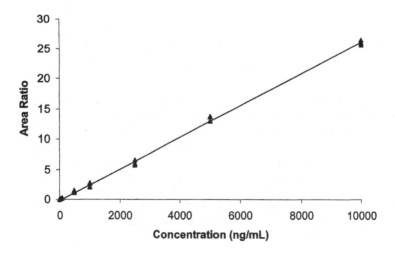

**Figure 9**   Typical quantitation using linear regression: phenytoin in rat plasma prepared using a liquid–liquid extraction as follows: 5–5000 ng/mL, 0.25 mL/min, 50/50 ACN/ $H_2O$, APCI negative ion PE Sciex API 3000.

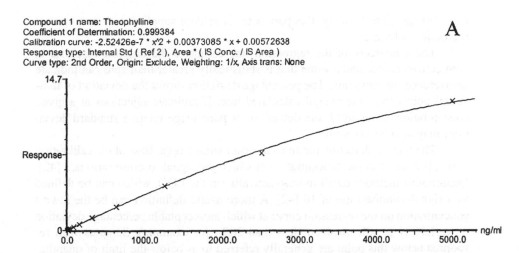

Compound 1 name: Theophylline
Coefficient of Determination: 0.999384
Calibration curve: -2.52426e-7 * x'2 + 0.00373085 * x + 0.00572638
Response type: Internal Std ( Ref 2 ), Area * ( IS Conc. / IS Area )
Curve type: 2nd Order, Origin: Exclude, Weighting: 1/x, Axis trans: None

**A**

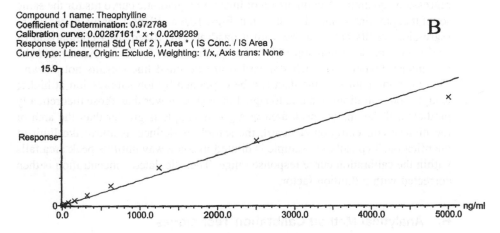

Compound 1 name: Theophylline
Coefficient of Determination: 0.972788
Calibration curve: 0.00287161 * x + 0.0209289
Response type: Internal Std ( Ref 2 ), Area * ( IS Conc. / IS Area )
Curve type: Linear, Origin: Exclude, Weighting: 1/x, Axis trans: None

**B**

**Figure 10**  Calibration curves for theophylline using (A) quadratic and (B) linear-regression fits. Standard concentrations from 5 to 5000 ng/mL. Standards prepared in rat plasma using acetonitrile protein precipitation. Analysis using Micromass Quattro II, 3-μL injection with 70/30 ACN/0.1% acetic acid, flow rate = 0.25 mL/min, MetaChem Phenyl 3, 5 μm, 2.0 × 50 mm.

the estimate. Unfortunately, this parameter is seldom supplied with commercial regression software.

The $y$ intercept of the regression line estimates the baseline response at zero concentration, and a value that is statistically greater than zero can indicate interferences or cross talk. The percentage deviation shows the deviation of individual points from the overall calculated line. If replicate injections at a given concentration are assayed, the deviation, or percentage relative standard deviation, may also be assessed.

The limit of detection for an analytical method, regardless of the calibration curve, is defined as the concentration at which the signal-to-noise ratio is 3 [42]. Quantitative methods dwell more practically on the LOQ, which can be defined as a signal-to-noise ratio of 10 [42]. A more useful definition can be the lowest concentration on the regression curve at which an acceptable percentage deviation or variability is still obtained (less than 20%, for example). Samples with responses below this point are generally referred to as below the limit of quantitation. At higher concentrations, the curve may deviate from a linear response due to factors influencing ionization. Samples with concentrations so high that the response deviates from linearity are referred to as above the limit of linearity. This problem can often be compensated for by the use of quadratic fits for the regression equations. A comparison of linear and quadratic curve fits for the same set of theophylline standards is shown in Figs. 10A and 10B. Clearly, better linear regression results are obtained with a quadratic fit, based on a comparison of coefficients of determination (0.999 for quadratic vs. 0.973 for linear) and the percentage deviation of each standard from the fitted line (results not shown). The quadratic nature of the data can be explained by ion suppression at higher analyte concentrations, leading to signal intensities lower than those theoretically predicted. If the analyte peak area of a given sample is greater than the area of the highest concentration standard, the sample is defined as above the limit of quantitation. Typically the sample is diluted in some way until its peak area falls within the calibration curve response range. The calculated concentration is then corrected with a dilution factor.

## H.  Analytical Method Calibration Techniques

The most common means of obtaining unknown sample concentration values is the construction of a standard curve, or calibration curve concurrent with sample preparation and assay. Although the ruggedness of LC/MS has increased greatly with improved source design and better sample preparation methods, peak area response invariably fluctuates from day to day by 50% or more. Use of a few selected standards that bracket the expected concentration range, rather than an entire series of standards, is generally not utilized. The stability of the LC/MS instrument cannot compare to UV or fluorescence; thus the need for running a calibration curve with each assay run. Because chromatographic run times are generally in the range of a few minutes, the additional work required by the calibration technique occurs primarily during the sample preparation stage.

Instrument response can be calibrated by the addition of an internal standard, which will minimize variability in both ionization processes and sample preparation. The ideal properties of the internal standard, as well as additional requirements for analytical methods, are discussed in Chapter 6.

If blank matrix is unavailable, or the number of samples is extremely small ($N < 10$), the method of standard additions can be used for quantitation [2,42]. In this method, the sample is separated into several individual aliquots, each of which is spiked with an increasing concentration of analyte standard solution. A plot of peak area as a function of sample plus standard can be generated and the sample concentration calculated from the $y$ intercept. Another less frequently used approach is normalization of sample area to area of a standard peak.

## I. Sources of Nonlinear Response versus Concentration Calibration

In general, nonlinear response is observed at the higher concentrations within a calibration curve. The most common cause is ion suppression, particularly for electrospray. In fact, the dynamic range of electrospray is considered less than that of APCI due to more frequent ion suppression effects. The simplest solution to this problem is to reinject at a lower volume to extend the linear range. Another approach is to use a quadratic curve to fit the data points. Although some analytical chemists shy away from quadratic curves, their usage should not be outlawed as long as thoughtful consideration is made with respect to assay ruggedness. One area for caution when using quadratic fits occurs when multiple analytes coelute during multicomponent quantitation. This situation is often encountered in cassette dosing (Chapter 11), where high standards containing multiple coeluting components will demonstrate nonlinearity. If a quadratic fit is used for each of these several component calibration curves, and not all of the same components are similarly represented in samples from dosed animals (due to lower bioavailability), then the curvilinear calibration will underrepresent the response of these components, thereby resulting in quantitation error. In electrospray, multimer formation at high concentrations can cause nonlinearity. If multimer formation occurs, the individual analyte molecules will not form a 1:1 correlation with each analyte detected as an $(M + H)^+$ or $(M - H)^-$, but will participate in competing processes and thus decrease the overall signal for monomers.

## J. Consequences of Pushing the Dynamic Range

Any LC/MS instrument will, at some point, reach the upper limit of linear response for a particular analyte molecules. If the dynamic range is "overextended," inaccuracies at high and low concentrations can result. Inappropriate limit of quantitation concentrations can be dramatically affected by injector carryover, which is usually not assessed for every compound assay conducted at the discovery stage. High concentrations can also be over- or underestimated and contribute to carryover of lower concentration samples. These problems will often only be discovered when the assay is validated for development purposes or is transferred to another laboratory.

## V.  TECHNIQUES FOR GETTING THE BEST QUANTITATION LIMITS

### A.  Sample Preparation Techniques

#### 1.  Concentration

One approach for improving detection limits in analytical methods is to concentrate the sample. If, for example, a 200-μL aliquot can be extracted and suspended in 50 μL, a fourfold improvement in sensitivity is achieved, unless ion suppression or another source of nonlinearity occurs. Because the remaining matrix components are also concentrated, the ion suppression threshold can be reached rapidly. In drug discovery programs, however, sample volumes are often limited to 50 or 100 μL, so this approach may not be practical.

### B.  Chromatographic Techniques

#### 1.  Injection Size

The volume of samples and standards injected can be tailored to suit linear or nonlinear response as well as sensitivity requirements. In situations where sensitivity is mass dependent (e.g., APCI or some electrospray situations), greater detector response can be attained by increasing the injection volume. This approach will provide a proportional increase in response until ion suppression begins to take effect, in which case response can actually decrease. Injection volumes for 2.0-mm columns are generally 1–5 μL. Greater volumes can overload the column, leading to peak fronting and tailing as well as nonlinear response at higher concentrations. However, if acceptable sensitivity is not obtained for a 2-μL injection, the volume can be increased to 5 μL as long as ion suppression or column overload does not occur. Attention should be paid to the precision and size of the injector syringe because reproducibility can suffer at extremely small volumes if proportionately large syringes are used.

In principle, electrospray ionization is a concentration-sensitive technique. This means that the signal response should be independent of the injection volume. It is often true, however, that the chromatographic separation concentrates the analyte at the head of the column and some mass-sensitive response is observed. Atmospheric pressure chemical ionization, on the other hand, is a mass-sensitive technique and should display a response proportional to increases in injection volume. The practical outcome of this is that a proportional increase in signal using either type of ionization can be obtained by injecting more sample if sensitivity is not limited by ion suppression. If ion suppression effects predominate, smaller injection volumes can actually increase signal intensities. Injection volumes as large as 50 to 100 µL can be used on analytical scale (4 and 4.6 mm i.d.). An article that considers injection volume and signal intensity has recently been published [34].

## 2. Column Dimensions

Column diameter can be adjusted to meet the sensitivity and dynamic range needs of the quantitative method. Decreasing the bore size from 3.0 to 2.1 while decreasing the flow rate may also show a "concentration" effect on band width and thus peak area response [43]. As shown in the equation below, if the diameter of the column is decreased from 3.0 to 2.1, the peak area response should double as follows:

$$\frac{\text{Area}_1}{\text{Area}_2} = \left(\frac{\text{Diameter}_2}{\text{Diameter}_1}\right)^2 = \left(\frac{3.0}{2.1}\right)^2 = 2.0.$$

This equation is applicable only if the extracolumn volume of the system remains small and chromatographic band broadening is not significant.

## 3. Chromatographic Efficiency

As with any LC separation, improving the efficiency of an LC/MS method will allow for greater detection sensitivity by narrowing the chromatographic band. This is sometimes difficult to accomplish in practice due to the limitations placed on LC/MS mobile-phase components and additives. An alternative approach to achieving narrower chromatographic bands is through the use of step gradients, which can easily lead to apparent efficiency increases of 10-fold or more.

## C. Mobile-Phase Properties

### 1. Volatility and Surface Tension

Mobile-phase volatility is required in API LC/MS due to the need to produce gas-phase ions, whether through electrospray or chemical ionization. Atmospheric pressure chemical ionization can require higher solvent gas-phase volatility than electrospray due to its ionization mechanism. Low-surface-tension solvents also perform better due to improved nebulization properties [44–46]. Solvents such as methanol, acetonitrile, and 2-propanol possess both the necessary volatility and low dielectric constants (which translate to low surface tension) to be practical for LC/MS. Surfactants increase surface tension and disrupt the nebulization process. Sample cleanup or dilution to remove surfactant effects are sometimes unavoidable.

## 2. Conductivity

Appropriate mobile-phase liquid conductivity is a necessary consideration in electrospray. Pure benzene, carbon tetrachloride, and hexane possess insufficient conductivity to form charged droplets and must be mixed with polar solvents before ion formation will occur. Typically this concern exists only when normal-phase or chiral separations are conducted. Trifluoroacetic acid is sometimes added to increase mobile-phase conductivity but its detrimental effects via ion pairing and surface tension increases can cause signal suppression [47]. An alternative approach to electrospray is to chose atmospheric pressure, which through a corona discharge, can produce a stable beam under normal-phase conditions.

## 3. Ionic Strength

Ionic strength effects via addition of salts in the mobile phase or sample have been examined by Constantopoulos et al. [19]. In general, dramatic decreases in signal intensity are observed only when the salt concentration reaches 0.1 M. Salts contained within the sample should be chromatographically separated. Excessively high salt content samples can be chromatographed with a divert mechanism to direct the LC flow to waste during the initial portion of the chromatographic run. In this manner, source cleanliness can be more easily maintained because salt buildup occurs rapidly, leading to decreased sensitivity with time.

## D. Mass Spectral Techniques

Provided all other parameters such as cone/orifice voltage or source temperature have been optimized, mass spectrometric sensitivity can be optimized by dwell time, ion energy, and resolution adjustment. *Dwell time* is defined as the amount of time a precursor–product ion pair is monitored before the quadrupoles switch to the next channel. *Ion energy* is determined by the difference between the ion's "starting potential" from any entrance (RF) lens and the first quadrupole multiplied by the number of charges on the molecule. The number of cycles in the quadrupole field the ion experiences will be directly affected by the ion energy. Increasing the ion energy will increase the differences in energy between ions and distort peak shape, thereby decreasing resolution. *Resolution* is ultimately related to the speed with which the ions pass through the quadrupole analyzer.

Resolution will increase as ion speed through the quadrupole decreases, but signal intensity will also decrease.

To maximize sensitivity, relatively straightforward modifications to the mass spectral method can be utilized. First, the dwell time can be increased as long as adequate chromatographic sampling (in the range of 1 to 2 points per sec) is maintained. The effect of dwell time on signal intensity has been discussed in Chapter 3. Figure 11 shows the effect of changing resolution settings on signal intensity when other variables are held constant. If resolution is decreased, signal

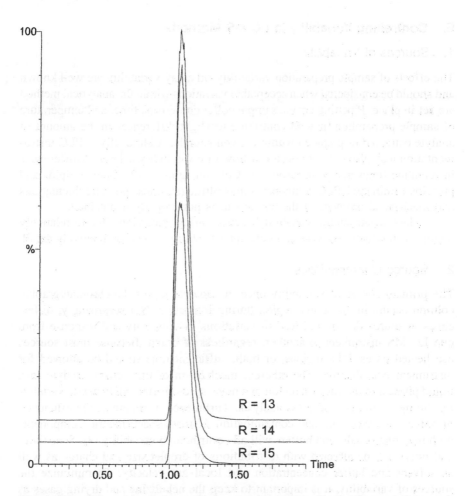

**Figure 11**   Effect of resolution settings on peak intensity, all other factors being equal. Compound analyzed on Micromass Quattro II, 3-μL injection with 70/30 ACN/0.1% acetic acid, flow rate = 0.25 mL/min, MetaChem Phenyl 3, 5 μm, 2.0 × 50 mm.

intensity will increase for a finite range of resolution values. Although the three chromatographic traces shown are offset, the improvement in signal intensity when the resolution decreases from index 15 to 14 (mass spectral peak width decreases from approximately 0.7 to 1.2 Da) is greater than the improvement observed when the resolution is changed from an index of 14 to 13 (mass spectral peak width decreases from approximately 1.2 to 1.8 Da). This type of variability is highly dependent on the instrument used. For lower resolution settings, ion energy can also be increased to improve signal intensity.

## E.  Controlling Variability in LC/MS Methods

### 1.  Sources of Variability

The effects of sample preparation variability on assay variability are well known and should be considered when acceptable variations within the analytical method are set in place. Pipetting errors, sample collection errors, time, and temperature of sample preparation may all contribute to slight differences in the amount of analyte extracted or prepared within a given sample. Additionally, HPLC instrumentation may also exhibit injector or flow rate variability leading to differences in retention times and peak responses. Column aging and buildup of lipids and proteins within the HPLC components may ultimately cause pressure fluctuations and mechanical problems if the instrument is not properly maintained.

Mass spectrometric variability occurs during ionization. Due to relatively mature technology, analyzer and detector variability should be relatively small.

### 2.  Source Characteristics

The primary source of variability once the analyte leaves the chromatographic column occurs in the source region during ionization. Not surprisingly, differences in source design will lead to variations in sensitivity and response from one LC/MS instrument to another, regardless of brand. Because most sources use heated gases, inlet regions, or both, sufficient time should be allowed for instrument equilibration. The inherent mechanism of transferring analyte in a liquid phase into the vacuum region of a mass spectrometer will generate variability during ionization and mass analysis. Turbulence, variation in the efficiency of solvent vaporization, and contamination of mass spectrometric components with nonvolatile salts and buffers will all contribute to variability [8]. Sensitivity and noise will be affected with distributions of droplet size and charge as well as solvent and buffer concentration and local gas velocity. To minimize the sources of variability, it is important to keep the nebulizing and drying gases as stable as possible during instrument operation.

In electrospray, corona discharge can occur, particularly in negative ion mode at very high capillary voltages. Signal variability and instability will result.

Addition of a scavenger gas such as $SF_6$ or a chlorinated solvent to the source region will prevent corona discharge [48,49], but is generally less practical than simply lowering the capillary voltage.

## 3. Choosing Internal Standards

On a routine basis, much of the previously discussed variability is easily surmounted by the use of an internal standard. By addition of a unique molecule that is structurally similar to the analyte, both sample preparation and instrumental variability effects can be minimized. Ideally, the internal standard is the analyte molecule that has been labeled with stable isotopes, such as deuterium or $^{13}C$. However, because this is both cost and time prohibitive at the discovery stage, the internal standard should at least possess the same ionizing groups as the analyte. Polarity switching or the need for a different mode of ionization (electrospray vs. APCI) in order to detect the internal standard will lead to selection of a different internal standard! The internal standard should also exhibit a chromatographic capacity factor ($k'$) similar to or identical with the analyte such that chromatographic run times are not unduly lengthened. The mass of the internal standard should be separated from the analyte of interest by greater than 3 Da so that naturally occurring isotopes of the parent analyte do not interfere with the response of the internal standard. An equal amount of internal standard added to each sample can be used to normalize for intersample variability caused by ion suppression as well as the injector. As is discussed in Chapter 6, the internal standard should also possess extraction characteristics similar to those of analytes to account for variability during sample preparation. If known, the purity of the internal standard, whether structural analog or stable isotope labeled, should be reflected in calculation of quantitative results.

## 4. Dimers, Clusters, and Adducts—Na$^+$, K$^+$, and NH$_4^+$

Additional variability during LC/MS is introduced if the analyte displays a tendency to form dimers or higher multimers either in solution or gas phase. Because a finite number of analyte molecules are ionized, formation of multimers rather than singly charged species will lead to a decrease in signal for the monitored singly charged channel. This phenomenon will be concentration dependent, with aggregate formation of analyte $(M_n + H)^+$ predominating at higher concentrations and contributing to nonlinear responses. The presence of multiply charged clusters indicates that a solution-phase process is occurring, which can be suppressed by the addition of a small amount of acid or base. Cluster formation of the analyte with solvent molecules may be unavoidable but can usually be minimized by increasing cone or orifice voltage, although increased analyte in-source fragmentation could result.

Cation or anion adducts will cause the same undesirable effects as clusters or multimers, although this type of adduct may not be as easily removed. Certain types of molecules are more efficiently cationized than protonated, and cations may be deliberately added to the sample preparation to promote cationization. Some types of sample cleanup techniques will help minimize ion adducts but, unfortunately, cannot completely eliminate them.

### 5. Multiply Charged Species

As discussed in Chapter 3, the ion formation mechanisms differ greatly for APCI and electrospray. By the very nature of ionization, multiply charged species formed in the gas phase are observed primarily by electrospray. In some cases, these multiply charged ions, particularly for proteins and peptides, can provide structural confirmation of molecular weight [1]. Most instrument manufacturers provide software that allows computation of molecular weight based on deconvolution of the analyte ion charge distribution. Quantitation using a multiply charged ion is more difficult due to interinjection variability in charge-state distribution. These multiply charged ions are often collisionally activated during ionization. A shift toward lower charge state will be observed at higher analyte concentrations [50]. The sample pH and ionic strength also affect this distribution. Because a finite number of charges reside on the droplet surface, at higher analyte molecule populations the ratio of charges/ion will drop. This situation can be improved by limiting the linear range of the analytical assay.

## VI. SUMMARY

The extensive discussion in this chapter can be distilled into two salient points: (1) Avoid contaminating the LC/MS with nonvolatile materials and (2) use the capability of the LC column and mobile phase to provide as much separation of the analyte from the void volume and other sample components as possible. Following these two recommendations will perhaps not guarantee success but will make its likelihood much greater. Unfortunately, many of the choices and finer points of LC/MS method development will depend on the molecule(s) to be determined. The suggestions offered in this chapter should provide a useful starting point for the analyst.

## REFERENCES

1.  WMA Niessen. Liquid Chromatography/Mass Spectrometry. 2nd ed. New York: Marcel Dekker, 1998.

2. R Willoughby. A Global View of LC/MS: How to Solve Your Most Challenging Analytical Problems. Pittsburgh, PA: Global View, 1998.
3. BS Larsen. Mass Spectrometry of Biological Materials. 2nd ed. New York: Marcel Dekker, 1998.
4. RB Cole. Electrospray Ionization Mass Spectrometry. New York: Wiley, 1997.
5. AG Harrison. Chemical Ionization Mass Spectrometry. 2nd ed. Boca Raton, FL: CRC Press, 1992.
6. AL Yergey, CG Edmonds, IAS Lewis, ML Vestal. Liquid Chromatography/Mass Spectrometry: Techniques and Applications. New York: Plenum, 1989.
7. MA Brown, ed. Liquid Chromatography/Mass Spectrometry: Applications in Agricultural, Pharmaceutical, and Environmental Chemistry. Vol. 420. Washington DC, ACS Symposium Series, 1990.
8. BA Thomson. Atmospheric pressure ionization and liquid chromatography/mass spectrometry—together at last. J Am Soc Mass Spectrom 9:187–193, 1998.
9. E Brewer, J Henion. Atmospheric pressure ionization LC/MS/MS techniques for drug disposition studies. J Pharm Sci 87(4):395–400, 1998.
10. TR Covey, ED Lee, AP Bruins, JD Henion. Liquid chromatography/mass spectrometry. Anal Chem 58:1451A–1461A, 1986.
11. P Kebarle, LK Tang. From ions in solution to ions in the gas phase: the mechanism of electrospray mass spectrometry. Anal Chem 65(22):972A–986A, 1993.
12. AP Bruins. Liquid chromatography–mass spectrometry with ion spray and electrospray interfaces in pharmaceutical and biomedical research. J Chromatogr 554(1–2):39–46, 1991.
13. AP Bruins. Mass spectrometry with ion sources operating at atmospheric pressure. Mass Spectrom Rev 10:53–77, 1991.
14. LR Snyder, JJ Kirkland, JL Glajch. Practical HPLC Method Development. 2nd ed. New York: Wiley, 1997.
15. JM Miller. Chromatography: Concepts and Contrasts. New York: Wiley, 1988.
16. JC Giddings. Dynamics of Chromatography. Part 1. New York: Marcel Dekker, 1965.
17. G Zweig, J Sherma, eds. In RN Gupta, I Sunshine, eds. CRC Handbook of Chromatography. Boca Raton, FL:CRC Press, 1981.
18. CF Poole, SA Schuette. Contemporary Practice of Chromatography. Amsterdam, Elsevier, 1984.
19. TL Constantopoulos, GS Jackson, CG Enke. Effects of salt concentration on analyte response using electrospray ionization mass spectrometry. J Am Soc Mass Spectrom 10:625–634, 1999.
20. PR Tiller, IM Mutton, SJ Lane, CD Bevan. Immobilized human serum albumin: liquid chromatography/mass spectrometry as a method of determining drug-protein binding. Rapid Commun Mass Spectrom 9:261–263, 1995.
21. D Temesi, B Law. The effect of LC eluent composition on MS responses using electrospray ionization. LC/GC 17(7):626–632, 1999.
22. I Hayati, AI Bailey, TF Tadros. Mechanism of stable jet formation in electrohydrodynamic atomization. Nature 319:41, 1986.
23. I Hayati, AI Bailey, TF Tadros. Investigations into the mechanisms of electrohydrodynamic spraying of liquids 1: effect of electric field and the environment on pendant

drops and factors affecting the formation of stable jets and atonization. J Colloid Interface Sci 117:205–221, 1987.

24. I Hayati, AI Bailey, TF Tadros. Investigations into the mechanisms of electrohydro-dynamic spraying of liquids II:mechanism of stable jet formation and electrical forces acting on a liquid cone. J Colloid Interface Sci 117:222–230, 1987.

25. WH Schaefer, F Dixon Jr. Effect of high-performance liquid chromatography mobile phase components on sensitivity in negative atmospheric pressure chemical ioniza-tion liquid chromatography-mass spectrometry. J Am Soc Mass Spectrom 7:1059–1069, 1996.

26. CS Wu. Handbook of Size Exclusion Chromatography. Portland, OR: Book News, Inc., 1995.

27. WW Yau, JJ Kirkland, DD Bly. Modern Size-Exclusion Liquid Chromatography. New York: Wiley-Interscience, 1979.

28. DJ Aaserud, L Prokal, WJ Simonsick. Gel permeation chromatography coupled to Fourier transform mass spectrometry for polymer characterization. Anal Chem 71: 4793–4799, 1999.

29. JT Eleveld, HA Claessens, JL Ammerdorfer, AM van Herk, CA Cramers. Evaluation of mixed-mode stationary phases in liquid chromatography for the separation of charged and uncharged oligomer-like model compounds. J Chromatogr A 677:211–227, 1994.

30. MP Henry. In: MTW Hearn, ed. HPLC of Proteins and Peptides and Polynucleotides. New York:VCH 1991, Chapter 5.

31. DR Nau. In: MTW Hearn, ed. HPLC of Proteins and Peptides and Polynucleotides New York:VCH 1991, Chapter 11.

32. JJ Kirkland, JL Galjch, RD Farlee. Synthesis and characterization of highly stable bonded phases for high-performance liquid chromatography column packings. Anal Chem 61:2–11, 1988.

33. J Keever, RD Voyksner, KL Tyczkowska. Quantitative determination of ceftiofur in milk by liquid chromatography–electrospray mass spectrometry. J Chromatogr A 794:57–62, 1998.

34. EK Kindt, DT Rossi, K Gueneva-Boucheva, H Hallak. Quantitative method for bio-markers of collagen degradation using LC/MS/MS. Anal Biochem 283:71–76, 2000.

35. KH Bauer, TP Knepper, A Maes, V Shatz, M Voihsel. Analysis of polar organic micropollutants in water with ion-chromatography–electrospray mass spectrometry. J Chromatogr A 837:117–128, 1999.

36. R Anacardio, MG Cantalini, F DeAngelis, M Gentile. Quantification of S-carboxy-methyl-(R)-cysteine in human plasma by high-performance ion-exchange liquid chromatography/atmospheric pressure ionization mass spectrometry. J Mass Spec-trom 32:388–394, 1997.

37. SG Lias, JE Bartmess, JF Liebman, JL Holmes, RD Levin, GW Mallard. Gas-phase ion and neutral thermochemstry. J Phys Chem Ref Data 17(suppl. 1): 1988.

38. S Kawasaki, H Ueda, H Itoh, J Tadano. Screening of organophosphorus pesticides using liquid chromatography–atmospheric pressure chemical ionization mass spec-trometry. J Chromatogr 595:193–202, 1992.

39. H Itoh, S Kawasaki, J Tadano. Application of liquid chromatography-atmospheric-

pressure-chemical-ionization mass spectrometry to pesticide analysis. J Chromatogr A 754:61–76, 1996.

40. BA Thomson, DJ Douglas, JJ Corr, JW Hager, CL Jolliffe. Improved collisionally activated dissociation efficiency and mass resolution on a triple quadrupole mass spectrometer system. Anal Chem 67:1696–1704, 1995.

41. DT Rossi. A simplified method for evaluating sampling error in chromatographic data acquisition. J Chrom Sci 26:101–105, 1988.

42. DA Skoog, FJ Holler, TA Niemann. Principles of Instrumental Analysis. 5th ed. Philadelphia: Saunders College, 1998, p 13.

43. DT Rossi, BA Phillips, JR Baldwin, PK Narang. Improved methodology for subnanogram quantitation of doxorubicin and its 13-hydroxymetabolite in biological fluids by liquid chromatography. Anal Chim Acta 211:59–68, 1993.

44. ML Vestal. In: KG Standing, W Ens, eds. Methods and Mechanisms for Producing Ions from Large Molecules. New York: Plenum, 1991 p 157.

45. S Zhou, M Hamburger. Effects of solvent composition on molecular ion response in electrospray mass spectrometry: investigation of the ionization processes. Rapid Commun Mass Spectrom 9:1516–1521, 1995.

46. RP Schneider, MJ Lynch, JF Ericson, HG Fouda. Electrospray ionization mass spectrometry of semduracin and other polyether ionophores. Anal Chem 63:1789–1794, 1991.

47. J Esrhraghi, SK Chowdhury. Factors affecting electrospray ionization of effluents containing trifluoroacetic acid for high-performance liquid chromatography/mass spectrometry. Anal Chem 65:3528–3533, 1993.

48. MG Ikanomou, AT Blades, PJ Kebarle. Electrospray mass spectrometry of methanol and water solutions suppression of electric discharge with $SF_6$ gas. J Am Soc Mass Spectrom 2:497–505, 1991.

49. FM Wampler, AT Blades, PJ Kebarle. Negative ion electrospray mass spectrometry of nucleotides: ionization from water solution with $SF_6$ discharge suppression. J Am Soc Mass Spectrom 4:289–295, 1993.

50. G Wang, RB Cole. Mechanistic interpretation of the dependence of charge state distributions on analyte concentrations in electrospray ionization mass spectrometry. Anal Chem 67:2892–2900, 1995.

# 6

# Sample Preparation and Handling for LC/MS in Drug Discovery

**David T. Rossi**
*Pfizer Global Research and Development, Ann Arbor, Michigan*

> Take a method and try it. If it fails, admit it frankly and try another. But by all means try something.
>
> *—Franklin Delano Roosevelt*

## I. WHY SHOULD SAMPLE PREPARATION BE DONE?

There are several important reasons to include sample preparation in modern liquid chromatography/tandem mass spectrometry (LC/MS/MS) methods. These include the elimination of matrix components from the sample; reduction of ion suppression; and, sometimes, improved sample utilization.

Once an analytical method has been established, it is desirable that method performance remains reasonably consistent over time. If a number of samples need to be assayed to answer a particular scientific question, then the results supplied by the method should be relatively free from systematic error and the relative error should be, to some extent, characterized and consistent. From the simplest and most direct perspective, sample preparation is used to ensure that a method maintains certain basic elements of ruggedness and consistency.

One example of the necessity of a sample preparation step involves LC/MS work with blood plasma. It is widely known that because of the large amounts of protein present in plasma, conventional high-performance liquid chromatography (HPLC) columns will not tolerate the direct introduction of plasma. Depending on the sample injection volume, a conventional HPLC column will clog and cease to function after only a few injections. This degree of ruggedness is generally unacceptable for any application. An appropriate extraction technique, aimed at removing most (>99%) of the protein from a sample, allows hundreds or thousands of samples to be injected and assayed [1], thereby generating sample assay data that could answer many types of scientific questions.

A more subtle example of the need for sample pretreatment involves the handling of samples from in vitro screening studies. These experiments, described in Chapters 8 and 9, typically involve the assay of drug compounds in a buffer matrix of high (50–200 mM) ionic strength [2]. Although the HPLC system would not be seriously affected by direct introduction of these high-salt samples, the suppression of ionization in the ion source of the mass spectrometer would severely limit the instrumental sensitivity and dynamic range, most likely limiting the value of the resulting data. Additionally, the cumulative effect of salt introduced from the sample could eventually play havoc with the mass spectrometer, causing the ion source and possibly even the vacuum region to become contaminated and requiring frequent cleaning and maintenance. In this situation, it is feasible to perform a simple dilution on samples prior to introduction [3] or to shunt rapidly eluting salt components to waste by use of a switching valve before the instrument ion source [4,5].

A second and higher level reason for utilizing sample preparation involves the value added to the information obtained from some pharmacological experiments. Specifically, drugs in biological samples are, to some extent, bound to proteins and/or tissues. Certain approaches to sample preparation, specifically ultrafiltration and in vivo microdialysis (Chapter 12) can be used to process biological samples in a way that the free (unbound) portion of the drug is separated from the protein-bound portion. In this way, the free concentration of drug can be readily quantified. This type of information is valuable whenever free drug fraction is of interest [6,7].

## II.  SAMPLE PREPARATION TECHNIQUES
## FOR DRUG DISCOVERY

Sample preparation for drug discovery applications represent a collection of many empirical techniques that are rapidly evolving, as necessity demands. As applied to biological and in vitro samples, these techniques have the goal of removing lipids, salts, cellular components, or proteins from the samples before injection into the LC/MS. Table 1 lists the major classifications of sample preparation, along with a ranking of their relative effectiveness for removing the major unwanted matrix components and their relative ease of use. The techniques are listed from least effective to most effective at removing matrix components and from easiest to use (direct injection) to most elaborate (solid-phase extraction). There is a direct trade-off between ease of use and effectiveness in sample prepa-

**Table 1** Relative Effectiveness and Ease of Use of Common Sample Preparation Approaches for Eliminating Matrix Components

| | Relative effectiveness | | | | Ease of use |
|---|---|---|---|---|---|
| | Lipids | Salts | Cellular components | Proteins | |
| Direct inject | − | + | − | − | +++ |
| Protein precipitation (96) | − | − | +++ | +++ | ++ |
| Ultrafiltration | − | − | +++ | +++ | + |
| On-line SPE | +/− | ++ | + | +/− | + |
| LLE (96) | +/− | +++ | +++ | +++ | +/− |
| SPE (96) | ++ | ++ | +++ | +++ | − |

Note: +++, very effective; +/−, somewhat effective; −, not effective.

ration. Certain techniques are very good at particular tasks (such as protein precipitation for removal of proteins and cellular components), but perhaps not as good in general. With the emphasis on greater efficiency, many of these approaches have become automated or semiautomated. Thus, there has been greater emphasis on the 96-well formats: an approach that can lend itself more readily to automated workstations. This format has been used successfully for protein precipitation, liquid–liquid extraction, and solid-phase extraction [8–10]. Ultrafiltration (UF) in a 96-well format is also being evaluated and shows some potential, but products and applications are not yet fully developed. Automated techniques for sample preparation and each of the sample preparation techniques listed in Table 1 are described below.

## A. Direct Injection

Shown pictorially in Fig. 1, direct injection is conceptually simple and efficient. A sample is obtained, diluted (optional), and injected onto the chromatographic column. As depicted, both analytes and matrix components are injected and it is from this feature that problems can arise. Samples containing appreciable amounts of protein and cellular components are not always suitable for direct injection and samples containing high concentrations of salt can also be problematic, as described above. Direct injection is most suitable for clean matrices, such as single- or multiple-component solutions and admixtures. If samples can be diluted fivefold or more, direct injection can be applied to samples from in vitro experiments as well.

One effective approach for reducing the contamination associated with samples having high salt content has been to install an automated divert valve

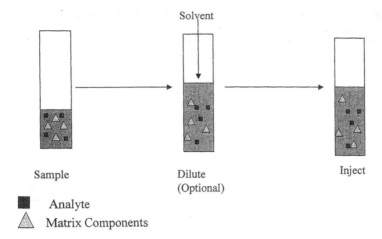

Sample               Dilute              Inject
                     (Optional)

■    Analyte
△    Matrix Components

**Figure 1**   The direct injection process, including the optional dilution step.

between the HPLC column and the mass spectrometer ion source. This configuration, depicted in Fig. 2, allows the divert valve to be automatically controlled by timed contact closures from the pump or autosampler. From injection to some point before analyte elution, the column effluent containing the highest concentration of salt is directed to waste. At some time immediately prior to the elution of the analytes, the effluent is directed to the ion source for detection. This simple approach helps to keep the mass spectrometer ion source and focusing quadrupole cleaner longer.

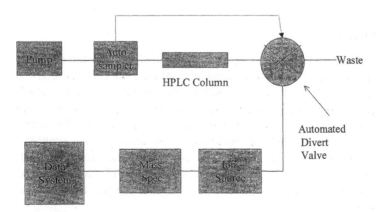

**Figure 2**   The inclusion of an automated divert valve before the mass spectrometer ion source.

Two recently introduced direct injection techniques have been developed to deal with the special problems posed by biological samples. These techniques involve the use of restricted access media (RAM) [11,12] and turbulent flow chromatography [13] and are described later.

## B.  Protein Precipitation

Figure 3 depicts the essential elements of the protein precipitation experiment. An exact volume of denaturing solvent (often two parts of denaturant to one part of sample, volume to volume) is added to a sample containing protein and/or cellular components. Protein denaturation is facilitated by vortexing, and centrifugation, allowing collection of the denatured protein into a pellet at the bottom of the vessel. This process is essentially a phase separation. The denaturation process causes a disabling of the protein's ability to bind analyte molecules. Analytes are typically released from the proteins and remain in the supernatant liquid, although some analyte molecules are occluded and are dragged to the bottom with the protein pellet. Samples are centrifuged and the phases can be separated by siphoning off the clear supernatant, typically with a liquid-transferring workstation. Occasionally, as a time-saving measure, the supernatant is injected directly, without a prior phase separation [8].

The most common LC/MS-compatible protein precipitating solvents and

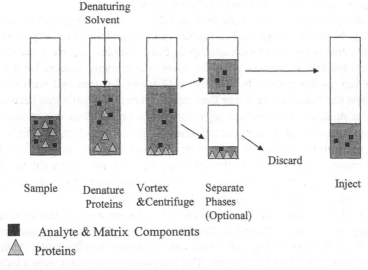

**Figure 3**   The protein precipitation process, including the steps of denaturation, centrifugation, phase separation, and injection.

**Table 2**  The Effectiveness of Common LC/MS-Compatible Solvents for Plasma
Protein Precipitation

| Solvent | Supernatant pH | % Protein removed (ratio of solvent to plasma) | |
|---|---|---|---|
| | | 1 | 2 |
| Methanol | 8.5–9.5 | 73.4 | 98.7 |
| Ethanol | 9–10 | 91.4 | 98.3 |
| Acetone | 9–10 | 96.2 | 99.4 |
| Acetonitrile | 8.5–9.5 | 97.2 | 99.7 |
| ZnSO₄: acetonitrile | | | |

*Source*: Ref. 1.

agents have been tabulated in Table 2. Simple organic solvents such as methanol and acetonitrile are effective at removing more than 98% of proteins from mammalian blood plasma when used in ratios of 2:1 or greater. The combination precipitant containing $ZnSO_4$ and acetonitrile (1:1) has been reported to be effective at providing very clean supernatants [14]. Acidic solvents such as perchloric acid, trichloroacetic acid, and trifluoroacetic acid are effective at denaturing and removing proteins from samples but are incompatible with LC/MS because they produce massive ion suppression and unreliable quantitation.

Although precipitation is simple and useful as a sample preparation tool, it leaves a lot to be desired in terms of cleanliness and ruggedness. Protein precipitation is ineffective at removing salts and lipids from samples, and these inevitably end up on the HPLC column or in the mass spectrometer source. Because of lipid buildup on the column head, narrow (2.1 and 1.0 mm and narrower) HPLC columns can become clogged or their retention characteristics can become altered after a number of precipitant injections. It is highly desirable to use a guard column when using protein precipitation to process biological samples. Despite this precaution, the column head of an analytical column can still become corrupted because of lipids breaking through the guard column after a single run of several samples. Salt in the supernatant is usually not an insurmountable problem, and the use of a postcolumn divert valve can be effective.

Other problems associated with protein precipitation are chromatography problems associated with injecting strong solvents onto narrow bore columns and sample dilution. Chromatography can be altered if too large a volume of a strong solvent is injected onto an HPLC column. The injection solvent can create a local region of strong elution, thereby causing tailing or extensive peak broadening. These types of chromatographic problems can be minimized by ensuring that the organic composition of mobile phase is at least as strong as that of the injection solvent and by keeping the HPLC injection volume small (1 or 2 μL). Sample dilution arises from the requirement of a threefold dilution in the denaturation step. If maximum sensitivity is required for an analyte or group of analytes then protein precipitation is not the best approach.

## C. Ultrafiltration

Ultrafiltration, as depicted in Fig. 4, allows for a separation based on molecular size [15]. The central component in this approach is a molecular weight cutoff membrane, positioned in the bottom of a small cup containing the sample (donor). Upon sample introduction and centrifugation, this membrane allows molecules smaller than the molecular weight cutoff (typically 3, 10, or 30 kDa) to pass to a receiver container, while retaining larger molecular species. Small drug molecules that are not protein bound are permitted to pass through. Large plasma proteins and any drug molecules bound to them do not pass and are retained in the donor compartment. Of course, small matrix molecules such as inorganic salts, lipids, and water will readily pass through the membrane as well. This ultrafiltrate fluid in the receiver compartment is typically directly injected into the LC/MS system, without additional processing. Because of large concentrations of low-molecular-weight matrix components present in the ultrafiltrate, the use of a postcolumn divert valve is usually a good idea. Reasonable chromatographic

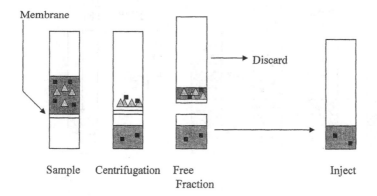

Sample    Centrifugation    Free    Discard    Inject
Fraction

Analyte & Matrix Components
Proteins

**Figure 4**  The ultrafiltration process, including a membrane separation of high-molecular-weight components.

capacity factors ($k'$ of 2 to 5) are also a good way to minimize ion suppression associated with the matrix components.

In principle, ultrafiltration can be an easy way to quantify free drug fraction present in plasma, serum, or other biological fluids. The approach is not without pitfalls, however, in that ion suppression, clogged membranes, and poor sensitivity due to extensively bound drugs can derail this type of assay. Still, ultrafiltration has a lot of untapped potential and could become pervasive as membranes are made more robust and adapted to high-throughput formats such as 96-well plates. A related sample preparation technique, microdialysis, is discussed in Chapter 12.

## D.  Liquid–Liquid Extraction

Liquid–liquid extraction (LLE) has been around for a long time and has been used extensively as an analytical sample pretreatment to remove unwanted matrix components [16,17]. It is based on the principles of differential solubility and partitioning equilibrium of analyte molecules between aqueous (the original sample) and organic phases. Depicted in Fig. 5, LLE initially involves pH adjustment of the sample with an appropriate buffer. This pH adjustment is intended to neutralize the molecule, making it more amenable to extraction. The next step is the addition of an immiscible organic extraction solvent, followed by agitation

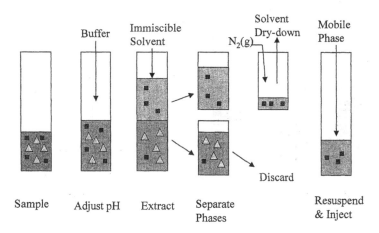

■  Analyte

△  Matrix Component

**Figure 5**  The liquid–liquid extraction process, including the steps of pH adjustment, extraction, phase separation, solvent evaporation, and resuspension.

(vortexing) to facilitate equilibrium partitioning of analyte molecules between phases. The phases are separated and the aqueous component is discarded. The organic phase is evaporated to dryness and resuspended with mobile phase or a similar solvent system and then injected onto the column. Back-extraction into a secondary aqueous phase, once common when UV absorbance was the end detection step, is unnecessary when the higher selectivity of tandem mass spectrometric detection is available.

Liquid–liquid extraction is a very good sample cleanup technique for non-polar or moderately polar analytes that can be deionized in solution using pH adjustment. The best recoveries are obtained when an excess of organic solvent (3- to 10-fold excess) is used. Method development is easy to implement, as the approach has a high probability of working the first time. The technique can, however, be labor intensive and it does not provide very good recoveries for highly polar or zwitterionic species. The adaptation of 96-well automation has gone a long way toward making LLE an effective tool for bioanalysis in drug discovery [9,18].

## E. Solid-Phase Extraction

Solid-phase extraction (SPE) (Fig. 6) is a miniature version of the liquid chromatography experiment. It was commercialized and introduced in the late 1970s. With the availability of prepacked cartridges, SPE first became popular in the mid-1980s [19]. It has been widely applied to sample cleanup in the drug discov-

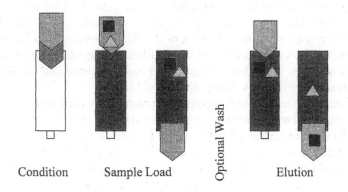

■ Analyte
△ Matrix Component

**Figure 6** The solid-phase extraction process, including column conditioning, sample loading, optional washing, and analyte elution.

ery and development areas. Excellent books and articles describing most theoretical and practical aspects of the technique have been recently published [20–22]. Several varieties of SPE can be utilized, such as reverse phase, ion exchange, and mixed mechanism retention, and are more practical for biological and aqueous samples and should be considered first.

The initial step in SPE is conditioning of the sorbent bed. For a reverse-phase sorbent such as octadecylsilane (ODS) this is typically done with a small volume of methanol or acetonitrile (2 µL/mg of sorbent) and water (2 µL/mg) or buffer. The sample is, again, buffered so that ionization of the analytes is minimized. The sample is loaded onto the sorbent and the sample solvent is pulled through by vacuum, positive pressure, or centrifugal force. Analyte and many matrix components partition with the sorbent and are retained. Some salts and matrix components are pulled through with the sample solvent. A series of sorbent wash steps are then implemented. These wash steps use a weak solvent (possibly 5% organic in water or buffer for a reverse-phase separation) to remove salt and some matrix components from the sorbent while leaving the analyte unaffected. Finally, an elution step uses a strong solvent (70 to 100% organic for normal phase separations) to remove the analyte molecules from the solid-phase sorbent. This eluate, containing analytes from the elution step, is collected and either injected directly or evaporated to dryness with a stream of $N_2(g)$ before resuspention and injection.

More recently, membrane disks [23–25] have been developed and are commonly utilized for sample preparation. Membrane disks contain chromatographic partitioning functionality similar to packed-bed sorbents, but use a thick ($\sim$0.2 mm) membrane to support the sorbent. Relative to packed-bed sorbents, membrane disks allow smaller solvent volumes ($\sim$5$\times$) to be used for conditioning, washing, and elution. For a 3-mm disk, volumes as small as 50 to 200 µL are effective, thereby allowing the omission of the final dry-down step for suitable elution solvents. Disk devices have not been without problems, however, in terms of clogging and flow difficulties. A number of iterations have had to take place before a practical device was finally developed [25].

An interesting and possibly unexpected feature associated with solid-phase extraction disks and other nonsilica sorbents is their lack of silanol activity. This functionality is an important characteristic of all silica-based materials, originating from acidity associated with Si-OH moieties in the silica beads. Endcapping of these silanol sites with methyl groups is moderately effective, but never eliminates all silanol activity. The practical significance of silanol activity is that, in addition to a partitioning mechanism, cationic species will be retained by a secondary ion-exchange interaction. This ion-exchange-mediated retention has proven useful in that it adds an additional dimension of retention and separation to the solid-phase extraction experiment. Great selectivity can be obtained by full utilization of silanol activity [26].

With the exception of immunoaffinity extraction, which is a specialized and elaborate sample-preparation approach [27,28], solid-phase extraction generally provides the cleanest extract of all sample-preparation techniques in terms of selectivity. The price paid for this performance is that method development is generally the most complex and time consuming [29–31]. Generic conditions for automated 96-well solid-phase greatly reduced the need for extraction method development and work for about 85% of the small organic molecule analytes typically encountered in drug discovery [32]. These approaches are described later.

## III. CHOOSING A SAMPLE PREPARATION APPROACH

When choosing a sample preparation approach for an LC/MS/MS discovery-phase experiment, at least five questions should be asked. These include What is the composition of the matrix? What is (are) the chemical structure(s) of the analyte(s)? How long will it take to develop the sample preparation? Can the procedure be automated? and How well does the procedure work? If each of these questions can be dealt with effectively, then, barring analyte instability or adhesion difficulties, a reasonably effective sample preparation can usually be quickly developed.

## A. What Is the Matrix?

Table 3 lists the most commonly encountered sample matrices in drug discovery (and, for that matter, drug development). These matrices represent samples har-

**Table 3**  Choosing the Appropriate Sample Preparation Approach

|                  | Salts | Lipids | Protein | Cellular components |
|------------------|-------|--------|---------|---------------------|
| Caco-2 buffer    | +++   | −      | −       | +/−                 |
| Perfusate buffer | +++   | −      | −       | −                   |
| Hepatocytes      | +     | +/−    | +       | +++                 |
| Plasma           | +     | +      | +++     | +/−                 |
| Urine            | +++   | +/−    | −       | −                   |
| Bile             | +++   | +/−    | −       | −                   |
| Brain tissue     | +     | +++    | ++      | +++                 |
| Brain dialysate  | ++    | ++     | −       | −                   |
| Soft tissue tumor| +     | +/−    | ++      | +++                 |

Note: +++, very abundant; +/−, somewhat abundant; −, not present.

vested from both in vitro and ex vivo experiments. Each of these sample matrices has been rated in terms of the extent to which it contains the four major matrix components: salts, lipids, proteins, and cellular components. The abundance of each of these components is a major consideration when selecting a sample preparation strategy.

For example, the Caco II experiment (Chapter 8) utilizes a high-salt buffer that can be, as described above, problematic when directly introduced into an LC/MS system. A sample preparation approach that addresses the high salt content and absence of protein, lipids, and cellular components is appropriate. Cross-referencing this information to the information presented in Table 1 suggests that a desalting sample preparation approach such as on-line solid-phase extraction (column switching), LLE, or SPE would be most appropriate. Direct injection is also a possible approach that could work in the short term, but by itself it does not address how to protect the instrument system from the salts which have been introduced. Some scientists who are unwilling to spend the extra time and effort required to do SPE or LLE on these samples have combined direct injection with a ballistic chromatographic gradient and a postcolumn divert valve to help eliminate sample salt from the system [33].

Considering plasma as an additional example, the major undesired component to be dealt with is the protein content. From Table 1, protein precipitation, UF, LLE, and SPE are the most effective at removing proteins from these samples. The choice between these techniques could be further narrowed by additional selection criteria such as the chemical structures of the analytes.

## B.  What Is the Analyte Structure?

The chemical structure of the analyte is another important consideration in designing a sample preparation method. The chemical structure gives clues as to the pH range over which the compound will be ionized or neutral, the relative hydrophobicity of the compound, and possible chemical reactivity. Using these clues, an educated guess as to how to achieve the best analyte recovery and the most direct sample preparation route, including pH adjustment and solvent choices, can be made.

The overriding principle in many extractions of aqueous samples is for the analyte molecules to be as near neutral as possible by adjusting the pH of the sample matrix. This guiding principle applies to most liquid–liquid and reverse-phase solid-phase extractions. If a compound cannot be made electrically neutral through pH adjustment, then an alternate approach could be considered. A compound such as a zwitterion, which cannot be made electrically neutral could be isolated from the matrix using an approach which is insensitive to the ionization state of the molecule, such as direct injection, protein precipitation, microdialysis, or ultrafiltration.

It is also possible that a solid-phase extraction technique, such as ion ex-change, could be used to isolate an ionized analyte. This approach can be difficult to implement, however, if a number of structurally dissimilar analytes need to be isolated from a single sample. Reverse-phase solid-phase extraction using an octal or octadecyl sorbent offers a more generic approach that can even work for charged compounds. These sorbents have a sticky affinity for organic molecules and, more often than not, will give at least some retention if the loading and washing solvents have a low organic content.

At this point, a few specific chemical examples illustrating these points are in order. The compound in Fig. 7a contains several very polar functional groups, including a carboxylic acid, a thiourea, a methoxy, and an amide. The $pK_a$ for this compound is approximately 4, owing to the dissociation of the carboxylic acid. Although it could be possible to isolate this compound from plasma matrix with a simple liquid extraction, the large number of very polar groups suggests that the recovery for this approach would be marginal. In practice, with a number of conventional organic solvents, the liquid–liquid extraction recovery rarely ex-ceeded 10%. Protein precipitation was another option, but the assay sensitivity requirements precluded that approach. In the end, a reverse-phase solid-phase extraction proved to be the most viable approach. Plasma samples were adjusted to pH 2.5 with acetic acid and the analyte molecules were retained on a $C_{18}$

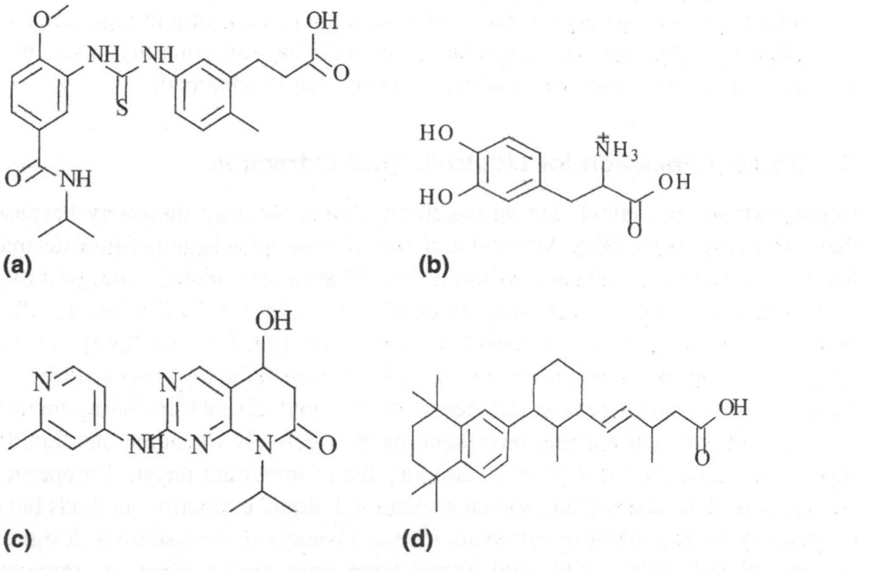

**Figure 7** Example compounds one (a), two (b), three (c), and four (d).

sorbent. Recoveries exceeding 60% were routinely obtained for this compound and numerous structural analogs.

A second example is the zwitterionic amino acid L-dopa, shown in Fig. 7b. This simple molecule is ionized over the entire pH range, with expected $pK_a$s of approximately 4 and 10. At pHs below 4, the compound has a net positive charge and above pH 10 a negative charge is present. Between pHs 4 and 10 both positive (amine) and negative (carboxylic acid) charges are present, giving a net neutral charge, but more importantly, making for a heavily ionized molecule that is poorly soluble in an organic phase. Again, depending on the matrix and assay requirements, either protein precipitation or solid-phase extraction would be most appropriate. Numerous literature procedures for SPE of L-dopa from many biological samples have been published.

A third example is shown in Fig. 7c. This compound contains a number of weakly basic functional groups, including aromatic and aliphatic amines and pyrimidines, which can be completely deprotonated above approximately pH 8. Because of the high proportion of nitrogen in this molecule (clog $P = 0.63$), it is fairly polar, yet can be readily extracted from samples with a simple liquid–liquid extraction using common polar organic solvents. Of course, solid-phase extraction could also work well, but would required some additional development time.

A fourth and final example, shown in Fig. 7d, illustrates a lipophilic compound with a single polar, ionizable functional group. After appropriate pH adjustment to protonate the acid group (below pH 4), the compound will extract well into most nonpolar organic solvents, including pentane or a mixture of alcohol and pentane. Solvent choice can be made on the basis of volatility or selectivity requirements and excellent analyte recovery can be obtained.

## C.  Solvent Selection for Liquid–Liquid Extraction

Liquid extraction methods are an excellent choice for drug discovery because they offer easy method development and effective sample cleanup. Some desired features of effective solvents in liquid–liquid extraction include immiscibility with water, polarity to match the analytes of interest, high volatility for easy drydown, moderate viscosity for effective volumetric transfer, and lower density than the aqueous phase so that the organic phase rises to the top. This last characteristic is not a requirement in that denser than water solvents, including methylene chloride and chloroform, have been used effectively for numerous liquid–liquid extractions. Table 4 gives a summary list of important physical properties for the most desirable organic solvents. Although liquid extraction methods have historically been difficult to automate, recent advances in workstations designed to manipulate liquids in a 96-well format have made highly effective semiautomated approaches possible [9,18].

**Table 4**  Important Physical Properties of Highly Desirable
Liquid–Liquid Extraction Solvents

|  | Boiling pt. (°C) | Density (g/ml) | Polarity |
|---|---|---|---|
| Methylethyl ketone | 80 | 0.80 | 4.7 |
| Ethyl acetate | 77 | 0.90 | 4.4 |
| MTBE +5% EtOH | 58 | 0.75 | 3.2 |
| Diethyl ether | 35 | 0.71 | 2.8 |
| MTBE | 55 | 0.74 | 2.5 |
| Butyl chloride | 78 | 0.89 | 1.0 |
| Pentane | 36 | 0.63 | 0.0 |

## D. How Long Will It Take to Develop the Sample Preparation Procedure?

The essence of bioanalytical chemistry for modern drug discovery is a combination of speed and accuracy. In keeping with these requirements, the sample preparation procedure must be developed in a short time, with a reasonably high certainty of success. Numerous modern sample preparation procedures utilize generic conditions that can be easily automated and will work a large percentage of the time. This is possible because method selectivity required for a liquid chromatography method with ultraviolet or fluorescence detection is not necessary when tandem mass spectrometric detection is utilized. This aspect of LC/MS, more than any other single aspect, has resulted in a significant decrease in method development time.

For a new class of compounds, with unfamiliar properties, several days or a week could be required to develop a quantitative method in the low nanogram-per-milliliter range. If a generic method has been established for a class of compounds, only 1 day is generally required to extend the method to additional compounds of the same class, even if multiple components are being determined. If all goes well, then study samples can be assayed on day two. Problematic compounds arise a small (5 to 10) percentage of the time, and additional time will be required for their method development.

## IV. CAN THE PROCEDURE BE AUTOMATED?

Numerous recent articles have demonstrated that automated sample preparation is highly desirable, not only because of time savings, but in order to relieve the tedium associated with processing large numbers of samples [34,35]. When sample numbers approach 100, it can be possible for an analytical chemist to save an hour or more on the sample preparation through the use of automation. It has been shown that the best way to incorporate automation into a sample preparation procedure is by developing the initial procedure using automation [29]. With the advent of dedicated sample preparation workstations for drug discovery support, this principle still applies.

Recently, automated sample preparation approaches utilizing parallel sample processing have been described for SPE [10], LLE [9], simple dilutions [11], and protein precipitation [8]. These procedures utilize commercially available workstations for liquid handling in a 96-well multichannel plate format. These workstations are evolving rapidly and are constantly gaining additional capabilities. A recent article has reviewed the most common types and describes the major advantages of each [36].

One especially facile workstation, the Quadra 96 (Tomtec), has proven to be a highly efficient 96-well liquid handler for SPE, LLE, and protein precipitation. Using this device as a liquid handler, aliquots of solvents can be manipulated and transferred, in parallel, to each of the wells in the plate in less than 1 min. One disadvantage of this device is the requirement for samples to be in a 96-well format at the start of the procedure. Once this requirement is met, however, and samples are in the correct format, parallel processing can proceed quickly in a semiautomated format, as shown in Fig. 8.

Using this workstation, a typical liquid–liquid extraction would conceptually proceed as follows [9]: (1) A 96-well plate containing samples, standards, and controls (25 to 100 $\mu$L) is placed in position 2 of the workstation. (2) To this plate internal standard is transferred (25 $\mu$L, position 6 of the workstation), followed by an optional buffer (position 5 of workstation) to adjust the sample pH (25 to 50 $\mu$L). An effort is made to keep the volume of the aqueous phase as small as possible in order to maximize the organic-to-aqueous ratio, thereby maximizing the recovery. (3) An organic solvent (position 4 of workstation) is delivered in parallel to each of the 96 wells. These wells can be deep-well vessels that have approximately a 1-mL volume, or they can be discrete tubes in a 96-well rack. Although the center-to-center distance for these discrete tubes is the same as that for a traditional 96-well rack, they are taller, thus holding a larger (1.2-mL) volume. This size tube is helpful when a large volume of organic (up to 800 $\mu$L) is desired. The plate is removed from the workstation, vortexed, and centrifuged and then placed back in the workstation. (4) Phase separation is conducted by the workstation, by having the 96-well transfer apparatus simultaneously dip into the top (organic) phase in each well and remove and transfer most of it (80–90%) to a clean 96-well plate located in position 1 of the workstation. Typically, 700 of the original 800-$\mu$L volume can be transferred effectively. (5) The plate containing the extracts is manually removed from the workstation and transferred to a 96-well dry-down apparatus, using heated nitrogen gas. (6)

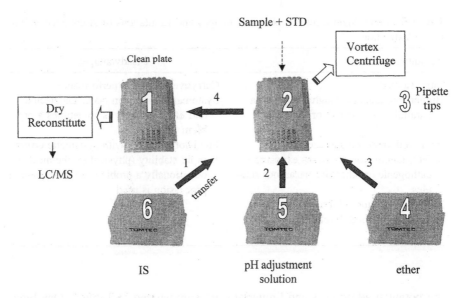

**Figure 8** A semiautomated 96-well liquid–liquid extraction using a Tomtec Quadra 96 liquid-handling workstation. The large numbers correspond to workstation stage positions; the small numbers correspond to individual steps in the procedure.

The dried residues are repositioned in the workstation and resuspended with 100 µL of mobile phase or a similar solvent. (7) After briefly vortexing, the completed plate is placed in the autosampler of the LC/MS instrument and between 1 and 10 µL of each utilized well is injected (unused wells can be skipped by the autosampler). Sample order can be easily randomized or otherwise sequenced by the programmability of the autosampler.

Using this procedure, a skilled analyst can prepare five to six plates (480 to 580 samples) in an 8-hr workday. A similar type of procedure has been used for 96-well SPE for plasma, serum, Caco, rat-intestinal perfusate, liver microsomes, and samples from several other matrices. Although more highly integrated plates, such as those consisting of 384 wells, could replace four of the 96-well plates, the extremely limited per-well sample volumes would make many common sample preparation techniques unfeasible.

## A. Advantages and Limitations of Automated SPE

In order to effectively apply automated solid-phase extraction it is again necessary that the advantages and limitations of the approach be understood. Some of

**Table 5** Some Significant Potential Advantages and Limitations of Automated Solid-Phase Extraction

| Advantages | Disadvantages |
|---|---|
| Time savings | Carryover can limit performance |
| Higher throughput through the use of a parallel processing algorithm | Systematic errors can occur undetected and error recovery is sometimes a problem |
| Improved precision and accuracy | Precision is worse with systematic errors |
| Safety; automation decreases exposure to pathogenic or otherwise hazardous samples | Sample stability (physical or chemical) is occasionally a problem when sequential procession is used |
| Reduced operator tedium | |
| Automated method development is possible | |

the potential advantages and limitations are summarized in Table 5. One long-standing advantage of automated SPE systems was that unattended operation and minimal operator intervention allowed for increased timesaving. Analysts could redirect their time to other tasks during the course of the automated solid-phase extraction.

An important advantage that has only been recently realized has been that automated systems are providing higher sample throughput than could be obtained from manual systems. This advantage has been made possible by utilizing the concept of *automated parallel sample processing*. With early, automated systems, individual samples were processed in series [37]. The next sample in the series was not started until the preceding sample had been completed or was well on its way. With serial sample processing, automated SPE systems were slower than manual systems, but because the workstation could operate continuously during the day, night, or weekend, timesavings were still achieved. Although not efficient in terms of processing time, this approach did prove to be effective and is still in use today. The fastest serial processing equipment currently extracts 25 to 50 samples/hr [36].

Around 1990, automated parallel processing solid-phase extraction was introduced and demonstrated to be practical [36]. Under this algorithm, numerous samples are extracted simultaneously. Although the equipment requirements for parallel processing can be more specialized, or at least more cleverly designed, the great payback occurs in terms of dramatically improved throughput. After parallel processing became commercially available in the form of the Zymark Rapid Trace [39], the speed of automated systems began to overtake that of manual approaches. As is described below, the fastest automation systems in existence today, including most 96-well microtiter plate systems, are parallel-processing systems. The fastest parallel-processing systems can achieve speeds of up to 400 samples/hr [36].

The ability of an automated SPE system to improve assay precision and accuracy is variable and depends on factors such as the consistency of analyte retention and the frequency of systematic errors, such as clogged cartridges and poor volume transfers. All things being equal, the best assay precision and accuracy is obtained by an expert, highly motivated analytical chemist who is not influenced by extraneous factors such as tedium. In terms of precision and accuracy, automated systems come second. Analysts with average skills and a normal propensity to boredom can be outperformed by automated systems when faced with moderate to high sample load [40]. Humanistic factors, such as safer handling of hazardous materials are also important advantages for automation. The ability to perform automated solid-phase extraction method development is an underutilized advantage of many modern workstations [29] and is described later.

The limitations of automated systems, although overshadowed by the advantages, are real and should be kept in mind. One practical limitation is analyte carryover. Carryover is dependent on many variables, including the particular apparatus being employed, the range of analyte levels, the adsorption properties of the analytes, the matrix, sensitivity requirements of the assay, the extent of flushing, tip changing, and similar operations. At best, carryover can limit the dynamic range of an assay by giving erroneously elevated analyte response at low levels. At worse, carryover can severely affect the precision and accuracy of an assay method and give falsely higher results. For a given apparatus, carryover that is acceptable for one application can be completely unacceptable for another application. For this reason, carryover should always be evaluated over a wide, realistic range when automated solid-phase extractions are planned. Extent of carryover ranging from 0.01 to 0.5% is typical, and the smaller the carryover, the better the assay performance.

An important feature now available on many liquid-handling workstations is liquid level sensing. By using the presence or absence of electrical conductance between different areas on the liquid transfer tips, liquid level sensing can detect whether liquid is present in a suitable form for transfer. If liquid is not present, the transfer tip can be repositioned to reattempt the transfer. This feature has practical significance when clots, flocculent, or inhomogeneities are present in samples. In many cases, the workstation can be programmed to make several attempts at a suitable sampling of liquid. Typically, if a suitable sampling of liquid cannot be obtained, the workstation will skip the sample and continue. This approach is helpful for decreasing the number of systematic errors made by sample inhomogeneities such as clots or protein globules in mammalian plasma, but is not as reliable as avoiding and working around the clot in a manual transfer.

A final disadvantage for some automated solid-phase extraction systems is the physical or chemical stability of the samples. This issue is most problematic when the serial-processing algorithm is being used with unstable samples. For example, if a drug substance is determined in plasma, with an overall processing time of 3 min per sample and 60 samples are to be processed serially, it is important to know if the drug is stable for 3 hr. In addition, the physical stability of plasma comes into play. If protein flocculent begins to occur in the plasma due to denaturization, the incidents of clogged solid-phase cartridges or wells will increase and the extraction failure rate can increase dramatically at the end of a run. Parallel processing algorithms are less susceptible to sample stability requirements because the time required to process all samples is usually brief (10 to 20 min) and because removal of the analyte from matrix can physically or chemically stabilize the sample. For this reason and because of higher throughput, parallel processing approaches to automated solid-phase extraction have advantages over serial-processing methods.

## B.  A Brief Review of Equipment for Automated SPE

An excellent review of hardware for automated solid-phase extraction has been recently published by Smith and Lloyd [36]. In this review, Smith and Lloyd present an overview of at least 18 different commercially available systems for automated SPE and discuss 13 important attributes, such as the degree of automation, type and quantity of work, cost, functionality, and so on. Their discussion is intended as a guide for selecting a suitable system and it is highly recommended for this purpose. We do not attempt to redo this excellent review. Instead, we have classified the available systems into three categories: (1) on-line SPE techniques, (2) discrete column workstations, and (3) 96-well workstations. Their descriptions, advantages, and limitations follow.

### 1.  On-Line SPE Approaches

Three recently introduced on-line extraction techniques have been applied to biological samples. These techniques involve direct injection with restricted access media [11,12], turbulent flow chromatography [13], and on-line solid-phase extraction.

The concept of RAM combines a hydrophilic external surface and a hydrophobic internal surface in silica particles with controlled pore sizes (Fig. 9). Large biopolymers, such as proteins, are prohibited from entering the pores of the packing and are not well retained by the column. Therapeutic drugs and other small molecules permeate the pores of the column packing material, where they partition and are retained. This approach is, in principle, a combination of size exclusion and partition chromatography. In practice, to obtain a reasonable

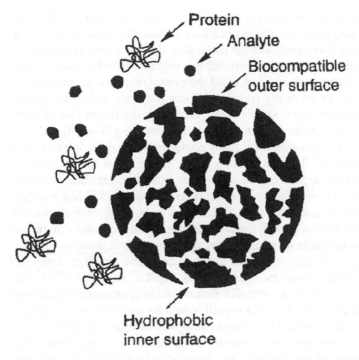

**Figure 9**   Use of a sorbent particle for restricted-access media chromatography. This media allows proteins and macromolecules to be excluded and elute in the solvent front, while small analyte molecules enter the pores and are retained.

amount of chromatographic efficiency, it is often necessary to perform a column or solvent switch with this approach, sometimes using a back flush setup. It occasionally takes a few days to develop suitable conditions, and sample volumes are limited to 10–50 µL. The RAM column typically requires 1.1 to 2 mL of solvent for washing after each injection and this can usually be done while analytes are eluting from a downstream analytical column. A limitation associated with this arrangement is that injection-to-injection cycle time can be long (8 to 15 min) and sample instability (both matrix and analyte) can be problematic when a large number of samples is involved. Injection-to-injection analyte carryover, although not well characterized, is also a potential problem.

Turbulent flow chromatography is a direct-inject sample-preparation technique that is accomplished on a special chromatography column. The technique has some applicability toward plasma and serum. The special column combines large particle (50 µm) and frit size (20 µm) with high flow rate (5 to 10 mL/min) to achieve eddies and nonlaminar flow. Under this arrangement, improved mass transfer and flow equilibration increases the analyte diffusion within the pores of the packing material. The net result is a separation of large biological matrix components from small analyte molecules. Sample pretreatment can be reduced to a simple centrifugation step, but the column is usually back-flushed and reequilibrated after each injection to rid the system of insoluble components. Method sensitivity can be nearly equal to off-line preparation (low nanogram-per-milliliter range for 50-μL sample size). There is a significant loss of chromatographic efficiency in the form of peak fronting or tailing (Fig. 10) and variable analyte recovery is prevalent. For many compounds, interinjection carryover (0.15 to 0.5%) seems to limit the dynamic range and utility of this technique.

The most widely applied on-line solid-phase extraction apparatus continues to be the Prospekt [41–49]. This device incorporates custom solid-phase extraction cartridges into an analytical-scale HPLC separation using three electrically actuated switching valves. Several hundred cartridges can be loaded into the instrument prior to initiation of a run. Under program control, samples in a 96-well format can be directly injected onto the head of the cartridge, undesired components washed to waste, and then analytes eluted to an in-line HPLC column, followed by detection. As with all column-switching arrangements, elution solvents are limited to those that are compatible with the downstream components and must be carefully selected to maintain acceptable chromatography. To circumvent this limitation, some investigators have gone so far as to propose eliminating the analytical column entirely [50]. Again, as with other serial processing techniques, sample stability must be considered.

There are advantages to on-line serial-processing approaches. Although the throughput is lower than pure, parallel-processing systems, they can gain back some lost efficiency by a direct link between the sample preparation and the downstream separation/detection and fewer liquid transfers are made. The trade-off to be made is that the downstream separation and detection need to be adapted to the on-line extraction through timing and solvent selection. This could lead to suboptimal performance.

## 2.   Discrete Column Workstations

One current arrangement for solid-phase extraction workstations is to use discrete SPE columns in commercially available syringe barrel sizes. Two systems that have successfully used this approach are the Rapid Trace (Zymark) and the Speed Wiz (Applied Separations). The biggest advantage of these systems is the wide selection of phase availability in the syringe barrel format. This advantage could diminish in time, as additional sorbent types become available for 96-well workstations. Another advantage is that each of these systems is, to some extent, a parallel-processing system, so at least some throughput advantages are enjoyed, relative to serial processing. In this regard, the Rapid Trace is a hybrid serial/

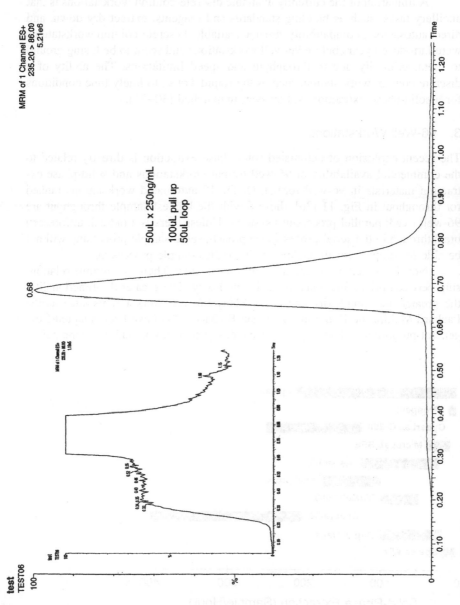

**Figure 10** Chromatogram for a typical turbulent-flow chromatography experiment (50-μL/L injection), showing fronting and tailing behavior. The numbers of theoretical plates (N) for this compound is ~350.

parallel-processing system; consisting of up to 10 modules operating in parallel, each module can process up to 10 samples serially.

A limitation of the currently available discrete-column workstations is that ancillary tasks, such as building standards and reagents, extract dry-down, and direct autosampler compatibility, are not available. Discrete-column workstations were introduced years before 96-well workstations and seem to be losing ground to them, primarily due to throughput and speed limitations. The ability of a discrete-column work station, such as the Rapid Trace, to finely tune conditions for a solid-phase extraction is, however, unmatched [30–31].

## 3. 96-Well Workstations

The recent explosion of automated solid-phase extraction is directly related to the commercial availability of 96-well format workstations and solid-phase extraction materials in 96-well format. Of the 10 automated workstations ranked for throughput in Fig. 11 [36], those 6 with the highest sample throughput are 96-well (read: parallel-processing) systems. Unless there is a radical, unforeseen breakthrough in the serial-processing approach, serial sample processing will not be able to compete for throughput with parallel sample processing.

Not all 96-well workstations were created equal. There is a definite relationship between price, functionality, and complexity. The greater the functionality, the greater the complexity and price. Many 96-well workstations such as the Packard Multiprobe II and the Beckman BioMec 2000 have been adapted from general-purpose use. These systems were designed as liquid handlers long before

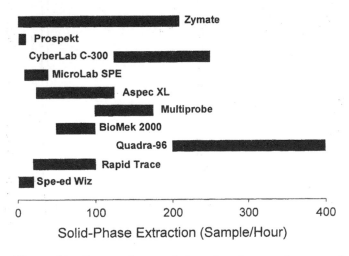

**Figure 11**  Comparative sample throughput for several automated solid-phase extraction workstations. (From Ref. 36.)

they were ever used for solid-phase extraction. As such, these automated work-stations have more flexibility and capabilities at the expense of efficiency. They allow for almost completely automated approaches while displaying greater facil-ity and lower throughput efficiency. The 96-well manifold, similar to that shown in Fig. 12, has been widely adapted to these workstations.

The 96-well systems are not the best available technology in terms of pro-ducing precisely tuned solid-phase extractions. Yet, because of the insatiable de-mand among some analytical groups for increased speed and throughput, 96-well workstations seem destined to gain popularity at the expense of other approaches until they are eventually supplanted by something better.

## C. How to Automate Solid-Phase Extraction

For an automated solid-phase extraction to be worthwhile, a minimum number of samples are required. In the past, this break-even number was a few hundred

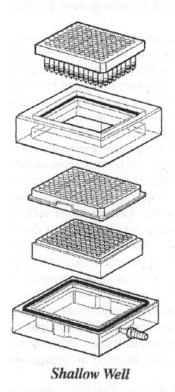

*Shallow Well*

**Figure 12**  Typical shallow-well 96-well format solid-phase extraction vacuum mani-fold adapted for use with liquid-handling workstations.

samples. The work required to automate an extraction demanded an assay of this many samples before a return was obtained on the time investment. As automated solid-phase extraction has become more commonplace and as better off-the-shelf solutions have become available, this break-even number has decreased. It now can be as low as 5 or 10 samples, depending on the system available and the analytical chemist's comfort level with automation.

There are few absolute rules in solid-phase extraction. One rule that stands out is that if a procedure is to ultimately be automated, it should be automated from the onset of method development, using the workstation on which it will be run. It is counterproductive for an extraction protocol to be developed manually then automated, as there are enough differences in pressures, flow rates, and solvent composition that transferring from manual to automation is like starting method development from scratch [40]. In addition, automated workstations have developed to the point where a number of efficiency advantages can be gained by doing the solid-phase extraction method development on the workstation [34]. Initial steps in automated solid-phase extraction, therefore, include selection of a workstation, based on assay requirements, on which to begin method development.

As the initial experiment, a solid-phase extraction method development paradigm might involve recovery evaluation at one concentration, in duplicate, on several different reverse-phase sorbents (eight samples plus four blank extracts for spiking as recovery standards). To achieve a first approximation of best case recovery, wash and elution solvents are chosen so that the analytes have the best chance of being retained and eluted. For a wash solvent in a reverse-phase extraction, 95:5 aqueous:organic could be chosen, and the converse (5:95 aqueous:organic) could be chosen as an elution solvent. Because the extraction selectivity is suboptimal at this juncture, the use of a highly selective detection method is desirable. The evaluation of recovery using nonmatrix samples is not recommended because interaction of analyte molecules with matrix components will affect recovery. On the basis of recovery, one or two sorbents could then be selected for further evaluation. An example of these experimental results is shown in Fig. 13. Several sorbent manufacturers can supply 96-well SPE method development kits. These kits are designed so that a different sorbent is located in each row of a single 96-well microtiter plate. Using one of these plates, a number of different sorbents can be evaluated in a single set of automated experiments.

Next, five or six wash solvents containing various amounts of organic (possibly 5, 10, 20, 30, 40, and 50%) are tested. This also requires duplicate determinations and recovery standards (18 samples). An example of these experimental results is shown in Fig. 14. From this plot, it is easy to select an appropriate wash solvent composition to maximize recovery.

This experiment could be followed by an evaluation of elution solvents at typical organic compositions of 70, 80, 90, and 100% (12 samples), with 70%

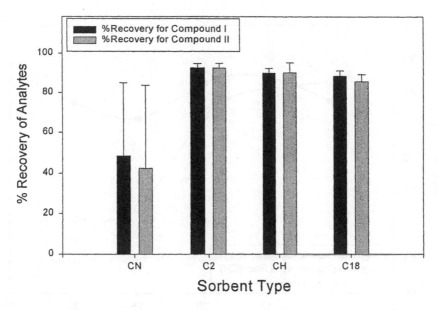

**Figure 13**   Solid-phase extraction recovery of two acidic compounds from four different solid-phase sorbents using a wash solvent of 5% acetonitrile and an elution solvent of 100% acetonitrile. The solid lines indicate the coefficient of variation associated with each determination. (From Ref. 31.)

as a practical limit for evaporating extracts in a reasonable time. If dry-down is to be omitted from the procedure, then this restriction need not apply.

A fourth experiment would be an evaluation of precision and recovery at one to four concentration levels ($n = 6$ to 24 plus a recovery standard), plus recovery of an internal standard ($n = 3$ plus recovery standard). Therefore, as a minimal method development exercise, 60–70 spiked samples would be prepared and extracted within 1 day. The experiments need to be performed sequentially because the results from each will impact how subsequent experiments are designed. Selectivity is assessed through the course of the method development. The analytical chemist with access to API LC/MS/MS will spend less time on solid-phase extraction selectivity development.

With many types of solid-phase extraction workstations, especially those with computer control of flow and pressure, some initial experience is required to select these parameters. After this experience has been gained, the same or similar settings can be used for a variety of applications without adjustment.

As mentioned earlier, carryover is a problem for automated SPE workstations and carryover performance should always be investigated. This set of

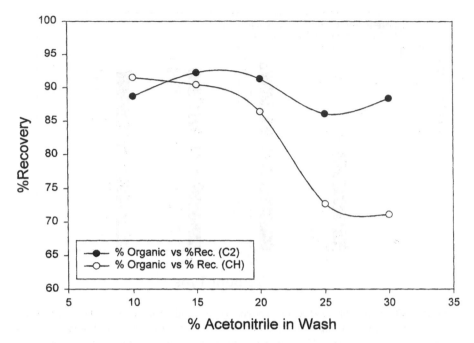

**Figure 14**  Solid-phase extraction recovery of a compound as functions of wash solvent concentration and sorbent. Solid dots (●) indicate recovery for $C_2$ sorbents, and hollow dots (○) indicate recovery for CH sorbent using 100% acetonitrile elution solvent. (From Ref. 31.)

experiments can be as simple as running a few matrix blanks in the same workstation positions after a few high-level samples or standards have been processed. If carryover is observed, then additional wash steps or transfer tip exchanges need to be added to the procedure and the carryover assessment needs to be repeated. If carryover cannot be eliminated in this way, then it could be possible to decrease carryover by selecting a more appropriate wash solvent for increased analyte solubility and decreased surface adsorption. For example, if the analytes are amine-containing compounds and demonstrate stickiness to surfaces, a mixture of methanol, water, and trace formic acid would be a better wash solvent choice than pure acetonitrile.

Carryover limits the performance and usefulness of some solid-phase extraction systems, automated or manual. Some degree of carryover is always present and can be observed if the detection technique is sensitive enough or the analyst looks hard enough. It is, therefore, important that realistic levels of carryover are assessed. If a method will be used to quantify an analyte over 1 to

1000 parts per billion, carryover from a 10 parts-per-million sample may not be reasonable if this level will not be encountered in sample assay work. Abnormally high analyte levels do occur sporadically in most analytical work and these occurrences must be recognized. The final options for eliminating carryover from automated methods are to deliberately limit the low-end dynamic range of the method so that false positives are not encountered or to avoid using an automated system for the extraction. If it is not feasible to increase the limit of quantitation, a different option must be found.

## D. A Generic Approach to Automated Solid-Phase Extraction

Because of the widespread use of LC/MS/MS for drug discovery bioanalysis, there is currently less of a necessity for finely tuned solid-phase extractions than there once was in this area. Instead, generic solid-phase extraction conditions that can accommodate many different analyte structures using the same extraction conditions have become more interesting. To make a solid-phase sample preparation useful for LC/MS/MS it must remove as much of the sample salts as possible in order to reduce the effects of ion suppression [47,51–52] and it must remove as many nonvolatile matrix components as possible so that the instrument ion source is not quickly fouled. Because the LC/MS/MS instrument is inherently so selective, added assay selectivity, per se, is no longer an objective of solid-phase extraction.

Examples of this approach has been reported by Janiszewski [10] and others [53] by utilizing extraction disks in a 96-well format to perform quick, automated solid-phase extractions under very simple wash and elution conditions. In one approach [10], a Tomtec Quadra 96-well workstation was used to perform the semiautomated solid-phase extraction with Empore $C_2$ extraction disks. The advantage of this piece of equipment is that it allows liquid to be transferred to or from all 96-wells simultaneously, thus giving the greatest throughput advantage.

The generic protocol for this type of extraction requires the sorbent to be conditioned with pure organic solvent and then water prior to sample loading. After the wells are processed with aliquots of wash solvent (typically water or buffer/organic mixture), a 96-well shallow plate is manually inserted into the collection position and the elution solvent (pure organic or organic with a small amount of acid or base) is added to each well. It is also possible to use multiple solvent elution steps to elute different classes of compounds separately. With extraction disks or small-mass (10-mg) packed-bed sorbents, the elution volume can be kept in the 75- to 150-μL range. As a simplification to the procedure, the eluate is not dried down. Rather, the contents of each well is diluted with a small volume of water or buffer to give an injection solvent with a composition (20 to 40% aqueous) that is compatible with the liquid chromatography mobile phase.

This approach provides minimal sample cleanup with little or no method development effort. The solid-phase extraction desalts and deproteinizes the samples: method selectivity is furnished by the LC/MS/MS system. The approach is capable of high throughput (up to 400 samples/hr) and appears to work a large percentage of the time, making it well suited for drug discovery support.

## V. EVALUATING SAMPLE PREPARATION PERFORMANCE

Analytical and bioanalytical methods for discovery can be evaluated and judged according to many of the same figures of merit used for good laboratory practices assays in support of toxicokinetic and clinical studies [54]. A list of typical parameters is given in Table 6. These parameters help to ensure data quality and build additional confidence in the analytical results. Because of the shortened time scale for discovery work, assays are necessarily developed and used over 2 or 3 days and then might never be used again. Also discovery experiments are often rudimentary in nature, yielding only preliminary drug information from a limited

**Table 6**  Useful Figures of Merit for Bioanalytical Methods

| Parameter | Routinely evaluated? | | Comment |
|---|---|---|---|
|  | Discovery phase | Development phase |  |
| Dynamic range | √ | √ | Based on adherence to linear or quadratic regression model. The correlation coefficient is useful only in that in measures scatter. Other parameters for goodness of fit are more useful. |
| Quantitation limit | √ | √ | Based on precision of replicates at the limit of quantitation. |
| Selectivity | √ | √ | Based on a combination blank matrix samples and chromatographic homogeneity in real samples. |
| Precision | √ | √ | Based on replicate spiked samples. |
| Accuracy |  | √ | Based on replicate quality controls. |
| Recovery |  | √ | Often not evaluated in discovery, >70% recovery is desirable for fully validated methods. |

number of samples. It is appropriate, therefore, to separate the analytical method figures of merit parameters that are essential (indicated by a check in Table 6) from those that are useful.

## A. Quantitation Limit, Dynamic Range, and Linearity

During method development and characterization, it is desirable to construct a calibration curve running from lowest to highest standard. This allows assessment of dynamic range; linearity; and, to some extent, quantitation limit. By assessing the chromatographic peak area counts and the signal-to-noise ($S/N$) ratio associated with the lowest standard a better assessment of quantitation limit can be gained. In practice, an $S/N$ of 10 suggests that a given quantitation limit is feasible. After the calibration curve has been constructed, the relative errors in the back-calculated standards are calculated for all standards. Relative errors of 15, 20, or even 30% in standards can be acceptable, depending on the degree of confidence required in the data. Deviations from linearity are usually acceptable, and linear or quadratic regression models have been used gainfully, along with unweighted or weighted regression lines (refer to Chapter 5). It is generally desirable to have as wide a calibration range as possible, as this will minimize the possibility of having some samples above the limit of quantitation. Two-and-one-half to 3 orders of magnitude in dynamic range are generally achievable.

## B. Assay Precision

Precision is a fundamental measure of assay performance and it should be assessed whenever possible [55]. This will allow a minimum understanding of how much confidence can be placed in the analytical data. In fully characterized assays designed to support definitive nonclinical and clinical studies, between-run and within-run assay precision of 15% or less (20% at the quantitation limit) is acceptable. Because of the shortened time scale for assay development and characterization in discovery these guidelines are excessively restrictive. Precision values of 20 to 30% are acceptable, as these are still less than intersubject variability associated with many nonclinical experiments. With this level of imprecision, data quality is sufficient to answer fundamental scientific questions such as relative levels of drug absorption, relative values of absolute bioavailability, and relative degree of drug clearance.

## C. Assay Selectivity

Because of the intrinsically selective nature of LC/MS/MS, assay selectivity is not as big an issue as it was in the past. Truly endogenous interferences are rare and can often be handled by modifying chromatographic selectivity or, as a last resort, the sample preparation procedure. A simple and effective tool for evaluating assay selectivity continues to be the evaluation of blank matrix samples from one or more sources. Another indication of selectivity issues can be the quality of a calibration curve construct. If an acceptable or high-quality calibration curve is produced (high degree of linearity, low scatter, and zero intercept) then it is probable that assay selectivity is under control. Nonzero intercepts, upward bending calibration curves, and high degrees of scatter can indicate selectivity problems or other issues [56–58].

The larger selectivity issue for LC/MS/MS arises not from the matrix, but from the compound itself. Instability of the compound, whether arising from chemical degradation or biotransformations, can lead to additional or split chromatographic peaks. These peaks are produced when chromatographically resolved degradation products or metabolites fragment in the source to recreate the parent drug. Once this *in-source fragmentation* occurs, the compound will behave exactly like the parent compound in the mass spectrometer because it *is* the parent compound. An example of this is given in Chapter 5.

Currently, there are no mass-spectrometric techniques to differentiate between original parent drug and that formed by in-source fragmentation. To circumvent this problem, it is advisable to have at least some chromatographic retention ($k'$ of 2 to 5) for each analyte to be quantified. Although metabolite/degradation product formation and retention cannot always be reliably predicted a priori, this precaution is likely to eliminate many selectivity problems caused by in-source fragmentation of unforeseen metabolites. A valuable tool for evaluating this type of assay selectivity is to observe the degree of chromatographic peak symmetry. Split peaks or shoulders often indicate that chemistry is occurring with the analyte(s).

## D.  Assay Accuracy

Accuracy is one of the most difficult assay performance parameters to estimate. The most readily accepted modern approach for estimating assay accuracy is to obtain (purchase or make) and assay replicate quality control samples containing the analyte(s) of interest. In this way, the relative error between theoretical and found concentrations of analyte can be calculated and used as an estimate of accuracy. This approach is constrained in that drug protein binding, drug lipid partitioning, or other synergistic drug/matrix interactions in real samples can limit the value of control-based accuracy estimates.

Because of the highly novel nature of new drug candidates, it is not possible to purchase controls containing these analytes. This leaves the possibility of preparing control samples. The rapid pace of drug discovery and the limited supply of untested and highly novel drug candidates (sometimes only a few milligrams are available for all testing) suggest that it is unrealistic to prepare quality controls for an assay that will be used only once. This means that many early in vitro and in vivo assays for drug screening will be run without an estimate of accuracy. Only at the latest discovery stages, when there are less candidates and any one candidate has a much higher chance for success, do quality-control-based accuracy assessments make sense. At later stages of discovery, when it is assessed, the accepted criterion for assay accuracy parallels that for precision: typically 20 to 30% relative errors.

## E. Extraction Recovery

As with accuracy, assay extraction recovery can only be estimated from control samples. Also, discovery phase analytical methods are, by constraint, less robust than those required for late-phase drug development work. An acceptable drug assay for discovery may need to be totally reworked to support drug development. For these reasons recovery is not often assessed during discovery phase analytical work.

At some point, however, assay recovery for an LC/MS method will be estimated and when this is done several important points are worth keeping in mind. Over the instrument linear range, mass spectral response is proportional to the number of ions which have entered the high vacuum region of the instrument. This number of ions is, in turn, dependent on how efficiently an analyte is ionized in the source. For a given analyte, the extent of ionization dictates its ability to compete with the matrix components present. As the amount of ionizable matrix components increases, the ability of an analyte to ionize, hence its apparent concentration, is diminished. This so-called *ionization suppression*, or *ion suppression*, can lead to nonquantitative behavior in LC/MS [51,52].

If not done carefully, ion suppression can severely distort the determination of assay recovery. For example, if the recovery of a liquid–liquid extraction is to be evaluated, the comparison of analyte peak response from a spiked, extracted matrix sample to the analyte peak response from a clean solution is not appropriate. The compound in clean solution will ionize much more readily than the compound in extracted matrix, so that when the ratio of response extracted to response nonextracted is calculated, an erroneously low estimate of recovery will result. If this type of recovery estimate is to be conducted, it is much better to extract an aliquot of blank matrix and spike the analyte into this residue for use as the recovery standard. In this way, the ionization environment of the recovery test sample and standard are approximately equal and a meaningful estimate of recovery can be made.

## F. Controlling Variability in Analytical Methods

It is valuable to understand the sources of uncertainty (variability) in analytical methods and in the sample preparation step in particular so that it can be managed and controlled accordingly. It has been well established by modern statistical theory that the overall random uncertainty of an analytical process is equal to the square root of the sum of the squares of individual standard deviations of various components in the process [59]. The equation for this is as follows:

$$S^2_{overall} = S^2_{aliquot1} + S^2_{aliquot2} + S^2_{sample\ prep} + S^2_{injection}$$
$$+ S^2_{ionization} + S^2_{fragmentation} + S^2_{detection}.$$

Here, the overall square of the standard deviation for the assay method equals the sum of the squares for volumetric transfer, sample preparation, injection, ionization, fragmentation, and detection. Under normal circumstances, it has been shown recently that if the proper precautions are taken, the standard deviations for volumetric transfers are small and well controlled, even when biological or fermentation media are involved [60]. The standard deviation for sample injection is typically less than 1% as well. Although the uncertainty associated with ion detection in mass spectrometry is dependent on a number of components, including the mechanism of detection and the signal-to-noise ratio, for $S/N$ ratios greater than 10 or 20, the random error associated with detection is also small. Likewise, fragmentation for the most common ion molecule processes can usually be controlled to within a small percentage. These terms in the above equation can be neglected, and the equation simplifies to the following:

$$S^2_{overall} = S^2_{sample\ prep} + S^2_{ionization}.$$

The limiting sources of uncertainty in LC/MS/MS methods are sample preparation and sample ionization. Under typical conditions, the random uncertainty associated with sample preparation is 5 to 15% relative standard deviation. This can increase to 25% or more when variables such as pH, solubility, or protein binding of the analyte within the matrix are not well controlled. If sample pH adjustment, for example, is poorly chosen during sample extraction, then the ionization state of the analyte molecules can be mixed. Small sample to sample differences in pH will then result in variable recovery, leading to high overall assay variability.

Analyte ionization can also be highly variable, ranging from 5 to 10% under the best circumstances and sometimes reaching 20% or more. There are several possible reasons for this higher variability, most of which are traceable to the sample preparation. One possible reason for highly variable ionization is source fouling. As a number of samples are passed through the ion source, matrix components can accumulate on or around the source cone orifice plate or RF lens. The deposition of debris on these components can cause a loss in ion formation over the course of a few dozen samples. Modern ion sources such as the Z-spray™ and the Sciex sources have greatly reduced this problem, but it seems unlikely that the problem will ever be completely eliminated. The undesired accumulation of nonvolatile matrix components is one of the perils associated with the interfacing of a condensed phase domain of samples with the vacuum phase domain of ion chemistry.

A second possible reason for ionization variability is the presence of semivolatile or volatile components in the chromatographic effluent. If these inorganic salts and organic matrix components coelute with the analyte molecule, they can compete with and curtail ionization. Effective deterrents to sample derived ionization suppression include ensuring that the sample preparation approach does an adequate job of sample cleanup, eliminating as much salt as possible, and providing adequate chromatographic separation of salts from the analytes. Longer retention times and ballistic gradients have been used for this purpose [33].

A better way to minimize the effects of this ion suppression is to have the luxury of a *stable-label internal standard*. The stable-label internal standard has the same structure as the analyte molecule, but certain atoms in the molecule will be replaced by nonradioactive isotopes, thus giving the stable-label compound a different molecular mass, but virtually identical properties in every other way. This stable-label internal standard will extract, chromatographically elute, and ionize in the same ways as the analyte. It will be distinguishable by $m/z$ in parent and, possibly, daughter ions depending on the position(s) of the isotopic label. Unfortunately, stable-label internal standards are not available for most early-phase drug discovery work because of the rapid time scale and the large number of compounds involved.

## G. Assessing Discovery Assay Performance—An Example

Figure 15 depicts a set of electrospray MRM chromatograms for a cassette consisting of seven compounds plus an internal standard (top chromatogram). These chromatograms are for the low standard in a curve that ranges from 2.5 to 1000 ng/mg in brain tissue. The compounds included in this assay were structurally similar to that depicted in Fig. 16, containing both positive and negative ionizing groups. For convenience, a positive ionization mode was selected for this experiment. The sample preparation included sonic homogenization of weighed amounts of brain tissue in saline, followed by liquid extraction using methyl-*tert*-butyl ether in a semiautomated 96-well format. Although the retention times are short, the capacity factors for analyte molecules were still acceptable (ranging from 1.2 to 4.2), owing to the small dimension analytical column used (3.0 × 1.0 mm). Analyte compounds were injected individually during method characterization to verify that interference, derived from common parent–daughter ion combinations, was not occurring. Liquid–liquid extraction often draws a large amount of lipids from brain tissue. Although these components do not interfere with the detection selectivity of the assay method, these materials can often build up on the head of the column. The use of a guard column or a large-diameter (3.0 mm diameter or greater) column will allow for a reasonable column life.

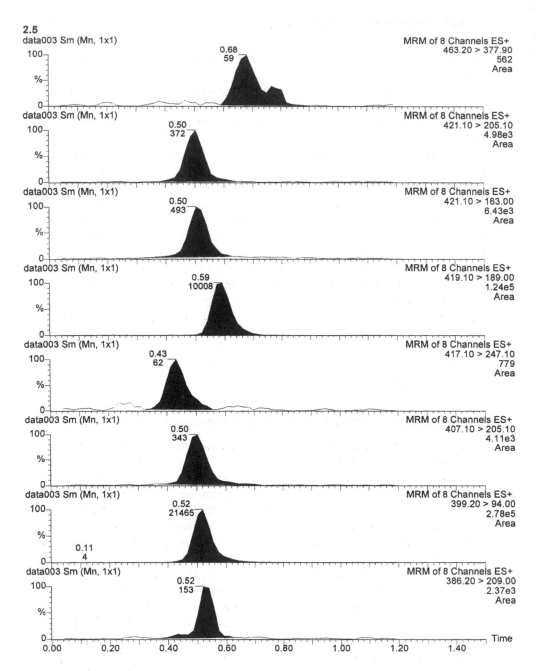

**Figure 15** Electrospray positive multiple-reaction monitoring (MRM) chromatograms for seven dosed compounds plus an analytical internal standard (top chromatogram). Compounds were extracted from brain tissue (2.5 ng/mL) by 96-well semiautomated liquid–liquid extraction.

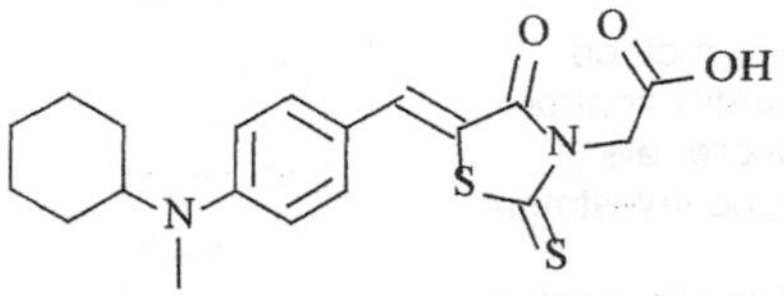

**Figure 16** Chemical structures of the compounds separated in Fig. 15.

Additionally, the quality of a calibration curve was assessed by evaluating the relative errors (RE) associated with back-calculated concentrations (C) for standards (%RE = 100 $\times$ ($C_{nominal}^I$ − $C_{calculated}^I$)). These errors are generally expected to be less than ±30% across the calibration range. Replicate values for imprecision were assessed at the high and low ends of the calibration range and were found to be less than ±30%. This figure of merit is often used to set the quantitation limit for the calibration range. That is, during a practice run, replicates standards were run at several levels (1.0, 2.5, 5.0 and 10 ng/ml). If a particular level fails to attain a relative standard deviation (RSD) of less than 30%, the quantitation limit is set one level higher. In this example, the RSD was less than 30% at the 2.5 ng/ml level, but approximately 50% at the 1 ng/ml level. Therefore, the 1 ng/ml level was not used for this assay method. In this facile way, a certain degree of assay characterization is quickly obtained for discovery stage assays. Using this method, an analytical chemist can readily quantify the levels of these drug candidates in rat brain tissue.

## VI. THE EVOLUTION OF ANALYTICAL METHODS

The evolution of a bioanalytical method through drug discovery and development roughly parallels the evolution and growth of our overall knowledge about the drug candidate, as outlined in Fig. 17. Initially the method is used in a drug candidate screening mode, where the questions asked include how well the drug is absorbed, distributed, metabolized or excreted relative to other candidates. It is a rudimentary method in that it conforms to minimal standards of precision, linearity, or quantitation limits. Some performance parameters, such as accuracy or recovery, may not be assessed. It requires a small investment in time (usually a day or 2 for development) and will exhibit the highest failure rate of any position in the process. A formal validation is not conducted and quality controls are usually not prepared because of a lack of availability of the compound.

If the drug candidate is chosen to progress to the prelead (preclinical) stage, the method evolves to a somewhat more robust state. Additional quality control is built into the method and the analyst, having used the method one or more

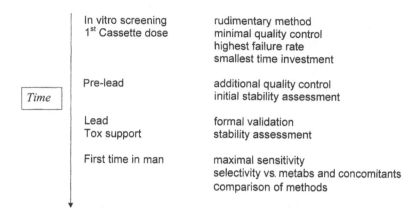

**Figure 17**  Bioanalytical methods evolve and improve as the drug discovery/development process progresses.

times, will have more confidence in it. Often, at this stage, quality control samples are prepared and used to assess assay accuracy and compound stability. Extensive assay validation is not yet conducted, however, as the viability of the drug candidate is still highly questionable.

If a compound is selected to enter preclinical development it is normally given significant resources and is expected to be around for an extended period of time. Now it is expected that the assay will be used more extensively to answer detailed questions about the drug with greater precision and confidence. The assay could be redeveloped and extensively characterized and tested so that it meets well-established guidelines for inter- and intraday precision and accuracy [54]. Other performance criteria, such as sample stability, are looked at as well. Sample preparation recovery could be defined in replicate. The assay is applied to large-scale, nonclinical studies in a variety of species.

As a compound proceeds into first-in-human clinical trials, the assay method reaches an apex in terms of performance. These methods require maximal sensitivity to be able to support dose escalation studies. The selectivity of the method is well established versus matrix components, concomitant medications, and metabolites. The assay is revalidated in the human matrices (plasma, serum, and urine) and will again meet well-established guidelines for inter- and intraday precision and accuracy. If significant changes are made to the method, a comparison of methods study will be conducted to understand the relative accuracy of the methods. After this benchmark, assay requirements, especially limit of quantitation, are usually less demanding and use of the assay becomes more routine, as it is applied again and again to additional clinical studies for pharmacokinetic support.

## VII. INTEGRATED SAMPLE HANDLING FOR DISCOVERY BIOANALYTICAL

One possible logistical roadblock to the widespread use of 96-well sample preparation workstations is the requirement that samples be in a 96-well format prior to the start of sample processing. Possible solutions to this include (1) manual transferal of samples, (2) the use of a liquid handling workstation to transfer samples, or (3) initial generation and delivery of samples in a 96-well format. Each of these approaches has been used to some extent, with the least desirable being manual transferal. The use of a liquid handling workstation, such as the Packard Multiprobe, has shown some utility in transferring samples from individual vials to a 96-well plate. Most recently, however, some discovery laboratories have settled on the latter approach: the initial generation and delivery of samples directly in a 96-well plate. Although not applicable to all types of sample matrices, most notably solids [61], this approach can streamline the sample preparation approach by eliminating one or more sample transfer steps and by allowing 96-well sample preparation to proceed more efficiently.

Figure 18 depicts an integrated sample/information flow for a typical discovery-phase ex vivo experiment. In this model, a discovery scientist designs

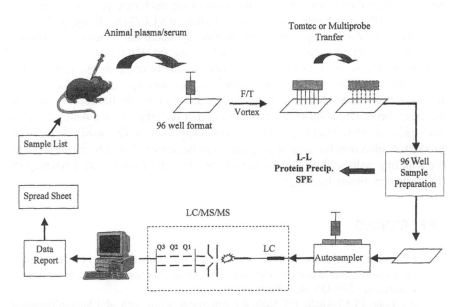

**Figure 18** Schematic representation of an integrated sample preparation/handling process for discovery bioanalytical. Electronic information flow regarding study design and sample collection moves in parallel with the physical handling of the samples.

a protocol and builds a sample list that is electronically available to all groups participating in the study. Animal modeling specialists execute the dosing protocol and collect and deliver samples to 96-well plates along a predefined fill scheme. A plate map delineating this fill scheme is available electronically. At this point, some of the sample locations in the 96-well plate remain vacant to accommodate standards and controls to be included at a later time. From this point onward, the samples remain in the 96-well format, although they may be frozen, thawed, centrifuged, automatically transferred, or otherwise processed by one of the approaches described earlier. After sample processing, a 96-well plate is delivered to an autosampler and injected into the LC/MS system for separation, detection, and quantitation. Quantitation results are reported in a format that has been previously defined (through the study protocol) and disseminated through electronic means. This approach offers a cogent and streamlined approach to sample collection and data handling for a majority of discovery phase experiments and serves to remove what many consider to be a bottleneck in the discovery process.

## VIII. SUMMARY

Sample preparation is an integral component of the analytical process, often providing the difference between success and failure in a LC/MS/MS approach. The selection of a sample preparation approach is dependent on a number of parameters, including the compound structure and the matrix. Each of the matrices encountered in discovery analytical chemistry contains components that require unique attention. The use of automation has become an important tool for increasing throughput in Discovery Bioanalysis. Although most of the recent attention has been directed at solid-phase extraction, other techniques have benefited from automation as well. Highly parallel processes appear to be the most efficient and productive direction for sample preparation in the future. The integration of sample handling with sample preparation has potential to improve the efficiency of the discovery bioanalytical process.

## REFERENCES

1.  J Blanchard. Evaluation of the relative efficacy of various techniques for deproteinizing plasma samples prior to high-performance liquid chromatographic analysis. J Chromatogr 226:455–460, 1981.
2.  AS Tang, PJ Chikhale, PK Shah, RT Borchardt. Utilization of a human intestinal epithelial cell culture system (Caco-2) for cytoprotective agents. Pharm Res 10: 1620–1626, 1993.
3.  H Lennernas. Human intestinal permeability. J Pharm Sci 87:403–410, 1998.

4. T Hagiwara, T Yasuno, K Funayama, SJ Suzuki. Determination of lycopene, alpha-carotene and beta-carotene in serum by liquid chromatography-atmospheric pressure chemical ionization mass spectrometry with selected-ion monitoring. J. Chromatogr B Biomed 708:67–73, 1998.
5. J Roboz, Q Yu, A Meng, R van Soest. On-line buffer removal and fraction selection in gradient capillary high-performance liquid chromatography prior to electrospray mass spectrometry of peptides and proteins. Rapid Commun Mass Spectrom 8:621–626, 1994.
6. AH Kumps. Therapeutic drug monitoring: a comprehensive and critical review of analytical methods for anticonvulsive drugs. J Neurol 228:1–16, 1982.
7. DS Hage, SA Tweed. Recent advances in chromatographic and electrophoretic methods for the study of drug-protein interactions. J Chromatogr B Biomed Sci Appl 699:499–525, 1997.
8. H Bi, G Pace, KL Hoffman, DT Rossi. Mixed-mechanism ionization to enhance sensitivity in atmospheric pressure ionization LC/MS. J Pharm Biomed Anal 22: 861–867, 2000.
9. N Zhang, KL Hoffman, W Li, DT Rossi. Semi-automated 96-well liquid liquid extraction for quantitation of drugs in biological samples. J Pharm Biomed Anal 22: 131–138, 1999.
10. J Janiszewski, P Schneider, K Hoffmaster, M Swyden, D Wells, H Fouda. Automating sample preparation using membrane microtiter extraction for bioanalytical mass spectrometry. Rapid Commun Mass Spectrom 11:1033–1037, 1997.
11. EA Hogendoorn, P van Zoonen, A Polettini, GM Bouland, M Montagna. The potential of restricted access media columns as applied in coupled-column LC/LC-TSP/MS/MS for the high-speed determination of target compounds in serum. Anal Chem 70:1362–1368, 1998.
12. SR Needham, MJ Cole, HG Fouda. Direct plasma injection for high-performance liquid chromatographic-mass spectrometric quantitation of the anxiolytic agent CP-93 393. J Chromatogr B Biomed Sci Appl 718:87–94, 1998.
13. J Ayrton, GJ Dear, WJ Leavens, DN Mallett, RS Plumb. The use of turbulent flow chromatography/mass spectrometry for the rapid, direct analysis of a novel pharmaceutical compound in plasma. Rapid Commun Mass Spectrom 11:1953–1958, 1997.
14. S Lam, L Boselli. HPLC of trazodone in serum after microscale protein precipitation. Biomed Chromatogr 1:177–179, 1986.
15. I Von Schwarzenfeld. Plasma protein binding: a review. Pharmazie 29:497–505, 1974.
16. DT Rossi, DS Wright. Analytical considerations for trace determinations of drugs in breast milk. J Pharm Biomed Anal 15:495–504, 1997.
17. RD McDowall. Sample preparation for biomedical analysis. J Chromatogr 492:3–58, 1989.
18. S Steinborner, J Henion. Semiautomated liquid-liquid sample preparation and high-throughput analysis of methotrexate and its metabolite 7-hydroxymethotrexate using the 96-well plate format. 46th ASMS Conference on Mass Spectrometry and Allied Topics, Orlando, FL, 1998, abstract 1436.
19. TR Krishnan, I Ibraham. Solid-phase extraction technique for the analysis of biological samples. J Pharm Biomed Anal 12:287–294, 1994.

20. J Scheurer, CM Moore. Solid-phase extraction of drugs from biological tissues. J Anal Toxicol 16:264–269, 1992.
21. RW Fedeniuk, PJ Shand. Theory and methodology of extraction from biomatrices. J Chromatogr A 812:3–15, 1998.
22. EM Thurman, MS Millsbook. Solid-phase extraction-principles and practices. Wiley, New York, 1998.
23. JP Franke, RA de Zeeuw. Solid-phase extraction procedures in systematic toxicological analysis. J Chromatogr B 713:51–59, 1998.
24. RS Plumb, RD Gray, AJ Harker, SJ Taylor. Use of reduced sorbent bed and disk membrane solid-phase extraction for the analysis of pharmaceutical compounds in biological fluids, with applications in the 96-well format. J Chromatogr B 687:457–461, 1996.
25. RS Plumb, RD Gray, CM Jones. HPLC assay for the sulphoxide metabolite of 2′-deoxy-3′-thiacytidine in human urine. J Chromatogr B 694:123–133, 1997.
26. DT Rossi, SK Overmyer, RC Lewis, PK Narang. Quantitation of a novel antiemetic (ADR-851) in plasma and urine by reverse-phase HPLC with fluorescence detection. J Chromatogr 566:257–265, 1991.
27. K Zheng, X Liang, DT Rossi, GD Nordblom, CM Barksdale, D Lubman. On-line analysis of affinity bound analytes by capillary LC-MS. Rapid Commun Mass Spectrom 14:261–269, 2000.
28. ML Nedved, S Habibi Goudarzi, B Ganem, JD Henion. Characterization of benzodiazepine combinatorial chemical libraries by on-line immunoaffinity extraction coupled column HPLC-ion spray mass spectrometry-tandem mass spectrometry. Anal Chem 68:4228–4236, 1996.
29. TD Parker, DT Rossi, DS Wright. Design and evaluation of an automated solid-phase extraction method-development system for use with biological fluids. Anal Chem 68:2437–2441, 1996.
30. TD Parker, N Surendran, BH Stewart, DT Rossi. Automated sample preparation for drugs in plasma using a solid-phase extraction workstation. J Pharm Biomed Anal 17:851–861, 1998.
31. H Huang, JR Kagel, DT Rossi. Automated solid-phase extraction workstations combined with quantitative bioanalytical LC/MS. J Pharm Biomed Anal 19:613–620, 1999.
32. YF Cheng, UD Neue, LJ Bean. Straightforward solid-phase extraction method for the determination of verapamil and its metabolite in plasma in a 96-well extraction plate. J Chromatogr 828:273–278, 1998.
33. JP Bulgarelli, SM Michael. Simultaneous analysis of multiple drug mixtures for the quantitation of in-vitro permeability through the CACO-2 model utilizing LC/MS/MS. 46th ASMS Conference on Mass Spectrometry and Allied Topics, Orlando, FL, 1998, abstract 14.
34. DT Rossi. Automating solid-phase extraction method-development for biological fluids: trends and applications in bioanalysis, LC-GC 17:S4–S8, 1999.
35. DT Rossi, N Zhang. Automated solid-phase extraction: current aspects and future prospects. J Chromatogr 2000.
36. GA Smith, TL Lloyd. Automated solid-phase extraction and sample preparation-finding the right solution for your laboratory. LC-GC S22–S31, 1998.

37. H Fouda, RP Schneider. Robotics for the bioanalytical laboratory: a flexible system for the analysis of drugs in biological fluids. TRAC 6:139–147, 1987.

38. EL Johnson, KL Hoffman, LA Pachla. Automated sample analysis from method development to sample analysis. In: JR Strimaitis, G Hawk, eds. Advances in Laboratory Automation Robotics. Hopkington, MA: Zymark, 1988, pp 111–122.

39. FX Diamond, WE Vickery, J de Kanel. Extraction of benzoylecgonine and opiates from urine samples using the Zymark rapid trace. J Anal Toxicol 20:587–591, 1997.

40. DT Rossi. Automating solid-phase extraction method-development for biological fluids Biotech Solutions 1:14–19, 1999.

41. JA Pascual, J Sanagustin. Fully automated analytical method for codeine quantification in human plasma using on-line solid-phase extraction and high-performance liquid chromatography with ultraviolet detection. J Chromatogr 724: 295–302, 1999.

42. M Hedenmo, BM Eriksson. On-line solid-phase extraction with automated cartridge exchange for liquid chromatographic determination of lipophilic antioxidants in plasma. J Chromatogr 692:161–166, 1995.

43. A Pastoris, L Cerutti, R Sacco, L De Vecchi, A Sbaffi. Automated analysis of urinary catecholamines by high-performance liquid chromatography and on-line sample pretreatment. J Chromatogr 664:287–293, 1995.

44. G Garcia Encina, R Farran, S Puig, MT Serafini, L Martinez. Automated high-performance liquid chromatographic assay for lesopitron, a novel anxiolytic, in human plasma using on-line solid-phase extraction. J Chromatogr 670:103–110, 1995.

45. H Svennberg, PO Lagerstrom. Evaluation of an on-line solid-phase extraction method for determination of almokalant, an antiarrhythmic drug, by liquid chromatography. J Chromatogr 689:371–377, 1997.

46. DA Mcloughlin, TV Olah, JD Gilbert. A direct technique for the simultaneous determination of 10 drug candidates in plasma by liquid chromatography-atmospheric pressure chemical ionization mass spectrometry interfaced to a Prospekt solid-phase extraction system. J Pharm Biomed Anal 15:1893–1901, 1997.

47. A Marchese, C McHugh, J Kehler, H Bi. Determination of Pranlukast and its metabolites in human plasma by LC/MS/MS with Prospekt on-line solid-phase extraction. J Mass Spectrom 33:1071–1079, 1998.

48. A Kurita, N Kaneda. HPLC method for the simultaneous determination of the camptothecin derivative irinotecan hydrochloride, CPT-11, and its metabolites SN-38 and SN-38 glucuronide in rat plasma with a fully automated on-line solid-phase extraction system. J Chromatogr 724:335–344, 1999.

49. F Beaudry, JC LeBlanc, M Coutu, NK Brown. In vivo pharmacokinetic screening in cassette dosing experiments: the use of on-line Prospekt liquid chromatography/ atmospheric pressure chemical ionization tandem mass spectrometry technology in drug discovery. Rapid Comm Mass Spectrom 12:1216–1222, 1998.

50. GD Bowers, CP Clegg, SC Hughes, AJ Harker, S Lambert. Automated SPE and tandem MS without HPLC columns for quantifying drugs at the picogram level. LC-GC 15:48–53, 1997.

51. TL Constantopoulos, GS Jackson, CG Enke. Effects of salt concentration on analyte response using electrospray ionization mass spectrometry. J Am Soc Mass Spectrom 10:625–634, 1999.

52.  A Apffel, S Fischer, G Goldberg, PC Goodley, FE Kuhlmann. Enhanced sensitivity for peptide mapping with electrospray liquid chromatography-mass spectrometry in the presence of signal suppression due to trifluoroacetic acid-containing mobile phases. J Chromatogr A 712:177–190, 1995.

53.  H Simpson, A Berthemy, D Buhrman, R Burton, J Newton, M Kealy, D Wells, D Wu. High throughput liquid chromatography/mass spectrometry bioanalysis using 96-well disk solid-phase extraction plate for the sample preparation. Rapid Commun Mass Spectrom 12:75–82, 1998.

54.  VP Shah, KK Midha, S Dighe, IJ McGilveray, JP Skelly, A Yacobi, T Layloff, CT Viswanathan, CE Cook, RD McDowall, KA Pittman, S Spector. Analytical methods validation: bioavailability, bioequivalence and pharmacokinetic studies. Pharm Res 9:588, 1992.

55.  HT Karnes, C March. Precision, accuracy and data acceptance criteria in biopharmaceutical analysis. Pharm Res 10:1420–1426, 1993.

56.  TV Olah, JD Gilbert, A Barrish, TF Greber, DA McLoughlin. A rapid and specific assay, based on liquid chromatography-atmospheric pressure chemical ionization mass spectrometry, for the determination of MK-434 and its metabolites in plasma. J Pharm Biomed Anal 12:705–712, 1994.

57.  HT Karnes, CJ March. Calibration and validation of linearity in chromatographic biopharmaceutical analysis. J Pharm Biomed Anal 9:911–918, 1991.

58.  HT Karnes, G Shiu, VP Shah. Validation of bioanalytical methods. Pharm Res 8:421–426, 1991.

59.  DG Peters, JM Hayes, GM Hieftje. A Brief Introduction to Modern Chemical Analysis. Philadelphia: Saunders, 1976.

60.  DC Harris. Quantitative Chemical Analysis. 3rd ed. New York: Freeman, 1991.

61.  N Zhang, K Rogers, K Gajda, JR Kagel, DT Rossi. Integrated sample collection and handling for drug discovery bioanalysis. J Pharm Biomed Anal 1999.

# 7

# Mass Spectrometry in Combinatorial Chemistry

**David T. Rossi and Michael J. Lovdahl**
*Pfizer Global Research and Development, Ann Arbor, Michigan*

> The race is not always to the swift, nor the battle to the strong, but that's
> the way to bet.
>
> *—Damon Runyon, attributed*

## I. WHAT IS COMBINATORIAL CHEMISTRY?

The synthesis of a chemical entity to produce a successful drug has historically involved isolation of the active compound from a biological source, followed by individual synthesis and screening of structural analogs to enhance the biopharmaceutical profile. This time-consuming process of serial synthesis (completing the synthesis of one compound prior to initiating that of the next) is slow and costly and requires a high degree of expertise by the chemist. Combinatorial chemistry is a recently developed alternative strategy that has become widely used to improve the throughput and decrease the expense of manual serial synthesis. Combinatorial chemistry is applied to both lead discovery, where the structures of potential leads have yet to be established, and to lead optimization, where the basic structural features of an active molecule have been identified and absorption/distribution/metabolism/elimination (ADME) performance must be improved before the compound can become a successful drug.

One technology that has contributed to the growth and development of combinatorial chemistry has been the recent development of automated liquid-handling workstations [1]. These devices, available from a number of laboratory automation companies, have facilitated the rapid manipulation of small volumes

(1 to 200 µL) of solvents, reagents, and product liquors in 96- or 384-well micro-titer-plate formats. The use of one of these liquid-handling devices can facilitate the rapid preparation of large numbers of structurally related compounds, either as mixtures in the same reaction vessel or individually by parallel semiautomated synthesis. When combined with resin-based reagents and catalysts, even separations and purifications can be automated. With combinatorial chemistry approaches and an automated liquid-handling workstation, hundreds to millions of compounds can be synthesized in the time required to make a dozen analogs by manual serial synthesis [2].

## A. Solid-Phase Combinatorial Synthesis

*Resins* have long been used for solid-phase peptide and oligonucleotide synthesis, but the idea of synthesizing small organic molecules on resins was not widely accepted until recently. This situation has changed and there has been increasing acceptance of resins as supports for the synthesis of small molecule libraries. Resins are typically used as a vehicle for the construction of small organic molecules either in a linear assembly of building blocks or in a convergent synthesis, using complex intermediates. These methods have been described in detail for the synthesis of many combinatorial libraries [3].

In effect, a starting component of the final molecule is attached to the resin in such a way that a reactive site is available for synthesis. In a typical linear automated synthesis scheme, the resin is added to a reactor constructed to have inlets for liquids and frits to prevent the solid support from being flushed from the reactor. The reagent, with appropriate protecting groups, is loaded into the reactor. The reaction is then run with agitation and heating as necessary. Upon completion of the reaction the spent reagent solution is flushed away, the newly created synthetic intermediate or molecule can be deprotected and washed prior to the next reaction step. The resin, with the synthetic molecule attached, is prevented from leaving the reactor by a frit. The complete molecule is synthesized by repeating the previous steps of adding reagent, deprotecting, and washing for each synthetic step. After the last synthesis cycle is finished, the final product can be cleaved from the resin or stored attached to the resin.

## II. HOW ARE LIBRARIES BUILT AND DECONVOLUTED?

A fundamental recipe for preparing combinatorial libraries is the randomization approach known as *portion mixing*. This technique is also called *split-and-mix* or *divide-and-combine* [4,5]. In this approach, solid-phase resins containing combinatorial library building blocks are portioned out, coupled, and combined to produce a combinatorial library of compounds. It is a way to synthesize a large number of compounds using just a few grams of resin. An example of this approach is shown in Fig. 1. Three monomer units, A, B, and C, attached to solid support are combined to form an ABC mixture. The mixture is divided into three batches and each batch is coupled to A, B or C, respectively, yielding nine compounds (three per batch). These are combined then divided into three new batches and each batch is again coupled to A, B, or C to yield 27 compounds. The three

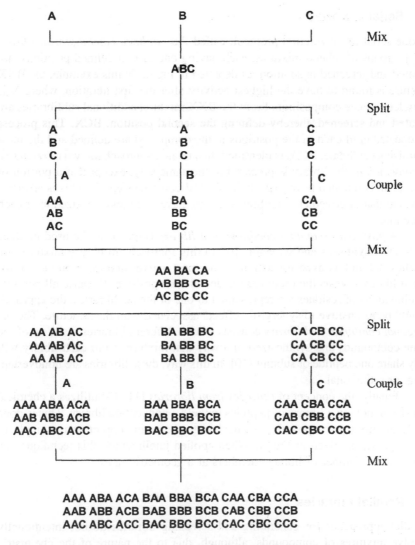

**Figure 1**   The split and mix combinatorial technique for three subunits, A, B, and C.

batches are combined into a single library containing all compounds. Larger libraries ($n^n$; e.g., 4 units = 256, 5 units = 3125, and 6 units = 46656) are produced using the same pattern, but require more steps. After synthesis, the compounds can be used directly (*parallel deconvolution*) while still attached to the resin support, or they can be cleaved from the resin into solution and used (*serial deconvolution*).

## A.  Serial Libraries

In one serial deconvolution protocol, called the *synthetic combinatorial library* [6,7], groups of related mixtures, each having one or two defined positions, are created and screened in solution, as depicted in Fig. 2. In this example, the BXX mixture is found to have the highest activity after the first iteration, where X is of indeterminate composition. From the BXX mixture, additional sublibraries are created and screened, thereby defining the second position, BCX. This process is repeated until each of the positions in the compound are defined and the most promising candidate, BCA, is identified. In a related approach known as *recursive deconvolution*, the library is prepared in the same way, except that a portion of each resin is retained prior to each divide-and-combine step [8]. This is advantageous in that intermediate sublibraries do not have to be resynthesized after each screening.

A *positional scanning combinatorial library* (Fig. 3) can be used in place of iterative synthesis and screening [9]. In this approach, all library mixtures are synthesized and assayed for activity. The most active mixture from each positional library reveals the preferred residue at each position. Because all possible combinations of residues are represented in one of the sublibraries, the approach should be insensitive to synergistic effects associated with the structure. The *orthogonal combinatorial library* consists of two different libraries made up of the same compounds and synthesized in a way that mixtures from each library will only share one peptide or subunit [10]. In this way, these libraries are nonoverlapping, or orthogonal.

Finally, the concept of *libraries from libraries* [11–13] utilizes a chemical transformation on a library of peptides to yield a nonpeptide library. This is done while the library compounds are attached to the resin support. An example of this approach is given in Fig. 4. When applied intelligently, this technique can increase the number of library members at a geometric rate.

## B.  Parallel Libraries

Parallel approaches for generating and screening libraries do not intentionally involve mixtures of compounds, although, due to the nature of the chemistry, mixtures are often present. Generally, the components of the library are synthe-

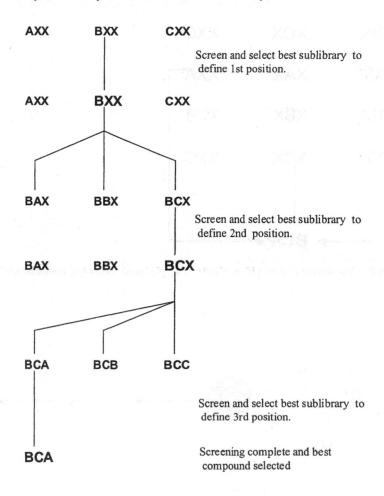

X = Randomized position carrying a mixture of A,B, and C

**Figure 2** The serial deconvolution protocol known as synthetic combinatorial library.

sized individually, either as an array of compounds in separate reaction vessels [14] or in a single reaction vessel with portion mixing. Although a well can contain a number of compounds, each resin bead contains only a single compound, an approach known as *"one- bead/one- compound"* [15]. Resin beads can be used directly in screening assays that involve antibody assay approaches [16–18]. When screening is conducted with a biological assay, orthogonal systems are usually used. This approach, illustrated in Fig. 5 and 6, allows a portion of the compound to be chemically released from the resin bead for biological

OXX
XOX
XXO
AXX
XAX
XXA
BXX
XBX
XXB
CXX
XCX
XXC
BCA

**Figure 3** The deconvolution protocol known as positional scanning combinatorial.

O
H
N
R2
N
H
R1
X
R4
N
N
N
R3
R1
R5

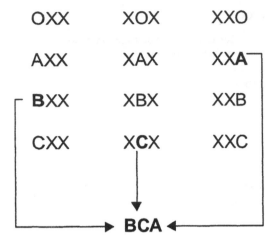

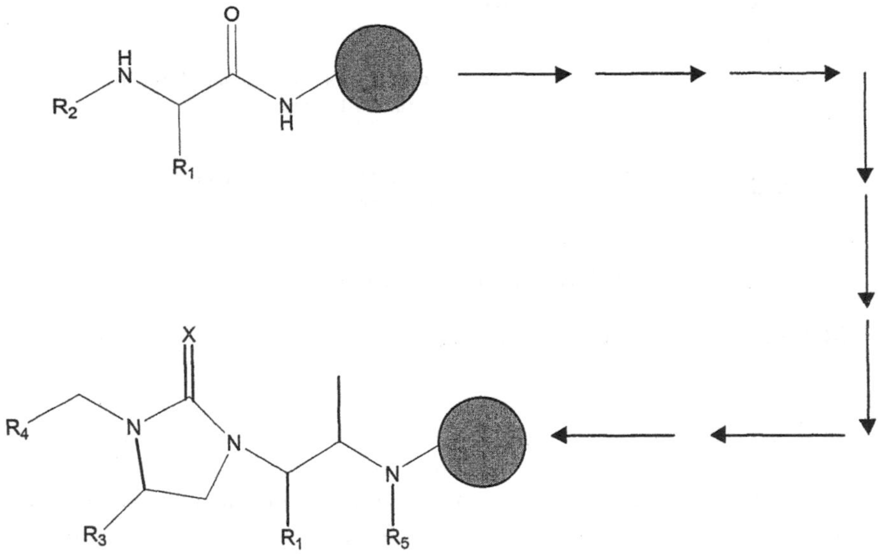

**Figure 4** The library from a library approach, showing that individual library components can be chemically modified to obtain new libraries.

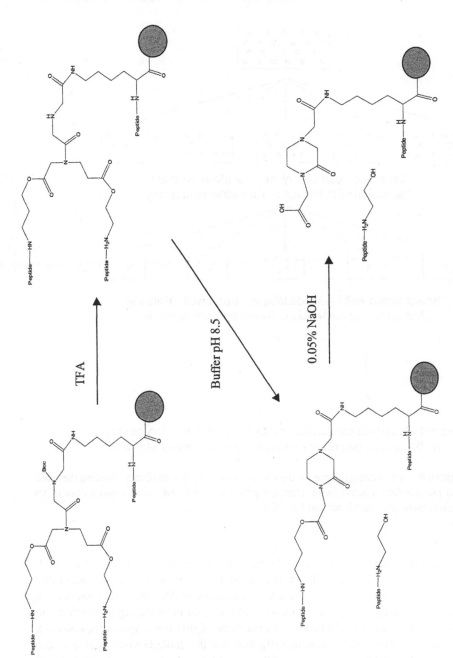

**Figure 5** The "one bead/one compound" approach showing a bead containing a peptide immobilized on several types of sites. Portions of the peptide can be made available for chemical analysis or screening by hydrolysis of the linkage under various conditions, thereby making a bead reusable.

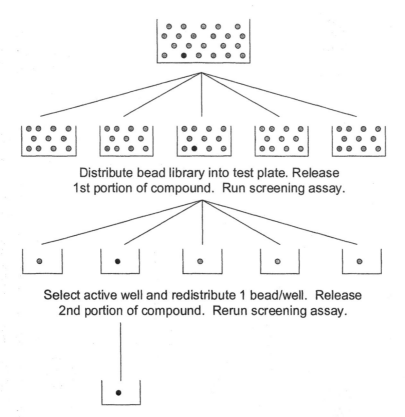

Distribute bead library into test plate. Release
1st portion of compound. Run screening assay.

Select active well and redistribute 1 bead/well. Release
2nd portion of compound. Rerun screening assay.

Select active well containing one bead. Determine the structure
of the active compound by sequencing or reading tag code.

**Figure 6**   A screening pattern for the one bead/one compound library showing the selection process for an active bead. Depending on the target and the size and nature of the library, more than one bead can be active.

assay. The beads from active wells are redistributed into individual wells with one bead per well. The wells are reassayed and active wells are then identified.

Identification of active library components is done by micro sequencing (peptide libraries) or by incorporation and recognition of the tag associated with each bead (nonpeptide libraries). An example of this latter approach, known as *encoded combinatorial synthesis* [19], may use the dialkylamine tagging system shown in Fig. 7. The amine group of the tag addition site is acylated with iminodiacetic anhydride to yield a pentafluorophenyl ester derivative. The active ester is reacted with a binary mixture of secondary alkylamines, leading to a tag polymer.

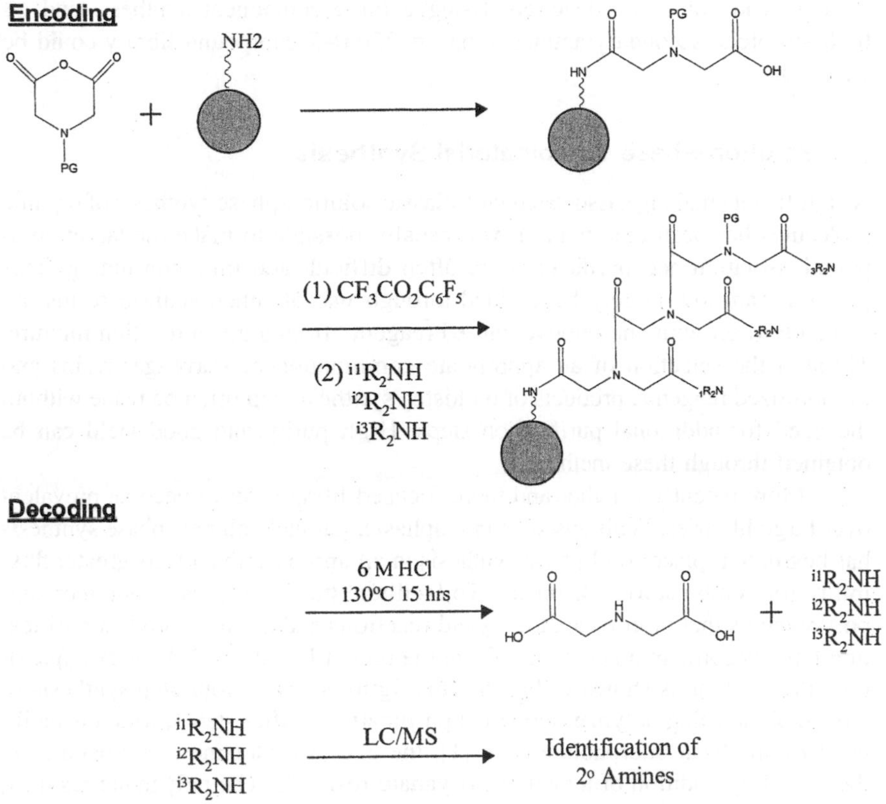

**Encoding**

**(1) $CF_3CO_2C_6F_5$**

**(2) $^{i1}R_2NH$**
   $^{i2}R_2NH$
   $^{i3}R_2NH$

**Decoding**

6 M HCl
130°C 15 hrs

$^{i1}R_2NH$
$^{i2}R_2NH$
$^{i3}R_2NH$

$^{i1}R_2NH$
$^{i2}R_2NH$  LC/MS  Identification of
$^{i3}R_2NH$         2° Amines

**Figure 7** Pictorial representation of an encoding/decoding scheme based on a polypeptide tag. The beads are encoded through the polymerization of secondary amines. The amines are later hydrolyzed and detected to decode the identity of the bead.

Application of this procedure at each step of a multistep synthesis either before or after addition of a ligand monomer, but prior to pooling the resin, allows the chemical reaction history of each bead to be labeled. Approximately 10% (~30 pmol) of the reactive sites of a bead are tagged, with the rest being available for ligand synthesis. It is essential that any encoding scheme be compatible with the specific combinatorial chemistry of the ligand synthesis.

During tag decoding, secondary amines are released from individually selected beads by acid hydrolysis. Because of the ability to rapidly determine molecular weight, electrospray LC/MS provides a reliable means of identification through tag recognition. The selected amines would form ammonium ions having distinctive molecular masses. Based on those amines present, the reaction history of any bead could be identified. Using a three-component synthesis and 63 building-block secondary amines, a $63^3 = 250,047$ compound library could be encoded.

## C. Solution-Phase Combinatorial Synthesis

A significant challenge associated with classic solution-phase synthesis of organic molecules has been that, while it was usually possible to make the target compound, isolation and purification are often difficult and time consuming. This problem can now largely be avoided through use of functionalized resins designed to react with and remove unused reagents from a given reaction mixture. Through the selection of an appropriate combination of scavenger resins and immobilized reagents, products of multistep synthesis can often be made without the need for additional purification steps. High purity and good yield can be obtained through these methods.

More recently, smaller and more focused libraries have become prevalent over large libraries. With this shift in emphasis, parallel solution-phase synthesis has begun to replace solid-phase synthesis in an approach that offers greater flexibility for combinatorial chemistry. To facilitate this shift, uses of polymer support reagents that offer clean, high-yield reactions and minimal product purification have become more common. From the recent literature [20], an example of how this is done is shown in Fig. 8. This figure shows a four-step synthesis of a pyrazolone using polymer-supported purifications. After step 1, which is facilitated by the basic morpholine resin [1], the excess hydrazine is removed from the vessel by addition of a methyl isocyanate resin [2]. The polyamine resin [3] is used to quench and remove acid impurities through salt formation. After each reaction, the resins are removed by simple filtration. Product is then isolated by solvent evaporation. This so-called polymer support quench method allows the insertion of a simple purification step between the traditional steps of a multistep synthesis, resulting in a cleaner product. Although the reagents are more expensive on a weight basis, omitting the chromatographic purification steps allows for overall cost savings and a procedure that is more amenable to automation.

Resins can also be used to hold and deliver reagents or catalysts to facilitate selective reaction chemistry. A review of the application of supported catalysts to organic synthesis has been recently published [21] and references many additional review articles covering polymer-based synthesis.

## D. Dynamic Combinatorial Libraries

In a *dynamic combinatorial library* (DCL) the components of the library are not chemically static, but can interconvert through either covalent or noncovalent means. This is shown conceptually in Fig. 9a. The advantage of a dynamic combi-

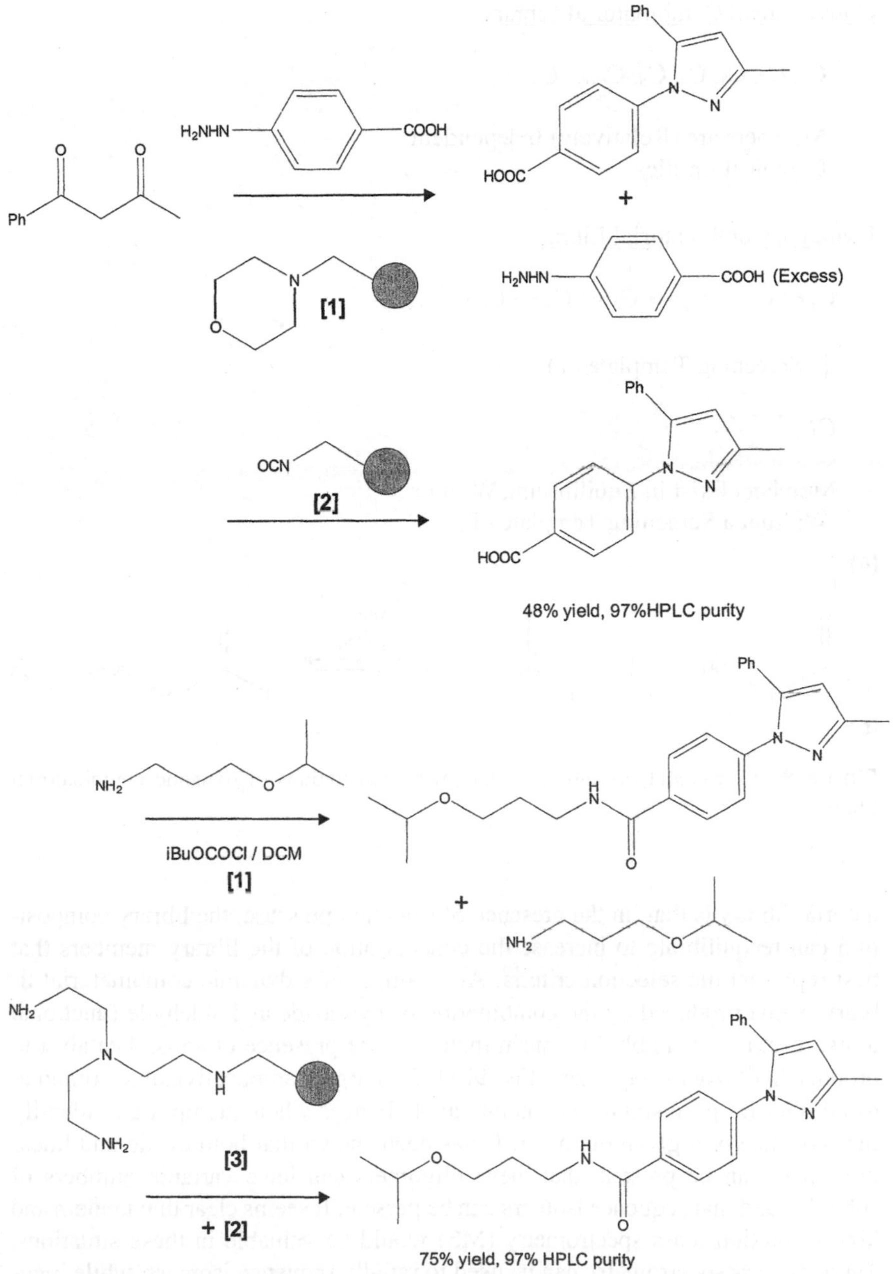

**Figure 8** A high-purity synthesis using functionalized resins to synthesize and purify the intermediate and final products.

Conventional Combinatorial Library

$C_1, C_2, C_3, C_4, C_5, C_6,....C_n$

Members are (Relatively) Independent
Chemical Entities

Dynamic Combinatorial Library

$C_1 \leftrightarrow C_2 \leftrightarrow C_3 \leftrightarrow C_4 \leftrightarrow C_5 \leftrightarrow C_6 \leftrightarrow ....C_n$

$\updownarrow$  Screening Template (T)

$C_{1-T}$

Members Exist in Equilibrium, With or
Without a Screening Template (T)

**(a)**

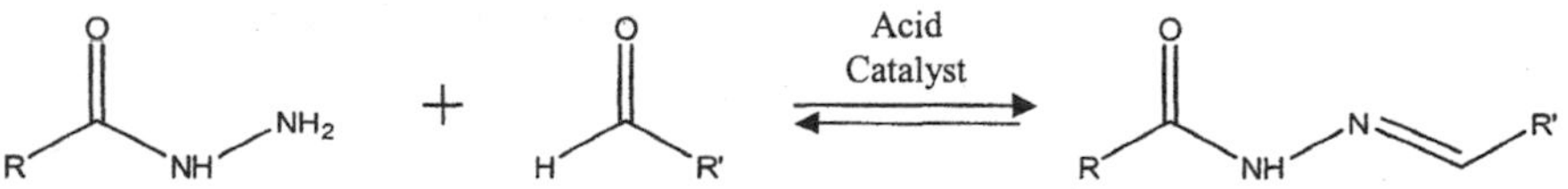

**(b)**

**Figure 9**  Conceptual (a) and chemical (b) representations of dynamic combinatorial libraries.

natorial library is that, in the presence of selection pressure, the library composition can reequilibrate to increase the concentration of the library members that best represent the selection criteria. An example of a dynamic combinatorial library is that produced by the combination of hydrazide and aldehyde functional units, reacting reversibly by transimination in the presence of an acid catalyst to produce hydrazones oligomers (Fig. 9b) [22]. Large, complex dynamic combinatorial libraries pose significant analytical challenges when attempting to identify the sequence of a given oligomer. It has been shown that both cyclic and linear oligomers can be present, that these oligomers can have variable numbers of subunits, and that sequence isomers can be present. It seems clear that tandem and high-resolution mass spectrometry (MS) would be valuable in these situations. Tandem mass spectrometry can be used to rapidly sequence isomers, while high-resolution mass spectrometry can be used to determine accurate mass information.

## E. The Advantages and Disadvantages of Combinatorial Synthesis

Parallel combinatorial synthesis has proven to be fast and conducive to automation. With either solid-phase or solution-phase synthesis/resin-supported reagents, excellent purity can be obtained for products, without additional purification. Many compounds can be produced in a short time period, using variations on a rationally designed molecular template. Structure–activity relationships can be developed and understood by combining these libraries with high-throughput screening (HTS). These structure–activity relationships can lead to better understanding of the disease-inhibition mechanism and can help to identify better ligand–receptor binding or new approaches to enzyme inhibition.

Combinatorial synthesis can also lead to many practical challenges because, with so many compounds being produced, the chemistry does not always proceed as predicted. The instability of some compounds can lead to unknown library components or the presence of large impurities that must be identified. As a result of the large numbers of compounds produced and the way in which they are created (as described earlier) it is also challenging to simply keep track of the compounds and where they are located in the library (hence the need for encoding and decoding strategies). Once some compounds have been made and found to be active in a HTS assay it is often challenging to simply identify the active compounds. These issues, combined with the small quantities of compounds available (300 to 500 pmol per bead) make for challenging analytical issues, many of which can be addressed through the use of atmospheric pressure ionization mass spectrometry.

## III. MASS SPECTROMETRY AS A TOOL FOR CHARACTERIZING COMBINATORIAL CANDIDATES

The relatively new combinatorial approaches described above are capable of producing vast numbers of complex samples that require fast and reliable analytical methods for composition information. Mass spectrometry is, in many respects, an ideal tool for characterizing the products of combinatorial synthesis and combinatorial libraries because it can provide both qualitative and quantitative information, including molecular weight, with great selectivity and good sensitivity. Generally, little or no method development time is required and it has a high success ratio when applied to compounds containing ionizable functional groups. High-throughput separation and detection approaches involving mass spectrometry are aimed at two distinct analytical challenges: (1) *synthesis control* and (2) *identification of active products from library screening*. The various ways to achieve these ends are described later.

## A. Synthesis Control

Synthesis control has two tasks directly associated with it. These are to identify or verify the identity of a combinatorial component and to determine the purity of the synthetic product. When characterizing a parallel library it is a relatively easy task to obtain a molecular weight from a small amount (femtomole) of compound and thereby obtain a crude identification of the product. This circumvents the need to perform more difficult NMR or IR spectral interpretation and sample introduction maybe performed by a simple flow-injection atmospheric pressure ionization (API)/MS system. Purity assessment is typically based on area percentage normalization of the total ion chromatogram, assuming equivalent ionization of impurities and parent compounds, or a secondary detector, such as UV.

## B. Direct Library Characterization

With flow-injection API mass spectrometry, test compounds can be rapidly characterized for molecular weight to confirm identity. To some extent, this approach can also be used to address the question of how pure a compound is [23,24]. The main advantage of this approach is to avoid injecting a poorly characterized compound onto a chromatographic stationary phase, as this can lead to difficulties. Generally, the flow-injection approach is used to confirm molecular identity of newly synthesized compounds by obtaining the molecular weight through the use of a relatively inexpensive single quadrupole mass spectrometer in full-scan mode. Compounds can be injected twice to obtain both positive and negative ion spectra or they can be sorted by ionization properties prior to injection [25]. This could require the selection of a mobile-phase pH that will readily allow formation of either positive or negative ions, depending on the class of compounds.

Because no separation is used, only crude information about the purity of the compounds can be obtained. For example, if unreacted, synthetic starting materials (Fig. 10b) are present in a sample of a combinatorial product (Fig. 10a), these can show up in the mass spectrum of the crude product (Fig. 10c). Because the starting materials are often structurally different from the finished product, a simple extraction can be used after the synthesis to remove much of the unused reactants and obtain a cleaner mass spectrum (Fig. 10d) for a solution-phase product. Conversely, washing the resin after solid-phase synthesis or using a scavenger resin in a solution-phase synthesis can also yield improved purity by removing these excess reactants. When direct flow injection is used to characterize combinatorial libraries, it is best to avoid dimethyl sulfoxide (DMSO) as a solvent, because it interferes with reliable ionization of the analytes.

---

**Figure 10** A crude combinatorial chemistry product (a) and unreacted starting material (b) are both present in the product well, exhibiting a combined mass spectrum (c). A cleaner spectrum (d) is obtained after solid-phase synthesis or by adding purification steps.

(a)

(b)

(c)

(d)

## C. Generic Separation and Detection Strategies

A more powerful strategy for library component characterization is to use a generic separation along with API mass spectrometry. This separation will usually consist of an HPLC separation, although supercritical-fluid [26], gas chromatographic [27], or capillary electrophoretic [28] separations have also been described. A typical generic separation for a single library component might involve a 3- to 5-min linear gradient from 5 to 95% acetonitrile, with heptafluorobutyric acid (HFBA) (0.01%) and isopropyl alcohol (1.0%) as aqueous component modifiers. Using a short (3 × 5-cm) HPLC column, this steep gradient will separate many synthetic impurities and starting materials from the final product, with a minimal investment in separation development. Generally, high success rates (90 to 95%) can be obtained with little or no method development. The mobile-phase components suggested here are compatible with either mass spectrometry or optical means of detection.

Primary detection is atmospheric pressure ionization mass spectrometry accompanied by a secondary detection. The purpose of the secondary detection is to quantify impurities, leading to a purity assessment based on peak height or area. It is generally accepted that a non-mass-spectrometric detector, such as UV absorbance detection (single or dual wavelengths at 220 or 254 nm), evaporative light-scattering detection, or laser light-scattering detection, is preferable for this task. Arguably, response from these instruments is thought to be more consistent than that for atmospheric pressure ionization. Dual-channel detection, although redundant, is useful if the response for one of the channels becomes saturated because it is possible that the response for the other channel will not be. Light-scattering detectors perform well when nonvolatile mobile-phase components are used, as described above. Absorbance detectors are unencumbered if UV interfering components such as HFBA are kept to a minimum.

In choosing between atmospheric pressure chemical ionization (APCI) and electrospray ionization, two considerations should be made. First, although some investigators consider APCI to have a slightly more universal nature, especially for polar, nonionic compounds, it is also true that the occurrence of in-source fragmentation is greater. This will lead to convoluted molecular weight information for a certain percentage of compounds. Electrospray ionization is sufficiently soft that many will choose it over APCI when consistent pseudomolecular ion formation (M + 1 or M − 1) is desired.

## D. High-Throughput Characterization

To improve the throughput for characterizing combinatorial libraries, automated approaches are being developed [29]. Some automated approaches include the combination of peak detection with preparative chromatography and fractionation

[30–33]. Compounds having less than a specified degree of purity (85% chromatographic peak area, for example) as specified by UV detection can be further purified by on-line preparative chromatography and fraction collection. These purifications are initiated by automated, real-time decisions that are based on signal intensities for target ions. Although some applications have included the purification of only a single component per well, it is possible in principle to purify multiple components per library well. High flow rates and short preparative columns have allowed cycle times for the preparative purification to be reduced (3 to 10 min), thereby allowing as many as 100 to 200 samples to be characterized and purified in a 24-hr period. It is also possible to run a number of these systems in parallel (so-called *parallel interface technology*) to further improve the throughput of library characterization. This approach uses a liquid-switching device or multiplexer to route multiple HPLC effluents to a single ion source, thereby improving the number of samples that can be run per instrument.

The data-controlled automation of LC/MS could also make it possible to perform the high-throughput sequencing of peptides [34] or oligonucleotides from pooled libraries. A constraint of this approach is vast amount of data that must be interpreted. Software-supported spectral interpretation must be improved before this approach can become practical.

## E. Library Characterization Using MALDI

Matrix-assisted laser-desorption ionization (MALDI) has been described in Chapter 3 as a technique that can provide mass spectral data for high-molecular-weight compounds such as peptides, nucleotides, and proteins. When combined with time-of-flight mass spectrometry, MALDI can provide improved mass resolution, largely because of the performance of the reflectron time-of-flight mass analyzer [25,29]. One area for the application of MALDI is for characterizing libraries of biopolymerlike drug candidates. This approach is useful because MALDI provides a ready means of sample introduction for these high-molecular-weight compounds. Perhaps a more important potential application for MALDI is with small molecule libraries, where the exact molecular weight can be determined to the 3rd or 4th decimal place ($\pm 10$ ppm mass accuracy), depending on the mass range to be covered. This makes a determination of an exact molecular formula possible in many cases.

One drawback for the use of MALDI with small molecules is that sample preparation is more intensive than that for electrospray because of the need for matrix addition. To some extent this sample preparation can, however, be automated. A more significant drawback is the presence of a low-mass-noise background arising from the matrix. This noise background can interfere with the direct assessment of the spectrum for the compound of interest. Considerable research efforts have been directed at approaches that eliminate the low-mass matrix background for MALDI [35,36]. A third drawback to MALDI is that coupling to a liquid-phase separation technique is not a routine option. This problem might be solved in the future as MALDI and micro-fabricated-chip-based separations are combined.

Because some components of small organic molecule libraries often lack ionizable groups, ionization tags can be introduced [37]. A photoclevable linker allows direct detection of each compound with MALDI, while the tag affords a mass shift that is useful in overcoming the signal overlap with matrix molecules. The tag generally does not interfere with the bioassay used for screening because different cleavage conditions are used. This approach has also been used successfully with electrospray mass spectrometry.

## IV. IDENTIFICATION OF ACTIVE PRODUCTS FROM SCREENING

### A. Synthetic Pooling

To facilitate the screening process, parallel libraries are occasionally mixed together after synthesis to create a pooled library. If an appropriate assay that is indicative of drug activity can be developed, *synthetic pooling* allows a more facile screening of large numbers of compounds by grouping or pooling them together [38]. When the drug activity screen is run, all compounds in the pool are tested simultaneously. In static combinatorial libraries, it is assumed that inactive components do not interfere or act synergistically with active components. If the active components can be identified they will provide a good starting place as a drug candidate.

When screening combinatorial libraries for activity, synthetically pooled samples or mixtures of substances are commonly encountered. The overall mixture activity against a specific target can be readily measured. It is not, however, a straightforward matter to identify all those components that give rise to activity. This process is referred to as *mixture deconvolution* and several approaches used to address this problem are described below.

### B. Bioassay-Guided Fractionation

*Bioassay-guided fractionation* (BGF) is an approach that has been employed for many years in the natural-product area to identify those components of a complex mixture that have a desired pharmacological effect [39,40]. A typical application of BGF is depicted in Fig. 11. A natural product, such as a plant component with reported anticancer activity, is extracted using solvent or fluid extraction. The

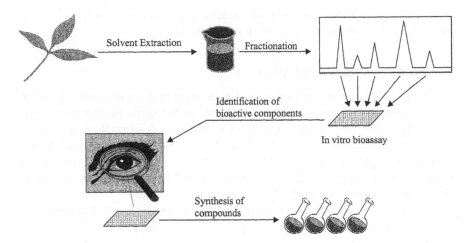

**Figure 11**   Conceptualized representation of bioassay-guided fractionation (BGF) as applied to isolation of a plant natural product. Application to a library component would start with the fractionation step.

extract is separated, often by HPLC, into individual components and the fraction containing each individual component is collected. Individual fractions are tested using an in vitro or in vivo bioassay. For anticancer properties, this test could consist of the rate or extent at which cancer cells are killed in vitro. Many other simple in vitro bioassays have been reported. Those fractions of the original sample showing biological activity are identified using structural elucidation approaches such as nuclear magnetic resonance (NMR) spectroscopy or mass spectrometry. Once identified, components of interest can be chemically synthesized and studied further.

This approach has also been proposed for combinatorial library screening because items in these libraries, even those that have been subject to cleanup, are seldom chemically pure. Rather, in addition to the target compound, each product contains structurally related substances, including unreacted starting materials or synthetic byproducts. Over time, even fairly pure compounds can degrade to form related substances. Libraries that have been isolated from natural sources present many of the same challenges that combinatorial libraries do, in that the impurities and related substances present in library samples can confuse the results from in vitro screens involving nonfractionated samples. Sometimes, to facilitate rapid screening of large libraries, several individual components of the library are pooled and their identities need to be established. Bioassay-guided fractionation provides a way to distinguish between active and inactive compounds. It can also be used as a follow-up screen once gross activity from a particular combinatorial sample has been established.

## C. Bioaffinity-Based Isolation

An interesting but underutilized approach to screening combinatorial libraries combines *bioaffinity-based isolation* and mass spectrometric identification. This approach requires the characterization of a drug to a macromolecular receptor such as an enzyme. This immobilized receptor can be conveniently packed into a chromatographic column [41], although configurations that do not use a column have also been reported [42]. An example of the latter, *pulsed ultrafiltration* (PUF) has been reported [43] and is illustrated in Fig. 12.

In PUF, the combinatorial library is simultaneously injected (pulsed) into

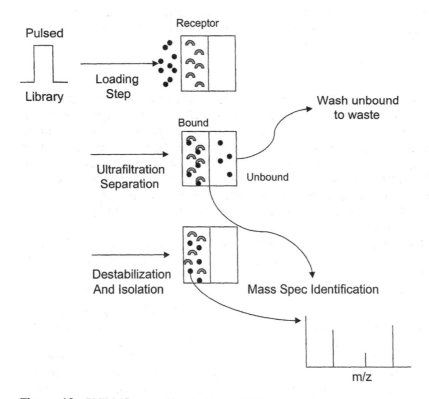

**Figure 12**  PUF-MS as used in combinatorial library screening is used to separate macromolecular bound library components from unbound ones. Mass spectrometry is used to identify the bound components after they have been released from the receptor.

an ultrafiltration cell containing a solution of the macromolecular receptor. Those components that have an affinity for the receptor are bound. The solution is subjected to ultrafiltration, facilitating the removal of the unbound, low-molecular-weight ligands from the system. Next, destabilizing conditions facilitate the release of the bound ligands from the receptor, thereby disrupting receptor–ligand binding and allowing subsequent API mass spectral detection and facilitating identification of the library components having affinity for the receptor. This can readily be accomplished through a pH change or addition of an organic solvent that produces a denaturation of the receptor protein. Subsequent downstream separation of those library components that have affinity for the receptor can be added to give additional selectivity.

The mass spectra of the retained components can be compared to spectra of the pure library components for identification. Alternatively, structural information can be obtained from product ion scans of the components of interest. Because numerous homologs in the library can have large affinities for the receptor, and because these homologs can be relatively similar in structure, it is sometimes impossible to definitively ascertain the exact structure of each "hit" using this approach.

One of the difficulties of this approach includes the time required to prepare and immobilize the receptor. If the receptor is produced in a way that allows appropriate ligand binding, ligand selectivity can be altered, resulting in erroneous compound selection results. It is also possible that the receptor for a signal-transduction pathway is not understood well enough for the receptor to be identified and cloned. If this is the case, the approach is not feasible.

### D. Ratio Encoding with Stable Isotopes to Identify Library Components

Mass spectrometry and isotope techniques can be used effectively for the encoding/decoding of pooled libraries. Two of these techniques are mass encoding [44] and *stable isotope encoding* [44,45]. Because stable isotope encoding is more direct and lends itself more readily to combinatorial library screening with mass spectrometry, it is discussed below.

A hypothetical combinatorial library, having been synthetically pooled, has undergone screening with an in vitro efficacy test, and five hits have been determined. The structures of these five compounds are shown in Table 1. Because of the close structural similarity of these five compounds (the molecular masses span a range of only 0.0364 Da) it is not feasible to distinguish between them by mass spectral differences alone. Also, because the positive ion moiety is contained in the piperazine ring, product ion spectra will be nearly identical. It is possible to distinguish between some of these compounds by either ultrahigh resolution mass spectrometry techniques, such as Fourier-transform ion cyclotron resonance (FTICR) mass spectrometry, but this would only differentiate those

**Table 1**   An Example of Isotope Encoding in a Combinatorial Library

| Compound | Molecular mass (Da) | Isotope encoding level (molar percentage of $m/z$ 298) |
|---|---|---|
| 1 | 292.1787 | |
| | | 0 |
| 2 | 292.1787 | |
| | | 12.5 |
| 3 | 292.2151 | |
| | | 25 |
| 4 | 292.2151 | |
| | | 37.5 |
| 5 | 292.2151 | |
| | | 50 |

components having different molecular masses. Compounds 1 or 2 could be distinguished from 3, 4, or 5, but not from each other.

During the solid-phase synthesis of these compounds, it is a relatively simple matter to incorporate deuterium labeling into the ethyl side chain of the phenyl ring to produce pentadeutero derivatives of the parent drug structures. If the pentadeutero derivatives are mixed with the unlabeled compounds in varying but unique molar ratios, then each compound could be identified by its unique isotopic ratio ($m/z$ 298 relative to $m/z$ 293). For the example shown in Table 1, the assigned isotopic encoding level varies from 0 (compound 1) to 50% (compound 5). In this way, each of these five components can quickly be identified by visual inspection of their mass spectra, as shown in Fig. 13. Ratio encoding requires additional synthesis and the molar ratio mixing of each library compound and its label, but once the library is prepared, compounds can be pooled, tested, and then rapidly identified. In this way, significant timesaving in library screening can be obtained.

## E.  Target and Ligand Identification and Characterization

Much drug discovery effort focuses on *proteomics*: understanding the functionality of proteins defined by the genetic code. Enzymatic proteins will bind small molecules and catalyze their chemical manipulation so that biochemical signals (signal transduction) are propagated and produce a physiological response. Ligands that bind to specific sites on proteins and inhibit their function can be used to regulate undesired physiological responses, and this provides the basic role for many therapeutic drugs. Because of its intrinsically soft ionization, leading to a direct stoichiometry of the protein and ligand, electrospray mass spectrometry has the potential to facilitate the identification and characterization of some of these molecular targets [46].

Ligands that bind to protein targets are readily studied by mass spectrometry as the protein/ligand complex survives transfer to the gas phase. Because of the softness of the electrospray ionization, weak, noncovalent complexes as well as strong covalent ones can be observed. One example of this approach has been developed for the screening of peptide libraries with carbonic anhydrase II [47]. The high mass resolution and fragmentation capabilities of electrospray FTICR mass spectrometry facilitate the formation of intact, noncovalent complexes in the gas phase and allow identification of free tight binding inhibitors after dissociation of the complexes. The best amino acid residues for ligand-to-protein binding can be identified.

A second example of target identification is the detection of oligonucleotide:ligand complexes by electrospray ionization mass spectrometry [48]. It has been shown that selective oligonucleotide:ligand binding can inhibit the action of a helicase protein and disrupt viral replication. Potential antiviral drugs would have the characteristic of binding to the oligonucleotide, creating an oligonucleotide:ligand complex that can be directly observed by electrospray mass spectrometry. An example of this approach, shown in Fig. 14, uses direct infusion of a buffered solution containing oligonucleotide and potential ligand into the electrospray ion source.

The uncomplexed oligonucleotide carries multiple negative charges, as does the oligonucleotide:ligand complex. For the uncomplexed oligonucleotide negative-8 charge state, a cluster with an $m/z$ of 1606 is observed and for the negative-7 charge state, a cluster with an $m/z$ of 1835 is present in the spectrum. When a ligand is present in solution at a sufficient concentration, an oligonucleotide:ligand complex is observed at slightly higher $m/z$. The exact $m/z$ observed will indicate the stoichiometry of the complex. In the example shown in Fig. 14, with distamycin A as the ligand, it appears that the stoichiometry is 1:2. Other ligands would generate different complex stoichiometries. Those ligands that do not bind or bind very weakly would not be observed in this mass range. By using a range of ligand concentrations, binding constants can be calculated. With this approach, up to 200 potential ligands can be evaluated per day.

## F. Fourier-Transform Mass Spectrometry

For combinatorial libraries where multiple components reside in a single well, it is usually necessary to identify those components that bind to a synthetic drug receptor. Although many forms of mass spectrometry can be used, one of the most powerful approaches is that of FTICR mass spectrometry. This mass spectral technique, described in Chapter 3, offers extended capabilities over other mass spectrometry approaches, including selective ion trapping ($MS^n$ capabilities), high resolution (resolution of 50,000, e.g., 25-fold better than for a quadrupole), and high mass accuracy over a broad mass range [49,50].

Because of these powerful qualitative capabilities, FTICR mass spectrometry is especially useful for characterization of DCLs. Because the composition of the DCL cannot be statically defined, except when in equilibrium with the receptor, it is essential to be able to identify the components at this stage. Using FTICR MS, identification can be done by a combination of exact mass determination and tandem-in-time mass spectrometry to yield parent/daughter ion structural information. Although FTICR is an expensive option for combinatorial library characterization, it can provide the most direct information for ligand identification in libraries.

## V. FUTURE MASS SPECTROMETRY NEEDS IN COMBINATORIAL CHEMISTRY

Mass spectrometry has been used to make significant contributions to improved efficiency in combinatorial chemistry. Additional improvements in certain areas

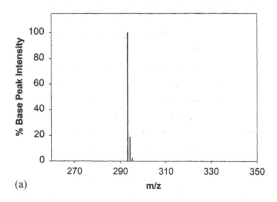

(a)

**Figure 13**  Positive ion electrospray mass spectra of compounds 1 through 5 (a–e) from Table 1. The spectra reflect varying ratios of isotope encoding at $m/z$ 298 using a penta-deutero analog. The parent masses are observable at $m/z$ 293.

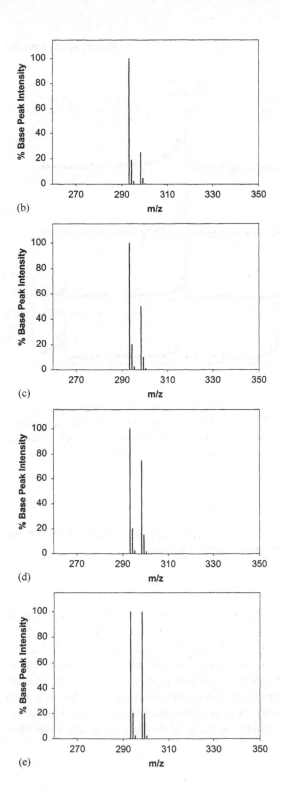

(b)

(c)

(d)

(e)

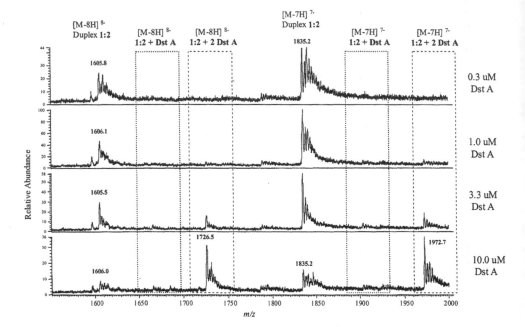

**Figure 14** Electrospray mass spectra at a fixed concentration of oligonucleotide and four concentrations of distamycin. The spectra show the uncomplexed oligonucleotide at both the −7 and −8 charge states and the appearance of the distamycin:oligonucleotide complex at higher concentrations of the ligand (distamycin = DstA).

will make the capabilities of combinatorial chemistry even better. A few of the possible areas for improvement are suggested below.

## A. Distinguishing Isomers by MS

One of the intrinsic limitations of mass spectrometry is its inability to routinely distinguish between positional isomers. If fragmentation chemistry is favorable, positional isomers can be elucidated with a fair degree of certainty, but this only occurs in a small number of the cases. One approach to this problem is to combine LC/MS with on-line NMR techniques [23]. On-line NMR, although expensive and less sensitive, provides rich structural information and is a good compliment to mass spectrometry. It is possible that this approach will be further refined and utilized in the future, as the reliability improves and equipment costs decline.

## B. Structural Elucidation Software

Part of the evolution of structural elucidation in drug metabolism (Chap. 9) has been the creation of specialized software capable of automatically identifying metabolites and interpreting mass spectra. Although this software is in its infancy, the rationale behind the approach is sound and, once evolved enough to be highly reliable, the software will be useful. It is anticipated that similar software would be created for the prediction of products, impurities, and related substances in the products of combinatorial synthesis. This goal is more challenging, however, in that unlike drug metabolism, where a limited number of clearance pathways are possible, combinatorial chemistry can be more diverse. Software for predicting oxidative and hydrolytic reactions for a wide variety of chemical structures will need to be developed before this goal can be attained.

## C. Data-Reduction Techniques

Parallel interface technology has increased the throughput of library characterization. As this throughput improves, the amount of data collected per unit time will dramatically increase as well. As this progression occurs, the reduction and interpretation of data becomes the rate-limiting step. One important improvement will, therefore, be the development of data-reduction techniques designed to sift through an ever-increasing mountain of data, make real-time decisions, and direct the next experiment. Some progress has been made in this area for high-throughput characterization of libraries, but additional progress is needed for data reduction in library screening.

## VI. CONCLUSION

As with many areas of drug discovery, mass spectrometry has become intertwined with combinatorial chemistry, especially in the areas of library characterization and screening. Some future advances in mass spectrometry will be driven by the needs of combinatorial chemistry, and mass spectrometry will, to some extent, regulate the progress in combinatorial chemistry.

## REFERENCES

1. GA Smith, TL Lloyd. Automated solid-phase extraction and sample preparation— finding the right solution for your laboratory. LC-GC Curr Trends Dev Sample Prep May: S22–S31, 1998.
2. EG Mata. Solid-phase and combinatorial synthesis in beta-lactam chemistry. Curr Pharm Des 5:955–964, 1999.

3.  WC Chan, PD White. Fmoc Solid-Phase Peptide Synthesis: a Practical Approach. Oxford, UK: Oxford Univ. Press, 2000.
4.  A Furka, F Sebestyen, M Asgedom, G Dibo. Cornicopia of peptides by synthesis. In: the 14th International Congress of Biochemistry. Prague 1988, Vol. 5, p 47.
5.  A Furka, F Sebestyen. General method for rapid synthesis of multicomponent peptide mixtures. Int J Pept Prot Res 37(6):487–493, 1991.
6.  R Houghten, C Pinilla, SE Blondelle, JR Appel, CT Dooley, JH Cuervo. Generation and use of synthetic peptide combinatorial libraries for basic research and drug discovery. Nature, 354:84–86, 1991.
7.  R Houghten, JR Appel, SE Blondelle, JH Cuervo, CT Dooley, C Pinilla. The use of synthetic peptide combinatorial libraries for the identification of bioactive peptides. Pept Res 5(6):351–358, 1992.
8.  E Erb, KD Janda, S Brenner. Recursive deconvolution of combinatorial chemical Libraries. Proc Natl Acad Sci USA 91:11422–11426, 1994.
9.  CT Dooley, RA Houghten. The use of positional scanning synthetic peptide combinatorial libraries for the rapid determination of opioid receptor ligands. Life Sci 52: (18):1509–1517, 1993.
10. B Deprez, X Williard, L Bourel, H Coste, F Hyafil, A Tartar. Orthogonal combinatorial chemical libraries. J Am Chem Soc 117:5405–5406, 1995.
11. B Domer, GM Husar, JM Ostresh, RA Houghten. The synthesis of peptidomimetic combinatorial libraries through successive amide alkylations. Bioorg Med Chem 4: 709–715, 1996.
12. A Nefzi, JM Ostresh, M Giulianotti, RA Houghten. Solid-phase synthesis of trisubstituted 2-imidazolidones and 2-imidazolidinethiones. J Comb Chem 1:195–198, 1999.
13. A Nefzi, NA Ong, MA Giulianotti, JM Ostresh, RA Houghten. Solid-phase synthesis of 1,4-benzothiazepin-5-one derivatives. Tetrahedron Lett 40:4939–4942, 1999.
14. MC Pirrung. Spatially addressable combinatorial libraries. Chem Rev 97:473–488, 1997.
15. KS Lam Michal Lebl, Krchák. The "one-bead-one-compound" combinatorial library method. Chem Rev 97:411–448, 1997.
16. V Krchnak et al. In: R Cortese, ed. Combinatorial Libraries: Synthesis, Screening and Application Potential. Berlin: de Gruyter, 1996, p 27.
17. CL Chen et al. In: JN Abelson, ed. Methods in Enzymology: Combinatorial Chemistry San Diego: Academic Press, 1996, p 211.
18. JJ Burbaum, MHJ Ohlmeyer, JC Reader, I Henderson, LW Dillard, G Li, TL Randle, NH Sigal, D Chelsky, JJ Baldwin. A paradigm for drug discovery employing encoded combinatorial libraries. Proc Natl Acad Sci USA 92:6027–6031, 1995.
19. S Brenner, RA Lerner. Encoded Combinatorial Chemistry. Proc Natl Acad Sci USA 89:5181–5183, 1992.
20. RJ Booth, JC Hodges. Polymer supported quenching reagents for parallel purification. Am Chem Soc 119:4882, 1997.
21. YR de Miguel. Supported catalysts and their applications in synthetic organic chemistry. J Chem Soc Perkin Trans 1:4213–4221, 2000.
22. SA Poulsen, PJ Gates, GRL Cousins, JKM Sanders. Electrospray ionization Fourier-transform ion cyclotron resonance mass spectrometry of dynamic combinatorial libraries. Rapid Commun Mass Spectrom 14:44–48, 2000.

23. BD Dulery, J Verne-Mismer, E Wolf, C Kugel, L Van Hijfte. Analyses of compound libraries obtained by high-throughput parallel synthesis: strategy of quality control by high-performance liquid chromatography, mass spectrometry and nuclear magnetic resonance techniques. J Chromatogr B. Biomed Sci Appl 725(1):39–47, 1999.

24. WL Fitch. Analytical methods for quality control of combinatorial libraries. Mol Divers 4(1):39–45, 1998–1999.

25. G Siuzdak, JK Lewis. Applications of mass spectrometry in combinatorial chemistry. Biotechnol Bioeng 61(2):127–134, 1998.

26. TA Berger, WH Wilson. High-speed screening of combinatorial libraries by gradient packed-column supercritical fluid chromatography. J Biochem Biophys Methods 43: 127–134, 2000.

27. C Chen, LA Ahlberg Randell, RB Miller, AD Jones, MJ Kurth. The solid-phase combinatorial synthesis of β-thioketones. Tetrahedron 53:6595–609, 1997.

28. YM Dunayevskiy, P Vouros, EA Wintner, GW Shipps, T Carell, J Rebek Jr. Application of capillary electrophoresis–electrospray ionization mass spectrometry in the determination of molecular diversity. Proc Natl Acad Sci USA 93:6152–6157, 1996.

29. RD Sussmuth, G. Jung. Impact of mass spectrometry on combinatorial chemistry. J Chromatogr B Biomed Sci Appl 725(1):49–65, 1999.

30. T Wang, J Cohen, DB Kassel, L Zeng. A multiple electrospray interface for parallel mass spectrometric analysis of compound libraries. Comb Chem High Throughput Screen 2(6):327–334, 1999.

31. L Zeng, L Burton, K Yung, B Shushan DB Kassel. Automated analytical/preparative HPLC-MS system for the rapid characterization and purification of compound libraries. J Chromatogr 794:3–13, 1998.

32. L Zeng, DB Kassel. Development of a fully automated parallel HPLC/MS system for the analytical characterization and preparation of combinatorial libraries. Anal Chem 70:4380–4388, 1998.

33. G Hegy, E Gorlach, R Richmond, F Bitsch. High throughput electrospray mass spectrometry of combinatorial chemistry racks with automated contamination surveillance and results reporting. Rapid Commun Mass Spectrom 10:1894–1900, 1996.

34. DC Stahl, KM Swiderek, MT Davis TD Lee. Data-controlled automation of liquid chromatography tandem mass spectrometry analysis of peptide mixtures. J Am Soc Mass Spectrom 7:532–540, 1996.

35. DR Owens, B Bothner, O Phung, K Harris, G Siuzdak. Aspects of oligonucleotide and peptide sequencing with MALDI and electrospray mass spectrometry. Bioorg Med Chem 6(9):1547–1554, 1998.

36. J Wei, JM Buriak, G Siuzdak. Desorption-ionization mass spectrometry on porous silicon. Nature 399:243–246, 1999.

37. MR Carrasco, MC Fitzgerald, Y Oda, SBH Kent. Direct monitoring of organic reactions on polymeric supports. Tetrahedron Lett 38(36):6331–6334, 1997.

38. DC Schriemer, O Hindsgaul. Deconvolution approaches in screening compound mixtures. Comb Chem High Throughput Screen 1:155–170, 1998.

39. KH Hollenbeak, FJ Schmitz. Aplysinopsin:antineoplastic tryptophan derivative from the marine sponge verongia spengelli. Lloydia 40:479–481, 1977.

40. M Zhu, NG Bowery, PM Greengrass, JD Phillipson. Application of radioligand receptor binding assays in the search for CNS active principles from Chinese medicinal plants. J Ethnopharmacol 54:153–164, 1996.

41. ML Nedved, S Habibi-Goudarzi, B Ganem, JD Henion. Characterization of benzodiazepine combinatorial chemical libraries by on-line immunoaffinity extraction, coupled column HPLC-ion spray mass spectrometry-tandem mass spectrometry. Anal Chem 68(23):4228–4236, 1996.

42. JE Bruce, GA Anderson, R Chen, X Cheng, DC Gale, SA Hofstadler, BL Schwartz, RD Smith. Bio-affinity characterization mass spectrometry. Rapid Commun Mass Spectrom 9(8):644–650, 1995.

43. RB van Breeman, CR Huang, D Nikolic, CP Woodbury, YZ Zhao, DL Venton. Pulsed ultrafiltration mass specrtometry: a new method for screening combinatorial libraries. Anal Chem 69:2159–2164, 1997.

44. HM Geysen, CD Wagner, WM Bodnar, CJ Markworth, GJ Parke, FJ Schoenen, DS Wagner. Isotope or mass encoding of combinatorial libraries. Chem Biol 3(8):679–688, 1996.

45. DS Wagner, CJ Markworth, CD Wagner, FJ Schoenen, CE Rewerts, BK Kay, HM Geysen. Ratio encoding combinatorial libraries with stable isotopes and their utility in pharmaceutical research. Comb Chem High Throughput Screen 1(3):143–153, 1998.

46. JA Loo, DE DeJohn, P Du, TI Stevenson, RR Ogorzalek-Loo. Application of mass sectrometry for target identification and characterization. Med Res Rev 19:307–319, 1999.

47. J Gao, X Chen, R Chen, GB Sigal, JE Bruce, BL Schwartz, SA Hofstadler, GA Anderson, RD Smith, GM Whiteside. Screening derivatized peptide libraries for tight binding inhibitors to carbonic anhydrase II by electrospray ionization-mass spectrometry. J Med Chem 39:1949–1955, 1996.

48. MJ Greig, JM Robinson. Detection of oligonucleotide:ligand complexes by ESI/MS as a component of high throughput screening. J Biomolec Screen 5(6):441–453, 2000.

49. SA Poulsen, PJ Gates, GR Cousins, JK Sanders. Electrospray ionization Fourier-transform ion cyclotron resonance mass spectrometry of dynamic combinatorial libraries. Rapid Commun Mass Spectrom 14(1):44–48, 2000.

50. V Swali, GJ Langley, M Bradley. Mass spectrometric analysis in combinatorial chemistry. Curr Opin Chem Biol 3(3):337–341, 1999.

# 8

# Drug Hydrophobicity and Transport

**Barbra H. Stewart, Steven M. Michael, and Narayanan Surendran**
*Pfizer Global Research and Development, Ann Arbor, Michigan*

> Good ideas are not adopted automatically. They must be driven into practice
> with courageous patience.
>
> —*Admiral H. Rickover*

## I. SIGNIFICANCE OF DRUG TRANSPORT IN DRUG DISCOVERY AND DEVELOPMENT

The goal of pharmaceutical therapeutics is to deliver the drug to its target site(s) within the body at concentrations adequate to obtain a positive pharmacological effect. Drug transport across biological membranes is a primary determinant of the drug concentration that will result at the target site. Table 1 provides examples of membrane barriers that must be traversed in order to reach target sites. The first three examples in Table 1 assume that the drug has been given by the intravenous route of administration and, hence, is immediately available to the systemic circulation. Under these conditions, the drug must still cross membranes of the tumor cell, virus, or blood–brain barrier to reach its site of action. When the more desirable alternative routes of administration are applied, additional membrane barriers are encountered, such as the epithelial cells of the gastrointestinal tract or the epidermal cells of the skin's stratum corneum. Although this chapter is devoted largely to the study of intestinal absorption, many of the principles and applications can be extended to transport phenomena in other tissues and targets.

**Table 1**  Examples of Pharmacological Targets or Routes of Administration and the Membrane Barriers That a Drug Must Traverse to Reach the Sites of Action

| Target | Membrane barrier |
|---|---|
| Chemotherapeutic (e.g., cyclin-dependent kinase) | Tumor cell membrane |
| Anti viral (e.g., HIV protease) | Viral cell membrane |
| CNS receptors (e.g., N-Methyl-D-aspartic acid) | Endothelial cells of the blood–brain barrier |
| Any drug after oral administration | Epithelial cells of the intestine |
| Any drug after transdermal administration | Epidermal cells |

## A.  Principles of Drug Transport

### 1.  What Is Permeability?

Membrane permeability is defined as a *rate* of transfer or translocation as follows:

$$P = D \times PC_m/h, \tag{1}$$

where

$D$ = diffusivity, $PC_m$ = the membrane partition coefficient, and $h$ = membrane thickness; units are in distance/time. Figure 1 depicts the various pathways by

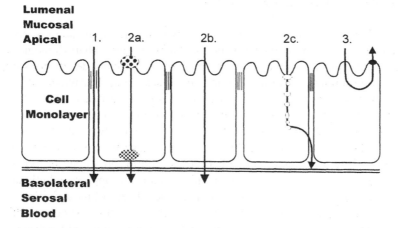

**Figure 1**  Parallel pathways of intestinal absorption: (1) paracellular pathway, (2) transcellular pathway, (2a) carrier-mediated, (2b) passive diffusion, (2c) receptor-mediated endocytosis, and (3) mediated efflux pathway. (From Ref. 30.)

which molecules may cross a membrane barrier, such as the small intestine. Depending on the functional and physicochemical properties of the molecule, one pathway generally dominates, although cases of mixed or parallel transport mechanisms are not uncommon [1].

## 2. What Is Permeability Not?

Permeability is not, for example, intestinal absorption, but a *component* of it. For freely soluble, nonmetabolized compounds, the fraction of dose absorbed is proportional to permeability as follows:

$$F_a = 1 - \exp[(-2P_{\text{eff}}/r)t_{\text{res}}], \tag{2}$$

where

$P_{\text{eff}}$ = effective permeability, $r$ = intestinal radius, and $t_{\text{res}}$ = residence time in the small intestine [2]. The critical assumptions in this relationship are (1) that the drug is in solution and hence available for transport across the intestinal membrane and (2) the drug is chemically stable in the gastrointestinal tract and not a substrate for proteolytic or metabolic enzymes in the intestinal lumen or gut mucosa. These are highly simplifying assumptions. The inappropriate use of Eq. (2) to calculate absorption for compounds that are limited by solubility, instability, or metabolism can lead to seriously erroneous and misleading conclusions.

For compounds limited by solubility but not metabolized, the fraction of dose absorbed is described as follows:

$$\begin{aligned} A_n &= (P_{\text{eff}}/r)t_{\text{res}} \\ D_o &= (M_o/V_o)/C_s \\ F_a &= 2A_n/D_o, \end{aligned} \tag{3}$$

where

$A_n$ = absorption number, $D_o$ = dose number, $V_o$ = gastrointestinal volume, $M_o$ = dose, and $C_s$ = aqueous solubility [3]. This relationship takes into account the limitation of low solubility, although not whether the molecule is unstable or a substrate for metabolic enzymes. In particular, $D_o$ expresses an important variable, the so-called dose-to-solubility ratio [4]. As this ratio increases, the fraction of dose that is *not* in solution also increases; hence, as $D_o$ is increased, $F_a$ must decrease since only drug that is dissolved is available for absorption.

Permeability is also not bioavailability, but a component of it. Permeability relates to absorption vide supra and absorption is related to bioavailability ($F$) as follows:

$$F = F_a(1 - F_g)(1 - F_h), \tag{4}$$

where

$F_a$ = fraction absorbed; $F_g$ = fraction extracted by gut; and $F_h$ = fraction extracted by liver, which is further defined by the following:

$$F_h = Q_H/(Q_H + f_{ub}Cl_{int}),$$

where

$Q_H$ = flow rate of blood to the liver, $f_{ub}$ = free fraction of drug (unbound to plasma proteins), and $Cl_{int}$ = the intrinsic clearance. This relationship describes, in quantitative terms, the resistance in series to an orally administered drug reaching the systemic circulation (i.e., a first-pass effect). As discussed above, both aqueous solubility and membrane permeability affect absorption ($F_a$). The fraction extracted by the gut ($F_g$) formally takes into account metabolism of the molecule in the gut mucosa. The fraction extracted by the liver ($F_h$) is dependent on blood flow to the liver; plasma protein binding; and, through $Cl_{int}$, the capacity of liver enzymes to metabolize substrate.

With higher- throughput approaches becoming standard practice, this has led to the $N$-in-1, or cassette-dosing, approach [5]. In this method, multiple compounds ($N$) are given in one dose to an animal by both oral and intravenous approaches; from the resultant dose-normalized areas under the curve, bioavailability can be calculated. The $N$-in-1 approach was made possible by the exceedingly selective and sensitive methods permitted by liquid chromatography/mass spectrometry (LC/MS) and represented a formidable advance in biopharmaceutics. The *next* advance is to quantify $F_a$, $F_g$, and $F_h$ from in vitro methods and calculate bioavailability prior to in vivo dosing. This offers a particular advantage for prediction of drug performance in humans when human cells or tissues are utilized.

## B. Applications of Drug Transport

### 1. Small Polar Molecules

In general, one would anticipate that small, polar molecules are candidates for diffusion through the paracellular pathway (Fig. 1, Path 1). Molecular size is the most relevant parameter relating to rate of diffusion through the restricted paracellular route [6,7]. For drugs primarily absorbed via the paracellular pathway, permeation could be limited in the large intestine. Higher transepithelial electrical resistance in the colon reflects that this organ contains tighter epithelial cell monolayers than the more permeable small intestine.

Other polar molecules may be substrates for carrier-mediated transport (Fig. 1, Path 2a) by amino acid [8,9], peptide [10,11], bile acid [12], sugar [13], or other transporters in the cell membrane. Transport rates for carrier-mediated substrates follow classic Michaelis–Menten kinetics [14] as follows:

$$v_0 = (V_{max} \times [S])/([S] + K_m), \tag{5}$$

where

$V_0$ = initial-rate velocity (permeability), $V_{max}$ = maximum velocity, $[S]$ = substrate concentration, and $K_m$ = substrate concentration at one-half $V_{max}$.

This relationship is depicted in Fig. 2. The permeability of carrier-mediated molecules is related to the affinity constant for the transporter ($K_m$), the number and activity of transporters ($V_{max}$), as well as the substrate concentration. Thus, at low concentrations relative to $V_{max}$, permeability is linear with respect to concentration, while at concentrations approaching $V_{max}$, permeability has zero-order dependence or is referred to as "saturated." Implications for the intestinal absorption of a drug that has carrier-mediated transport are that absorption will be saturated at high doses and that absorption may be restricted to the small intestine where the nutrient transporters exist in abundance.

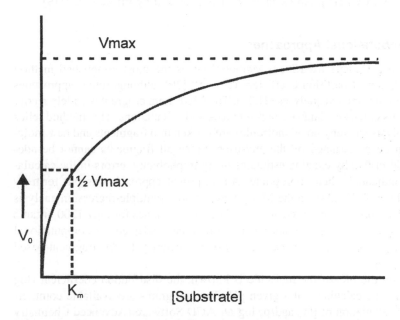

**Figure 2** The dependence on substrate concentration for permeability of a carrier-mediated substrate demonstrating saturable enzyme kinetics. The relationship is described mathematically in Eq. (5).

## 2.  Lipophilic Molecules

For the passive transport mechanism (Fig. 1, Path 2b), there is a linear dependence for permeability on diffusivity and the partition coefficient between aqueous and membrane compartments [Eq. (1)]. Diffusivity, which can be calculated from molar volume, generally varies within 1 order of magnitude, while partitioning can range over many orders of magnitude. Hence, for molecules with passive transport as the dominant pathway, determination of the partition coefficient can be a useful estimate of permeability.

## II.  LIPOPHILICITY MEASUREMENTS

Lipophilicity is the single most important physicochemical parameter in predicting and interpreting membrane transport. Intense study has provided a number of descriptors for hydrophobicity and membrane interaction, including partition and distribution coefficients using different organic phases, partitioning into liposomes [15], chromatography using retention on reverse-phase HPLC [16], immobilized artificial membranes (IAM) [17], and micellar electrokinetic capillary approaches (MECC) [18]. An excellent text is provided by Pliska et al. [19].

## A.  Computational Approaches

The Pomona College MedChem Methodology is the most established method for calculation of partition coefficient (clog $P$) [20], although other approaches also rely on fragmental analyses [21,22]. The MedChem program is widely recognized for its extensive database of references and calculations. This method relies on an analysis whereby novel molecules are broken into fragments and new molecules can be constructed by the end-user. Since all fragments cannot be adequately described by existing estimates of hydrophobicity, errors in the calculation are indicated when appropriate. An empirical approach was put forth by Moriguchi et al. [23,24]. In the Moriguchi method, multiple-regression analysis was used to correlate observed and calculated log $P$ values for over 1200 organic molecules. This method has not been commercially adapted for computational analysis but has been compared favorably in accuracy to the fragment-based methods [21].

When the p$K_a$ of the molecule is known, the distribution coefficient (log $D$) can also be calculated at a given pH. Two programs are available commercially for calculation of p$K_a$ and/or log $D$: ACD Software (Advanced Chemistry Development, Toronto, Ontario) and Pallas (CompuDrug, Budapest, Hungary). These programs are gaining widespread use in the pharmaceutical industry, but have been accepted only with qualification. It is largely agreed that either program is useful in estimating an initial value for $pK_a$ with subsequent experimental determination providing the definitive numbers.

## B. Oil:Water Partitioning (Shake-Flask Method)

The shake-flask method measures the equilibrium distribution of the solute between aqueous and organic phases. This method is considered definitive and large databases have been compiled. The solvent n-octanol has been most commonly used as the organic phase [25]. The ability of n-octanol to form hydrogen bonds is an advantage in that dimerization of the solute does not readily occur; however, this same ability is a disadvantage as a model for biological membranes since there is a dominant role for solvent cavity formation in the water:octanol system [26]. For structurally related series, there is a sigmoidal relationship between permeability and partition coefficient; however, for diverse compound sets there may be no simple correlation [27].

In the drug discovery arena of the 1990s, it was the hydrophobic new molecular entity (NME) that was of major concern. The shake-flask method is most reliable and precise; however, for very hydrophobic or hydrophilic solutes, this is not the case. For the highly hydrophobic solutes routinely encountered in drug discovery, either the volume of the aqueous phase will be miniscule or the amount of solute available for detection will be vanishingly small. These experimental challenges, along with low throughput capacity and potential for automation, have limited recent use of this method.

## C. Chromatographic Approaches

The chromatographic methods provide a much faster way to determine lipophilicity. The trade-off with higher throughput is that chromatographic approaches require correlation with reference compounds. Correlation should first be established between capacity factors and oil:water partition coefficients for reference compounds; subsequently, capacity factors for unknowns can be interpolated to obtain partition coefficients. Historically, chromatographic approaches have relied on UV or fluorescence detection, which limited throughput to injection of single compounds. The advent of HPLC/MS permits injection of multiple compounds because detection by molecular mass is more specific than generic UV or fluorescent techniques. So long as each compound is resolved by at least two mass units, the number injected is limited only by practical considerations such as relative retention times.

### 1. Reverse-Phase Chromatography

Reverse-phase (RP) methods were initially established on the basis of similarity between octanol:water partitioning and retention on an RP column for HPLC

[16,26,28]. Octanol-coated (ODS) columns are generally used. As noted vide supra, this method relies on correlation with reference compounds from the literature. References are ideally selected to span the range of lipophilicity estimated for the intended unknowns. Linearity should be demonstrated between the capacity factors ($k'_{ODS}$) and octanol:water partition coefficients before values for the unknown compounds can be determined by interpolation. The hydrogen-bonding character of the unknown should also be taken into account because inaccuracies can result from H-bonding of solute to residual silanol sites [29].

## 2.  Immobilized Artificial Membrane Chromatography

Immobilized Artificial Membrane (IAM) packings for HPLC provide a different environment from that of the hydrocarbon-based ODS columns [17,30–32]. In this model, IAMs are purified phospholipids that are covalently bonded to the silicon support. At this time, only IAM columns prepared from phosphatidylcholine are commercially available (Regis Technologies, Inc., Morton Grove, IL). The hypothesis is that the bonded phospholipid layer is akin to the cellular monolayer that represents a barrier to transport. Interaction of the solute with the phospholipid results in a capacity factor ($k'_{IAM}$) that is proportional to the membrane partition coefficient [$PC_m$ in Eq. (1)]. In this sense, the IAM approach does not attempt to correlate with literature values of oil:water partition coefficients, but seeks to establish a unique membrane interaction parameter.

   The physiological environment being modeled by the chromatographic interaction requires that solute partitioning is between an aqueous mobile phase and the stationary phase. For highly hydrophobic solutes, retention on IAM or RP columns can be extensive without addition of organic modifier to the mobile phase. For this approach, three or more organic:aqueous mobile-phase compositions are prepared and the capacity factors are determined in each composition. Capacity factors are plotted versus the reciprocal of mobile-phase composition, and the capacity factor in pure aqueous mobile phase ($k'_w$) is obtained by extrapolation to 0% organic.

   In addition to inducing a more ordered hydrocarbon region, a major distinction of IAM is the electrostatic component manifested by the polar headgroups [33,34]. Direct interaction between positively or negatively charged solutes with the negatively charged headgroups can result in $k'_{IAM}$ changes due to electrostatic attraction or repulsion. Thus, IAM may enhance discrimination of charged solutes due differences in lipophilicity.

## 3.  Micellar Electrokinetic Capillary Chromatography

Micellar electrokinetic capillary chromatography uses a totally aqueous buffer into which a surfactant in excess of its critical micelle concentration is introduced [18,35–37]. This approach makes use of solute separation via partitioning into micelles as opposed to separation on a bonded stationary phase. These features tend to minimize or eliminate major sources of error due to multiple injections to obtain $k'_w$ by extrapolation or laboratory-to-laboratory differences due to variation in column packing or aging. Micelles provide both electrostatic and hydrophobic sites of interaction, similar to those of the biomembrane being modeled. Sodium dodecyl sulfate is the most common surfactant, although others have been utilized. MECC is primarily used to resolve uncharged molecules, with migration times largely dependent on hydrophobic behavior. MECC, in which lysophospholipid was used, compared more favorably than IAM to liposomal partitioning [35].

## III. TRANSPORT MEASUREMENTS

There are a variety of in vitro, in situ, and in vivo techniques available for evaluating the intestinal permeability characteristics of NMEs in early discovery and preclinical settings. These models can be employed to evaluate the extent, rate, or mechanism of intestinal absorption of drug candidates. The choice of model(s) for a particular study largely depends on the nature of the problem an investigator is attempting to address, the stage of a compound or a series in the drug discovery/development process, and the various analytical tools available to an investigator. For example, in vitro cell monolayer models such as Caco-2 or MDCK (Madine Darby canine kidney) are useful in early drug discovery settings to rank order a series of compounds on their permeability characteristics. This information enables the synthetic chemists to incorporate the necessary structural features on the pharmacophore, eventually leading to NMEs with more favorable permeability characteristics. On the other hand, to examine the potential of NMEs to utilize a specific transport mechanism, cell lines transfected to overexpress transporters such as the oligopeptide transporter (e.g., CHO-hPEPT1) are an attractive choice. Ideally, the model(s) chosen should be an accurate reflection of the specific barriers to absorption relevant for that particular study. The ability to automate, and therefore increase, throughput would be an additional consideration if a particular technique were intended for use in early drug discovery. Regardless of the model that is being employed to evaluate the intestinal permeability of NMEs, the need for a sensitive, specific, and rapid analytical methodology cannot be overemphasized. Commonly, the availability or lack of analytical tools and techniques determines the types of models that an investigator may choose to employ for a particular study. The purpose of this section is to outline the types of in vitro, in situ, and in vivo models that are available, with particular emphasis on the type of information that can obtained from these models. Additionally, specific advantages or limitations of these models is highlighted wherever appropriate.

## A.  Subcellular Fractions

Subcellular fractions available for evaluating the permeability characteristics of NMEs include preparations of intestinal brush-border membrane vesicles (BBMV) and basolateral membrane vesicles (BLMV). Rat or rabbit tissue are most commonly used, although human tissue can also be obtained. Typically, the steps involved in the preparation of these systems include tissue homogenization (e.g., duodenum, jejunum, ileum, or colon), differential sedimentation, fractionation by density gradient centrifugation, and differential precipitation. Subsequently, subfractionation for either BBMV or BLMV can be performed. Before utilizing these preparations for studies, the fidelity is examined by analysis of marker enzymes for different membrane populations [38]. Tissue sources are generally from human or animal species such as rabbit, guinea pig, and in some cases rat. New molecular entities are characterized for extent of partitioning into the lipid bilayer of the cell membrane or kinetics of transport into the intravesicular space. The latter can be accomplished by manipulating the osmolality of the incubation medium and by demonstration of equilibrium uptake that is proportional to initial NME concentration. These systems can also be employed to elucidate specialized transport processes (i.e., active transport or facilitated diffusion) of various drugs, nutrients, and other endogenous compounds such as ß-lactam antibiotics, renin inhibitors, ACE inhibitors, amino acids, bile acids, monocarboxylic acids, sugars, and exogenous lipids [39–46].

The ability to examine transport as an apical (BBMV) or basolateral (BLMV) process is a significant advantage of these models. Additionally, the general availability of fresh tissue and well-characterized techniques for preparation and data interpretation from these models are advantages. The presence of hydrolytic enzymes such as in BBMV makes this a useful tool for evaluating drug stability and prodrug reconversion at the intestinal wall [47].

These models also suffer from distinct limitations. The ability to translate transport parameters in this model to in vivo rate or extent of absorption is limited. Contamination due to heterogeneity of tissue source or other membranes, as well as proper orientation after vesicle resealing, are additional issues. One of the major drawbacks of this technique is the need for a sensitive analytical methodology. Use of radiolabeled solutes is a viable alternative when available. In other cases, the availability of sensitive and specific analytical tools such as HPLC/MS/MS would be well suited. Depending on the method used to quench the uptake reaction, the matrix for assay may be a dilute buffer solution or a miscible mixture of buffer and organic solvent (e.g., acetonitrile).

## B.  Freshly Isolated Cells

Use of freshly isolated enterocytes to evaluate uptake of NMEs is well documented. Various laboratories [48–50] have established isolation techniques. Advantages to using this model include the availability of laboratory animals to isolate cells of interest. Limitations of this model include the restriction of evaluating only uptake of NMEs as opposed to bidirectional transport. The inability to store cell suspensions and stringency of viability maintenance during the course of the experiments are additional challenges. For drug assay, the matrix is typically a dilute buffer solution. In recent years, the use of this model has been limited, largely due to the availability of cell monolayer models that maintain polarity and offer the ability to examine absorptive and secretory transport mechanisms.

## C. Cell Monolayer Models

Cell monolayer models that represent the use of cells in continuous culture have gained immense popularity in recent years. Since fully differentiated intestinal cells do not thrive in primary culture, most cells used in monolayer preparations are of transformed origin. Caco-2 cells, derived from human colon adenocarcinoma, have the ability to differentiate spontaneously in culture when grown on permeable polycarbonate or nitrocellulose filters. Although these cells are polarized with distinct apical and basolateral domains and exhibit microvillar enzymes, nutrient transporters, ion conductance, and transepithelial electrical resistance, indicating formation of tight junctions, the cells are colonic in nature [51,52]. HT-29 is another cell type that is used in transport studies. These cells differentiate in the presence of galactose, express microvillar hydrolases, and display phenotypes resembling enterocytes and goblet cells [53,54]. Several subclones of Caco-2 and HT-29 cells have been developed in recent years that have either the capacity to synthesize drug-metabolizing cytochrome P450 enzymes or mucus-secreting components when differentiated [55,56]. MDCK cells are gaining popularity and have been used as an alternative to Caco-2 cells for examining the permeability characteristics of NMEs [57].

Some of the advantages of cell monolayer models include the ability to use human instead of animal cell types as well as the ability to perform cellular uptake and bidirectional cell transport studies for evaluation of absorptive and secretory processes. The potential for automation to achieve higher throughput in the early drug discovery setting is an added attraction. Regardless of the cell type used, the utility of these models in transport studies is based on the correlation between permeability properties determined in these models and those obtained in vivo, such as fraction of dose absorbed (Fig. 3). To date, numerous laboratories have established a correlation between apparent permeability coefficients ($P_{app}$) from Caco-2 or MDCK cells and in vivo fraction absorbed of drugs in solution [58–62]. Construction of correlation plots for known compounds or reference markers then provides an opportunity for interpolation of fraction absorbed for NMEs for which in vivo fraction absorbed is unknown.

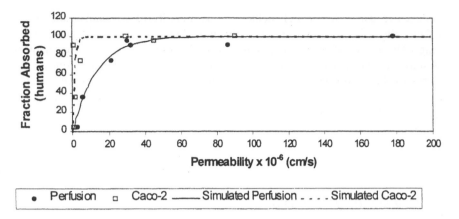

**Figure 3** The relationship between permeability determined using Caco-2 cell mono-layers or rat intestinal perfusion and the fraction of dose absorbed in humans. The relationship is described in Eq. (2). (From Ref. 61.)

These models, especially Caco-2 cells, suffer from certain limitations. First, these cells are homogenous and therefore do not represent the physiological environment of the intestinal barrier. Lack of mucus and low metabolic capacity are additional deviations from the in vivo state. Although these cells have endogenous nutrient and solute transporters such as those for bile acids, amino acids, and oligopeptides, qualitative and quantitative differences exist. Maintenance of cells in culture, routine seeding, and feeding are labor-intensive procedures requiring care over 21 days for Caco-2 cells and 4–7 days for MDCK cells. The availability of 3-day systems such as BIOCOAT for Caco-2 has been proposed to alleviate some of these concerns [63]. Probably the most important drawback of these models is laboratory-to-laboratory variability for reasons not limited to differences in cell maintenance techniques, passage number, choice of experimental devices, and choice of permeable support. Evaluation of NMEs with low solubility and/or significant nonspecific adsorption to devices used in these experiments is also problematic. The use of cosolvents or other modifiers and numerical corrections for nonspecific adsorption can be employed to circumvent or eliminate solubility and adsorption concerns [64]. With regard to data interpretation, it must be realized that the relationship between $P_{app}$ and fraction absorbed tends to be sigmoidal or hyperbolic depending on the series of compounds for evaluation in various laboratories. Therefore, while prediction accuracies are acceptable at the low and high ends of the permeability spectrum, interpretation in the hyper-variable region must be performed with caution and with supporting physico-chemical and structural information. Since Caco-2 cells are of colonic origin, inherently high transepithelial electrical resistance and low ion conductance may result in underestimated permeability for compounds that traverse the paracellular pathway (Fig. 1, Pathway 1). Therefore, caution should be exercised while examining the permeability of NMEs that are low molecular weight (<300 Da) and have negative log $D$ (distribution coefficient).

A novel approach whereby evaluation of permeability characteristics is coupled to the intrinsic biological activity of NMEs was demonstrated by Rubas et al. [65]. While this technique obviates the need for the development of a specific assay, the inability to accurately correlate the biological effects to the parent molecule must be borne in mind during data interpretation. Another interesting modification to evaluating membrane permeability is the use of artificial lipid bilayers in a high-throughput format as demonstrated by Kansy et al. [66]. Arguably, replacement of UV-based detection by the more sensitive and rapid HPLC/MS/MS will lead to a higher throughput process than originally proposed by these authors. To maximize the efficiency of initial screening of NMEs for permeability characteristics, the utility of the "cocktail" approach for screening mixtures from combinatorial libraries has been demonstrated [67]. The ability to use a sensitive and specific technique such as the HPLC/MS/MS makes this a reality. Sample preparation must take into account the relatively high-salt concentration of the buffer. This approach may work well for compounds that are transported by passive diffusion, but may confound results if there is saturation of transporters or metabolic enzymes in the transepithelial transport of NME mixtures.

## D. Intestinal Rings

The everted intestinal ring technique for the study of drug uptake has been commonly used for assessing the permeability characteristics of drug candidates [68]. Excised intestinal segments are first everted, and then the segment is cut transversely using a razor blade to obtain rings. Rings are typically incubated in oxygenated buffers containing the solute or NME of interest and placed in a shaking water bath, with provision for temperature and mixing control. Drug uptake can be determined as a function of concentration, time, or other variables. Uptake is quenched by rinsing with ice-cold buffer and then assayed using available analytical techniques.

Some of the advantages of everted intestinal rings include speed and simplicity of the method. The ability to obtain a large number of rings from the same animal or different segments of the intestine allow for sufficient control and evaluation of segmental differences in drug transport. The ability to control conditions such as temperature and buffer pH is an added advantage. Both passive diffusion and drug uptake using specialized mechanisms such as active transport or facilitative diffusion can be evaluated using this system [9,69,70].

The everted intestinal ring model also suffers from serious limitations. Assurance of viability during the course of the experiment is a concern. Deterioration of the epithelial membrane can occur for incubation periods exceeding 10–15 min. Therefore, poorly permeable markers such as mannitol, inulin, or PEG 4000 are routinely employed to assess barrier integrity. Additional complications in the interpretation of data include delineation of uptake versus nonspecific adsorption and elimination of intestinal metabolism as a confounding variable. Low drug accumulation is another concern that necessitates the use of a sensitive, specific analytical methodology in order to obtain reliable quantitation of cellular uptake. If radiolabeled compound is unavailable, solute assay may be difficult.

## E.  Isolated Intestinal Tissue

In this system, isolated intestinal segments from a suitable animal species (e.g., rabbit, rat, and guinea pig) are opened to form planar tissue sheets that can be mounted on diffusion chambers for conducting transport studies (Fig. 4). Generally, the diffusion cells have gas-lift systems that serve to oxygenate the tissue and to mix the buffers gently to minimize unstirred water layers. Transepithelial flux of NME can be evaluated using either mucosal or serosal chambers as donor solutions to determine absorptive or secretory fluxes, respectively. Similar to cell monolayer models, variables such as buffer pH, temperature, and concentration of NME can be controlled in this system. In cases where there may be interactions with the underlying musculature of the intestinal tissue, a process known as "stripping" is employed to remove the muscle layer. In addition to determining transport characteristic, ionic conductance or transepithelial electrical resistance can be monitored, thus providing a measurement of barrier property of the isolated tissue. Frequently, impermeable markers such as mannitol or PEG 4000 are used as internal references to monitor the barrier integrity of the tissue during the course of a study. This model has been used to investigate the mechanisms of transport of NMEs. Examples of NMEs that are transported by passive diffusion and other specialized mechanisms, such as active transport or facilitated diffusion, are abundant in the literature [71–73]. Isolated tissues have been used to characterize secretory processes in rat intestine, as well as the effect of different agents on the secretory transport of drugs and NMEs. Additionally, the ability to study different segments of the intestine offers an opportunity to compare segmental heterogeneity of NME transport.

Advantages of this model are the availability of tissue and the well-characterized techniques of tissue preparation and experimental conditions. Data interpretation is straightforward in most cases and the ability to discern different transport mechanisms and segmental differences in transport are noteworthy. Some of the limitations of this model are related to concerns of tissue viability during the course of the experiment due to a lack of blood supply as well as its potential

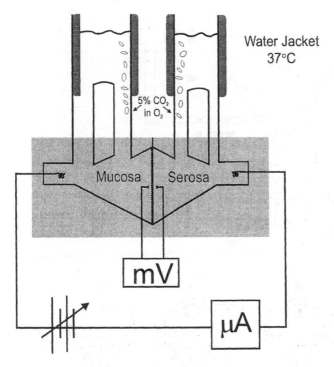

**Figure 4** Schematic representation of diffusion cell preparation for isolated tissue experiments. In this figure, $P_d$ represents the electrical potential difference (in millivolts) across the membrane and $I_{sc}$ represents the current flow; transepithelial resistance is $P_d$. (From Ref. 62.)

effect on the paracellular pathway. Additionally, metabolism of NMEs by the intestinal tissue may serve as a confounding variable during assessment of transport properties. Conversely, the presence of intestinal metabolic machinery in these tissues provides an opportunity to assess transport and metabolism coupling to reduce the rate and extent of transepithelial transport [74]. As with assay of drug levels from cell monolayer systems, sample preparation must take into account high salt contents in buffers.

## F.  Intestinal Perfusion

In situ perfusion of intestinal segments provides an excellent alternative to everted rings or isolated intestinal tissue for experimental determination of permeability [75]. The perfusion technique has been characterized extensively using rat as an animal model. A number of in situ flow patterns have been utilized to

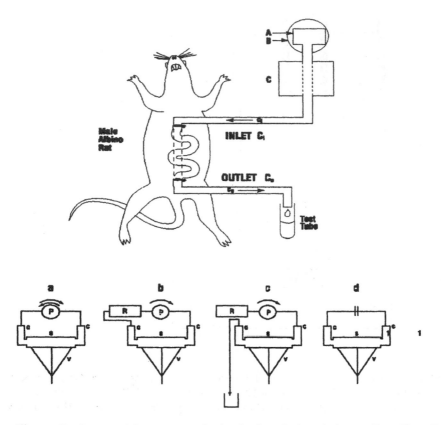

**Figure 5** Setup and flow patterns for in situ intestinal perfusion studies. (From Ref. 62.)

obtain permeability coefficients for NMEs (Fig. 5). In addition to evaluation of passive transport, this model can be used to examine specialized mechanisms of transport (i.e., active transport and facilitated diffusion). Correlation between permeability coefficients obtained by in situ perfusion and fraction absorbed in humans has been well established [2]. From an experimental point of view, there are several factors that must be borne in mind while using this model. First, the permeability coefficient obtained from this model is calculated from the disappearance of NME relative to its initial concentration when perfused to steady state. For well-absorbed compounds, aqueous resistance may dominate the overall resistance to permeability; in these cases, the perfusion flow rate controlling intestinal residence time should be adjusted to less than 15%. The viability of the intestinal segment during the experiment is generally improved relative to everted rings or isolated tissue. Exiting concentrations of NMEs must be corrected for water flux based on the dilution or concentration of nonabsorbable markers such as PEG 4000, or by gravimetric means. This correction is necessary to eliminate influences from secretion or absorption of water and also assess the barrier integrity of the isolated intestinal segment.

Some of the advantages of in situ intestinal perfusion are related to maintenance of physiological processes that are present in vivo. The presence of intact neural and endocrine input, intact blood, and lymphatic supply as well as other attributes, such as the presence of nutrient transporters and hydrolytic enzymes, are noteworthy. Because disappearance of the drug from the intestinal segment is measured, low concentrations are used to both minimize saturation of transport processes and drug interaction possibilities in a drug cassette. From an analytical point of view, assay sensitivity requirements are less stringent because loss of compound from the intestinal lumen is measured, not appearance of compound. Samples for drug assay have high salt content due to buffering. The ability to characterize regional dependence of NME permeability characteristics in the same animal is another advantage. An attractive advantage of this technique is that it has been adapted for human intestinal perfusion studies using an intubation procedure [76].

Limitations of the perfusion model are related to some basic assumptions made in developing the model. Because permeability coefficients are determined by loss of drug from the intestinal lumen, involvement of other confounding variables such as lumenal stability of NMEs must be borne in mind during data interpretation. This system is not well suited to evaluating efflux processes (i.e., secretion). Occasionally, anesthetics administered to rats prior to surgical procedures may influence the results. The inability to automate this model is probably the most serious limitation, especially at the early drug discovery stage, where higher throughput is necessary.

## G.  Transfected Cell Lines

In recent years, there has been increasing awareness regarding the importance of transporters in the absorption and disposition of NMEs. While the major portion of NMEs or marketed drugs traverse cell membranes by passive diffusion, there are numerous examples where the involvement of specialized transport mechanisms has been demonstrated. Examples include the role of oligopeptide transporters in the intestinal absorption of β-lactam antibiotics, angiotensin-converting enzyme (ACE) inhibitors, and novel NMEs as well as the role of P-glycoprotein (P-gp) in the secretion of molecules into the intestine [11,77–79]. Transfection of cells with the transporter protein of interest has permitted the evaluation of precise cellular mechanisms of uptake and transport of NMEs. Transfected cell lines by definition are tailor-made to overexpress the protein of interest; therefore, data interpretation is largely unambiguous. Most studies using transfected cell models have evaluated the effect of NME on the uptake or transport of radiolabeled or fluorescent substrates (i.e., inhibition screens). Evaluation of transport mechanisms of NMEs can also be accomplished provided sensitive, specific, and rapid methods of assay such as HPLC/MS/MS are available. Drug samples for HPLC/MS/MS assay would be typical of those obtained from subcellular (IIIA) or cellular (IIIB) suspensions (i.e., dilute buffer solutions or aqueous: organic mixtures).

## H.  Computational Approaches

Given the large numbers of NMEs emanating from combinatorial chemistry and the mass-screen mindset, there is a tremendous interest and demand for "in silico" predictions of NME permeability. Recent work by Waterbeemd and Palm has demonstrated the utility of these approaches to estimate the membrane permeability of NMEs [72,80]. The work of Waterbeemd et al. revolved around estimation of Caco-2 permeability using molecular descriptors (e.g., log $P$, molecular weight, $pK_a$, surface area, and molar volume). Palm et al. [81] have explored the use of polar surface area (PSA) to estimate fraction of dose absorbed in humans across different structural series. Regardless of the specific approach, these techniques are being explored widely and can be considered to be in the beginning stages. To have an impact in early drug discovery, it is necessary to have a robust, computationally nonintensive and validated model that can be used to screen NMEs on a routine basis across a variety of structural series. One simple mnemonic that has gained popularity in recent years is the Pfizer Rule of Five [82]. Rule of Five is used as an early alert system that is most ideally in place when NMEs are registered to an inventory or database system in a discovery organization. Rule of Five states that poor absorption is likely to occur when there are more than five H-bond donors, molecular weight >500 Da, clog $P > 5$, and the sum of N and O atoms is >10.

## IV.  BIOANALYTICAL SUPPORT OF DRUG TRANSPORT

There are several prominent experimental methodologies currently in use for evaluating and predicting the transport properties of new molecular entities in drug discovery. As in most sciences, the choice of experimental method or model to study drug transport is largely determined by the question the pharmaceutical scientist must address. Regardless of which model is utilized to investigate drug transport, there is a universal demand for fast and sensitive quantitative assay. Rapid, reliable determination of drug transport properties is one contributing piece of information toward the enhancement of bioavailability and, ultimately, probability for success as an NME progresses through the drug discovery process. The selectivity, sensitivity, and speed of HPLC/MS/MS make it an ideal analytical tool for the support of drug transport studies in drug discovery. Leveraging these characteristics of the HPLC/MS/MS instrument enable drug transport studies to impact the drug discovery process in a timely manner. In this section the analytical support of drug transport via mass spectrometry will be presented both from a practical application perspective, as well as a survey of current trends in the pharmaceutical industry.

## A. A Practical Approach

The analytical strategy for supporting drug transport studies is not unique. As in all analytical support, the application starts with defining the analytical challenge. The analytical scientist must know what question is to be answered by the data, what is the required throughput, and what turnaround time is required. A rational analytical strategy will consider all of these key factors and serve as the foundation for successful analytical methodologies to be developed and implemented.

## B. Defining the Analytical Challenge

The analytical scientist needs to determine what is required from the data to be delivered. The data must be able to appropriately address the question being investigated. Hence, care must be taken to ensure that the experimental design ensures that appropriately powered data will result from the assay. In conjunction with appropriately powering the data, the basic details of what must be quantified, from what matrices and what limits of quantitation are required must also be determined. Next, production considerations, i.e., throughput and cycle time, need to be factored into the decision-making process. When all of these factors are considered together an analytical approach can be planned to support the experimental investigation.

The various experimental models for in vitro and in situ transport studies result in samples that are largely similar from the analytical perspective. Both matrices are aqueous, buffered at pH 6.5, contain relatively large amount of buffer salts and very low amounts of protein. As a result, fairly generic analytical strategies can be put in place to support these studies. These samples are characterized by the analyte(s) of interest present in a high-salt-content aqueous matrix; organic solvent modifiers and other small molecules, i.e., reference compounds, are usually also present. The problems studied by transport experiments range from those requiring quantitation over 3 orders of magnitude to semiquantitative determination via comparison to a single known sample. The production requirements for analytical support of transport studies are rapid turnaround, typically 1 day to 1 week, with throughput demands varying from very low numbers of samples to support an individual study to high throughput required to support transport screening programs. While the analytical samples are similar across the variety of transport experimental models, the other program requirements often dictate the specifics of the analytical support required.

There are a variety of HPLC/MS/MS-based analytical programs that support transport studies across the pharmaceutical industry. All of these analytical groups take advantage of some or all of the speed, sensitivity, and selectivity characteristics of the mass spectrometer to meet the needs of their respective transport programs. Several groups [83,84] utilize gradient HPLC/MS/MS strategies. These groups report on transport samples that are pooled posttransport experiment and then directly injected into the HPLC/MS/MS system for assay. Pooling discreet transport experiment samples prior to assay enables simultaneous data acquisition for multiple experiments and, hence, greatly increases throughput capacity. The gradient HPLC is used to separate the analytes present from the experimental matrix, a fast on-line clean-up method, while the selectivity of tandem mass spectrometric assay (MS/MS) is used to discreetly quantify the analytes of interest. The ability of the mass spectrometer to separate multiple analytes in space versus temporal chromatographic separation is leveraged to result in both faster and higher throughput analytical support.

Other groups [85,86] report a similar analytical approach of gradient HPLC/MS assay of transport samples. These groups directly inject discreet transport samples onto a gradient HPLC to separate the analyte of interest from the matrix, followed by single-stage mass spectrometric assay. Again, this approach results in fast assay with high throughput capacity. This HPLC/MS approach is simplified relative to the HPLC/MS/MS analytical support strategy, as less method development is required prior to sample assay. Another group [87] reports direct injection assay of transport samples via an isocratic HPLC/MS methodology similar to those of Ackermann et al. and Wei et al. Regardless of variations in analytical methodologies, several of these groups report individual sample cycle times of a few minutes and throughput capacities of thousands of samples per week.

Analytical groups vary not only in methodology, but also in delivered product. Some analytical groups supporting transport studies deliver quantitative results from calibration curves of 10 standards over 3 orders of magnitude, while others are delivering relative ratios of an unknown sample to that of a known sample. This variation in the level of certainty of the delivered results is indicative of the diversity of transport experiments and what is done with the information generated by them.

The need for fast, reliable, and sensitive analytical support of drug transport programs is consistent throughout the pharmaceutical industry. Regardless of specific transport models and what the goals are for the information obtained, this need is widely being met by the application of mass spectrometry.

## V. CONCLUSIONS

The sensitivity and selectivity brought to drug discovery by the routine use of HPLC/MS/MS has revolutionized biopharmaceutical capabilities. This impact has been realized in the form of increased throughput and decreased cycle time. In a highly significant sense, these new analytical methods have enabled the introduction of absorption distribution metabolism excretion studies much earlier in discovery than previously possible. At one time the investment of lengthy assay development for transport studies on a discovery candidate was nearly unthinkable. With HPLC/MS/MS, this barrier has been removed. Prime ADME information can be provided at the earliest stages of discovery to aid in the selection of lead candidates.

## REFERENCES

1. N Jezyk, X Wu, C Li, BH Stewart, D Fleisher. The transport of pregabalin in rat intestine and Caco-2 monolayers. Pharm Res 16:519–526, 1999.
2. GL Amidon, PJ Sinko, D Fleisher. Estimating human oral fraction dose absorbed: a correlation using rat intestinal membrane permeability for passive and carrier-mediated compounds. Pharm Res 5:651–654, 1988.
3. GL Amidon, H Lennernas, VP Shah, JR Crison. A Theoretical basis for a biopharmaceutic drug classification: the correlation of in vitro drug product dissolution and in vivo bioavailability. Pharm Res 12:413–420, 1995.
4. JB Dressman, GL Amidon, D Fleisher. Absorption potential: estimating the fraction absorbed for orally administered compounds. J Pharm Sci 74:588–589, 1985.
5. J Berman, K Halm, K Adkison, J Shaffer, Simultaneous pharmacokinetic screening of a mixture of compounds in the dog using API LC/MS/MS analysis for increased throughput. J Med Chem 40(6):827–829, 1997.
6. Y-L He, S Murby, G Warhurst, L Gifford, D Walker, J Ayrton, R Eastmond, M Rowland. Species differences in size discrimination in the paracellular pathway reflected by oral bioavailability of poly(ethylene glycol) and D-peptides. J Pharm Sci 87:626–633, 1998.
7. JL Madara, K Dharmsathaphorm. Occluding junction structure–function relationships in a cultured epithelial monolayer. J Cell Biol 101:2124–2133, 1985.
8. BR Stevens, HJ Ross, EM Wright. Multiple transport pathways for neutral amino acids in rabbit jejunal brush border vesicles. J Membr Biol 66:213–225, 1982.
9. BH Stewart, AR Kugler, PR Thompson, HN Bockbrader. Saturable intestinal absorption of gabapentin is the underlying cause of dose disproportionality in humans. Pharm Res.
10. V Ganapathy, G Burckhardt, FH Leibach. Characteristics of glycylsarcosine transport in rabbit intestinal brush-border membrane vesicles. J Biol Chem 259:8954–8959, 1984.

11. N Surendran, K-MY Covitz, H-K Han, W Sadee, D-M Oh, GL Amidon, RM Williamson, CF Bigge, BH Stewart. Evidence for overlapping substrate specificity between large neutral amino acid (LNAA) and dipeptide (hPEPT1) transporters for PD 158473, an NMDA antagonist. Pharm Res 16:391–395.

12. IJ Hidalgo, RT Borchardt. Transport of bile acids in human intestinal epithelial cell line, Caco-2. Biochim Biophys Acta 1035:97–103, 1990.

13. CI Cheeseman. GLUT2 is the transporter for fructose across the rat intestinal basolateral membrane. Gastroenterology 105:1050–1056, 1993.

14. H Lodish, D Baltimore, A Berk, SL Zipursky, P Matsudaira, J Darnell. Molecular Cell Biology. 3rd ed. New York: Scientific American Books/Freeman, 1995, pp 81–83.

15. GV Betageri, A Nayernama, MJ Habib. Thermodynamics of partitioning of nonsteroidal anti-inflammatory drugs in the N-octanol/buffer and liposome systems. Int J Pharm Adv 1:310–319, 1996.

16. KB Sentell, JG Dorsey. Retention mechanisms in reversed-phase liquid chromatography: stationary-phase bonding density and partitioning. Anal Chem 61:930–934, 1989.

17. CY Yang, SJ Cai, H Liu, C Pidgeon. Immobilized artificial membranes—screens for drug membrane interactions. Adv Drug Del Rev 23:229–256, 1996.

18. BJ Herbert, JG Dorsey. n-Octanol-water partition coefficient estimation by micellar electrokinetic capillary chromatography. Anal Chem 67:744–749, 1995.

19. V Pliska, B Testa, H van de Waterbeemd. In Lipophilicity in Drug Action and Toxicology. New York: VCH, 1996.

20. AJ Leo. Hydrophobic parameter: measurement and calculation. In: JJ Langone, ed. Methods in Enzymology: Molecular Design and Modeling: Concepts and Applications, Part A, Proteins, Peptides and Enzymes. New York: Academic Press, 1991, vol 202A, pp 544–591.

21. RF Rekker, AM ter Laak, R Mannhold. On the reliability of calculated log P-values: Rekker, Hansch/Leo and Suzuki approach. Quant Struct–Act Relat 12:152–157, 1993.

22. T Suzuki, Y Kudo. Automatic log P estimation based on combined additive modeling methods. Comput-Aided Mol Design 4:155–198, 1990.

23. I Moriguchi, S Hirono, Q Liu, I Nakagome, Y Matsushita. Simple method of calculating octanol/water partition coefficient. Chem Pharm Bull 40:127–130, 1992.

24. I Moriguchi, S Hirono, Q Liu, I Nakagome, H Hirono. Comparison of reliability of log P values for drugs calculated by several methods. Chem Pharm Bull 42:976–978, 1994.

25. RN Smith, C Hansch, MM Ames. Selection of a reference partitioning system for drug design work. J Pharm Sci 64:599–606, 1975.

26. DA Paterson, RA Conradi, AR Hilgers, TJ Vidmar, PS Burton. A non-aqueous partitioning system for predicting the oral absorption potential of peptides. Quant Struct–Act Relat 13:4–10, 1994.

27. NFH Ho, JY Park, W Morozowich, WI Higuchi. Physical model approach to the design of drugs with improved intestinal absorption. In: EB Roche, ed. Design of Biopharmaceutical Properties through Prodrugs and Analogs. Washington, DC: APhA/APS, 1977, pp 136–277.

28.  R Kaliszan, A Kaliszan, IW Wainer. Deactivated hydrocarbonaceous silica and immobilized artificial membrane stationary phases in high-performance liquid chromatographic determination of hydrophobicities of organic bases: relationship to log P and CLOGP. J Pharm Biomed Anal 11:505–511, 1993.
29.  JE Haky, AM Young. Evaluation of a simple HPLC correlation method for the estimation of the octanol–water partition coefficients of organic compounds. J Liq Chromatogr 7:675–689, 1984.
30.  BH Stewart, OH Chan. Use of immobilized artificial membrane chromatography for drug transport applications. J Pharm Sci 87:1471–1478, 1998.
31.  S Ong, H Liu, X Qiu, G Bhat, C Pidgeon. Membrane partition coefficients chromatographically measured using immobilized artificial membrane surfaces. Anal Chem 67:755–762, 1995.
32.  C Pidgeon, S Ong, H Choi, H Liu. Preparation of mixed ligand immobilized artificial membranes for predicting drug binding to membranes. Anal Chem 66:2701–2709, 1994.
33.  F Barbato, MI La Rotonda, F Quaglia. Chromatographic indexes on immobilized artificial membranes for local anesthetics: relationships with activity data on closed sodium channels. Pharm Res 14:1699–1705, 1997.
34.  P Barton, AM Davis, DJ Webborn, PJH Webborn. Drug-phospholipid interactions. 2. Predicting the sites of drug distribution using n-octanol/water and membrane/water distribution coefficients. J Pharm Sci 86:1034–1039, 1997.
35.  JA Masucci, GW Caldwell, JP Foley. Comparison of the retention behavior of β-blockers using immobilized artificial membrane chromatography and lysophospholipid micellar electrokinetic chromatography. J Chromatogr A 810:95–103, 1998.
36.  M Hanna, V de Biasi, B Bond, C Salter, AJ Hutt, P Camilleri. Estimation of the partitioning characteristics of drugs: a comparison of a large and diverse drug series utilizing chromatographic and electrophoretic methodology. Anal Chem 70:2092–2099, 1998.
37.  S Yang, JG Bumgarner, LFR Kruk, MG Khaledi. Quantitative structure–activity relationships: studies with micellar electrokinetic chromatography; influence of surfactant type and mixed micelles on estimation of hydrophobicity and bioavailability. J Chromatogr A 721:323–335, 1996.
38.  E Kinne-Saffran, RKH Kinne. Membrane isolation: strategy, techniques, markers. In: S Fleischer, B Fleischer, eds. Biomembranes: Transport: Membrane Isolation and Characterization. New York: Academic Press, 1989, vol. 172, part 2, pp 3–17.
39.  K-I Inui, T Okano, H Maegawa, M Kato, M Takano, R Hori. H+ coupled transport of p.o. cephalosporins via dipeptide carriers in rabbit intestinal brush border membranes: diffecence of transport characteristics between cefixime and cephadrine. J Pharmacol Exp Ther 247:235–241, 1988.
40.  N Hasimoto, T Fujioka, T Toyoda, N Muranishi, K Hirano. Renin inhibitor: transport mechanism in rat small intestinal brush border membrane vesicles. Pharm Res 11:1448–1451, 1994.
41.  H Yuasa, D Fleisher, GL Amidon. Non competitive inhibition of cephadrine uptake by enalapril in rabbit intestinal brush-border membrane vesicles: an enalapril specific inhibitory binding site on the peptide carrier. J Pharmacol Exp Ther 269:1107–1111, 1994.

42. R Muniz, L Burguillo, JR del Castillo. Effect of starvation on neutral amino acid transport in isolated small intestinal cells from guinea pig. Pflugers Arch 423:59–66, 1993.

43. W Kramer, G Wess, G Neckerman, G Schubert, J Fink, R Girbig, U Gutjahr, S Kowalewski, K-H Baringhaus, G Boger, A Enhsen, E Falk, M Friedrich, H Glombik, A Hoffman, C Pitius, M Urman. Intestinal absorption of peptides by coupling to bile acids. J Biol Chem 269:10621–10627, 1994.

44. MT Simanjuntak, I Tamai, A Tsuji. Carrier mediated uptake of nicotinic acid by rat intestinal brush border membrane vesicles and relation to monocarboxylic acid transport. J Pharmacobiodyn 3:301–309, 1990.

45. CI Cheeseman. Role of intestinal basolateral membrane in absorption of nutrients. Am J Physiol 263:R482–R488, 1992.

46. F Bellemare, J Noel, C Malo. Characteristics of exogenous lipid uptake by renal and intestinal brush border membrane vesicles. Biochem Cell Biol 73:171–179, 1995.

47. CN TenHoor, BH Stewart. Reconversion of fosphenytoin in the presence of intestinal alkaline phosphatase. Pharm Res 12:1806–1809, 1995.

48. DD Harrison, HL Webster. The preparation of isolated intestinal crypt cells. Exp Cell Res 55:257–260, 1969.

49. HL Webster, DD Harrison. Enzymic activities during the transformation of crypt to columnar intestinal cells. Exp Cell Res 56:245–253, 1969.

50. MM Weiser. Intestinal epithelial cell surface membrane glycoprotein. Synthesis. I. An indicator of cellular differentiation. J Biol Chem 248:2536–2641, 1973.

51. M Pinto, S Robine-Leon, MD Appay, M Kedinger, N Triadou, E Dussaulx, B Lacroix, P Simon-Assman, K Haffen, J Fogh, A Zweibaum. Enterocyte-like differentiation and polirazation of the human colon carcinoma cell line Caco-2 in culture. Biol Cell 47:323–330, 1983.

52. E Grasset, M Pinto, E Dussaulx, A Zweibaum. JF Desjeux. Epithelial properties of human colonic carcinoma cell line Caco-2: electrical parameters. Am J Physiol 247:C260–C267, 1984.

53. A Zweibaum, M Pinto, G Chevalier, E Dussaulx, N Triadou, B Lacroix, K Haffen, JL Brun, M Rousset. Enterocytic differentiation of a sub-population of the human colon tumor cell line HT-29 selected for growth in sugar-free medium and inhibition by glucose. J Cell Physiol 122:21–29, 1985.

54. M Pinto, MD Appay, P Simon-Assman, G Chevalier, N Dracopoli, J Fogh, A Zweibaum. Enterocytic differentiation of cultured human cancer cells by replacement of glucose by galactose in the medium. Biol Cell 44:193–196, 1982.

55. P Schmiedlin-Ren, KE Thummel, JM Fisher, MF Paine, KS Lown, PB Watkins. Expression of enzymatically active CYP3A4 by Caco-2 cells grown on extracellular matrix-coated permeable supports in the presence of $1\alpha$, 25-dihydroxyvitamin D3. Mol Pharmacol 51:741–754, 1997.

56. C Huet, C Sahuquillo-Merino, E Coudrier, D Louvard. Absorptive and mucus-secreting subclones isolated from a multipotent intestinal cell line (HT-29) provide new models for cell polarity and terminal differentiation. J Cell Biol 105:345–357, 1987.

57. JD Irvine, L Takahashi, K Lockhart, J Cheong, JW Tolan, HE Selick, JR Grove. MDCK (Madin–Darby canine kidney) cells: a tool for membrane permeability screening. J Pharm Sci 88:28–32, 1999.

58. P Artursson, J Karlsson. Correlation between oral drug absorption in humans and apparent drug permeability coefficients in human intestinal epithelial (Caco-2) cells. Biochem Biophys Res Commun 175:880–885, 1991.

59. RA Conradi, KF Wilkinson, BD Rush, AR Hilgers, MJ Ruwart, PS Burton. In vitro/in vivo models for peptide oral absorption: comparison of Caco-2 cell permeability with rat intestinal absorption of renin inhibitory peptides. Pharm Res 10:1710–1714, 1993.

60. C-P Lee, DC Chiossone, IJ Hidalgo, PL Smith. Comparison of in vitro permeabilities of series of benzodiazepines and correlation with in vivo absorption. Pharm Res 10: S177, 1993.

61. BH Stewart, OH Chan, RH Lu, EL Reyner, HL Schmid, HW Hamilton, BA Steinbaugh, MD Taylor. Comparison of intestinal permeabilities determined in multiple in vitro and in situ models: relationship to absorption in humans. Pharm Res 12:693–699, 1995.

62. BH Stewart, OH Chan, N Jezyk, D Fleisher. Discrimination between drug candidates using models for evaluation of intestinal absorption. Adv Drug Delivery Rev 23: 27–45, 1997.

63. S Chong, SA Dando, RA Morrison. Evaluation of BIOCOAT intestinal epithelium differentiation environment (3-day cultured Caco-2 cells) as an absorption screening model with improved permeability. Pharm Res 14:1835–1837, 1997.

64. OH Chan, HL Schmid, B-S Kuo, DS Wright, W Howson, BH Stewart. Absorption of Cam-2445, an NK1 neurokinin receptor antagonist: in vivo, in situ and in vitro evaluations. J Pharm Sci 85:253–257, 1996.

65. W Rubas, MEM Cromwell, RJ Mrsny, G Ingle, KA Elias. An integrated method to determine epithelial transport and bioactivity of oral drug candidates in vitro. Pharm Res 13:23–26, 1996.

66. M Kansy, F Senner, K Gubernator. Physicochemical high throughput screening: parallel artificial membrane permeation assay in the description of passive absorption process. J Med Chem 41:1007–1010, 1998.

67. EW Taylor, JA Gibbons, RA Braeckman. Intestinal absorption screening of mixtures from combinatorial libraries in the Caco-2 model. Pharm Res 14:572–577, 1997.

68. I Osiecka, PA Porter, RT Borchadt, JA Fix, CR Gardner. In vitro drug absorption models I: brush border membrane vesicles, isolated mucosal cells and everted intestinal rings: characterization and salicylate accumulation. Pharm Res 2:284–293, 1985.

69. PA Porter, I Osiecka, RT Borchardt, JA Fix, L Frost, CR Gardner. In vitro drug absorption models II: salicylate, cefoxitin, α-methyl dopa, and theophylline uptake in cells and rings: correlation with in vivo bioavailability. Pharm Res 2:293–298, 1985.

70. V Mummaneni, JB Dressman. Intestinal uptake of cimetidine and ranitidine in rats. Pharm Res 11:1599–1604, 1994.

71. BJ Aungst, H Saitoh. Intestinal absorption barriers and transport mechanisms, including secretory transport, for a cyclic peptide, fibrinogen antagonist. Pharm Res 13:114–119, 1996.

72. K Palm, K Luthman, A-L Ungell, G Strandlund, P Artursson. Correlation of drug absorption with molecular surface properties. J Pharm Sci 85:32–39, 1996.

73. DC Dawson. Principles of membrane transport. In: SG Schultz, M Field, RA Frizzell, eds. Handbook of Physiology—The Gastrointestinal Tract. American Physiological Society, 1991, vol. IV, pp 1–44.

74. SM Rogers, DJ Black. The use of Ussing chambers for the study of intestinal metabolism in vitro. Prog Pharm Clin Pharm 7:43–53, 1989.

75. D Fleisher. Gastrointestinal transport of peptides: experimental systems. In: MD Taylor, GL Amidon, eds. Peptide-Based Drug Design. Washington DC: American Chemical Society, 1995, pp 500–523.

76. H Lennernas, O Ahrensedt, R Hallgre, L Knutson, M Ryde, LK Paalzow. Regional jejunal perfusion, a new in vivo approach to study oral drug absorption in man. Pharm Res 9:1243–1251, 1992.

77. A Tsuji. Intestinal absorption of β-lactam antibiotics. In: MD Taylor, GL Amidon, eds. Peptide-Based Drug Design. Washington DC: American Chemical Society, 1995, pp 101–134.

78. S Yee, GL Amidon. Oral absorption of angiotensin-converting enzyme inhibitors and peptide prodrugs. In: MD Taylor, GL Amidon, eds. Peptide-Based Drug Design. Washington DC: American Chemical Society, 1995, pp 135–148.

79. Y Zhang, LZ Benet, Characterization of P-glycoprotein mediated transport of K02, a novel vinylsulfone peptidomimetic cysteine protease inhibitor across MDR1-MDCK and Caco-2 cell monolayers. Pharm Res 15:1520–1524, 1998.

80. H Waterbeemd, G Camenisch, G Folkers, OA Raevsky. Estimation of Caco-2 cell permeability using calculated molecular descriptors. Quant Struc–Act Relat 15:480–490, 1996.

81. K Palm, P Stenberg, K Luthman, P Artursson. Polar molecular surface properties predict the intestinal absorption of drugs in humans. Pharm Res 14:568–571, 1997.

82. CA Lipinski, F Lombardo, BW Dominy, PJ Feeny. Experimental and computational approaches to estimate solubility and permeability in drug discovery and development settings. Adv Drug Del Rev 23:3–25, 1997.

83. JP Bulgarelli, SM Michael. Simultaneous analysis of multiple drug mixtures for the quantitation of in-vitro permeability through the Caco-2 model utilizing LC/MS/MS. Proceedings of the 46th American Society for Mass Spectrometry Conference, Orlando, FL, 1998, p 14.

84. SA Wring, LW Frick, D Morgan, JW Polli, JE Humphreys, C Rewerts, L Tyler, I Silver, J Salisbury, P Soo, DM Higton, C Sersbjit-Singh. Increased throughput of in vitro ADME screens in drug discovery: cassette techniques with HPLC-mass spectrometry. The Tenth International Symposium on Pharmaceutical and Biomedical Analysis, Washington, DC, 1999.

85. BL Ackermann, RE Stratford Jr., SA Castetter, BR Hanssen, KJ Ruterbories, DA Laska, TD Lindstrom. Enhanced throughput absorption screening by Caco-2 using fast gradient elution LC/MS with column-switching. Proceedings of the 47th American Society for Mass Spectrometry Conference, Dallas, TX, 1999.

86. R Wei, H Lee, LYT Li, JN Kryanos. Development of a high throughput analytical technique for compound permeability studies using Caco-2 Cells and HPLC/MS. Proceedings of the 47th American Society for Mass Spectrometry Conference, Dallas, TX, 1999.

87. DS Richards, HM Verrier, H Major, S Preece. Evaluation of an atmospheric pressure ionization ion source for high-throughput analysis of drugs directly from cell culture media at high ionic strength. Am Biotech Lab 17:58–60, 1999.

# 9

# The Mass Spectrometer in Drug Metabolism

**Michael W. Sinz**
*Bristol-Myers Squibb, Pharmaceutical Research Institute,*
*Wallingford, Connecticut*

**Terry Podoll**
*CEDRA Corporation, Austin, Texas*

> Under the most rigorously controlled conditions of pressure, temperature,
> humidity, and other variables, the organism will do as it damn well pleases.
> —*Unknown*

## I. INTRODUCTION

Prior to the 1990s, the properties of absorption, distribution, metabolism, and
elimination (ADME) were not of major concern during the process of drug
discovery. Drug discovery research at this time was dominated by the fields
of chemistry and pharmacology, where chemical synthesis (patents) and pharmacological activity were the determining factors on which compounds would
be selected for further development. During the 1990s, many pharmaceutical
companies began developing discovery programs that incorporated various
ADME properties, such as pharmacokinetics, metabolism, and biopharmaceutical properties (solubility/log *P*). As a result of better understanding the impact
that ADME can have on the discovery process and the alleviation of problems
later in development, many pharmaceutical companies incorporated procedures
to screen and select improved discovery candidates based on ADME properties. The following discussion highlights the importance of drug metabolism in

**271**

pharmaceutical drug discovery and development and its relationship to mass spectrometry.

Drug metabolism (xenobiotic biotransformation) is the process by which lipophilic foreign compounds are detoxified through biochemical or chemical reactions to hydrophilic metabolites that are eliminated directly as intact primary metabolites or after conjugation with endogenous cofactors. These are the processes by which the body removes foreign compounds (drugs or ingested natural products) in the most efficient and safe manner. It is well recognized that not all biotransformation reactions lead to nontoxic or pharmacologically inactive metabolites. Pharmacological activity (duration of activity), either through a lack of metabolism or metabolism to an active metabolite, and toxicity (as a result of metabolism to reactive metabolites) are two essential issues that should be evaluated in any drug discovery program. As shown in Fig. 1, many compounds have been found to generate reactive intermediates or pharmacologically active compounds through metabolism that have led to toxicity or prolonged efficacy, respectively. Several examples are illustrated throughout this chapter to emphasize the significance of particular metabolic reactions, either beneficial or harmful.

Drug metabolism as a science goes back over 200 years, when metabolites from ingested food products or early medicines were isolated from animal and human urine. Drug metabolism as a discipline really came together with the publication of *Detoxification Mechanisms* by R. T. Williams in 1947, who assimilated what was known at the time about the metabolism of drugs and organic compounds [1,2]. Over the centuries, the means of detecting and identifying xenobiotic metabolites has increased dramatically, culminating with the use of the mass spectrometer in almost every aspect of research in drug metabolism today. Gas chromatography mass spectrometry (GC/MS) was the first mass spectrometric method employed to separate complex mixtures of biological fluids and characterize both endogenous and xenobiotic metabolites found in blood, urine, bile, and feces. GC/MS instruments worked well for volatile compounds or compounds that could be derivatized to volatile compounds; however, it did not work well for compounds that could not be volatilized or would degrade under the conditions necessary to separate components. The advent of liquid chromatography mass spectrometry (LC/MS) provided a means to separate a wider variety of complex mixtures at room temperature without the need to volatilize compounds. Today, the most widely employed technique to study biotransformation reactions and products is the atmospheric pressure ionization (API) LC/MS. This chapter endeavors to describe the different types of biotransformation reactions as well as presents a discussion on how drug metabolism and the mass spectrometer are used in combination in the fast-paced areas of drug discovery and drug development.

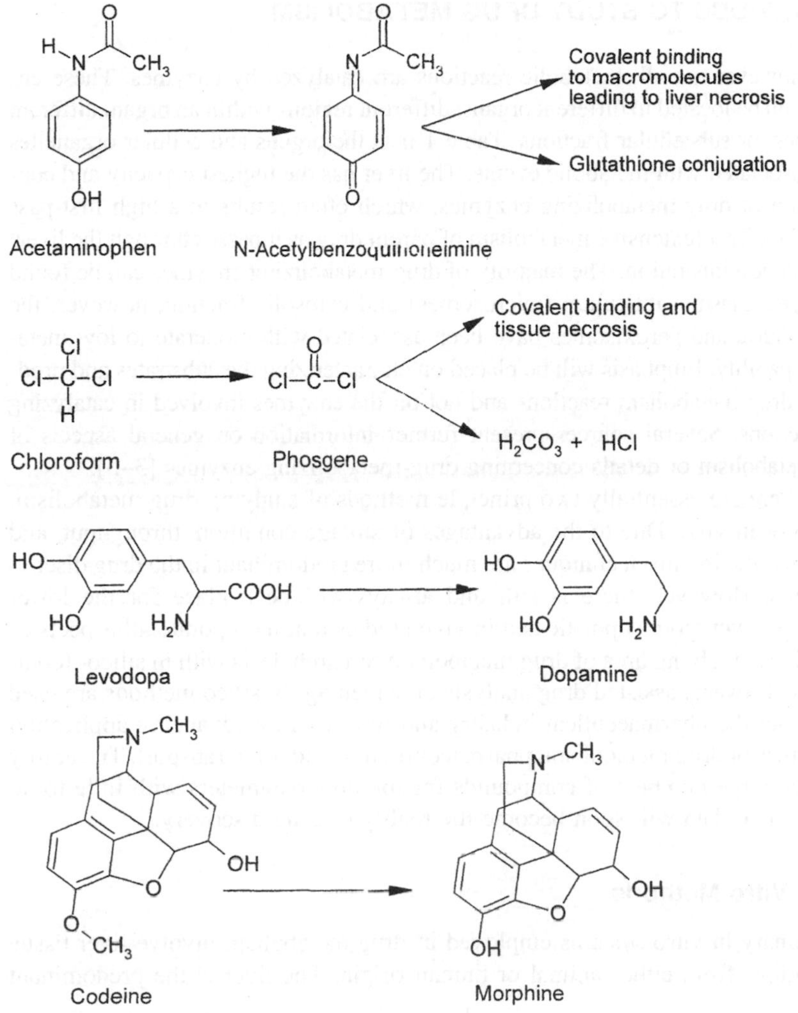

**Figure 1** Examples of compounds that are metabolized to toxic or active metabolites. Acetaminophen and chloroform are metabolized to reactive intermediates that lead to toxicity, whereas levodopa and codeine are metabolized to active metabolites.

## II.  METHODS TO STUDY DRUG METABOLISM

Most, however not all, metabolic reactions are catalyzed by enzymes. These enzymes can be located in different organs, different regions within an organ, different cell types, or subcellular fractions. Table 1 lists the organs and cellular organelles often associated with metabolic events. The liver has the highest capacity and concentration of drug-metabolizing enzymes, which often results in a high first-pass metabolic effect (extensive metabolism of parent drug as it passes through the liver) after oral administration. The majority of drug metabolizing enzymes can be found in the endoplasmic reticulum (microsomes) and cytosolic fraction; however, the mitochondria and peroxisomes have been associated with moderate to low metabolic capability. Emphasis will be placed on characterizing the substrates and products of drug metabolism reactions and not on the enzymes involved in catalyzing the reactions. Several sources present further information on general aspects of drug metabolism or details concerning drug-metabolizing enzymes [3–5].

There are essentially two principle methods of studying drug metabolism: in vitro or in vivo. Due to the advantages of storage condition, throughput, and flexibility, the in vitro techniques are much more predominant in the drug discovery arena. However, there is still and always will be a place for the lower throughput, yet more sophisticated in vivo studies that incorporate all aspects of ADME. An evolving area of drug metabolism research deals with in silico- (computer or software) assisted drug analysis or screening. In silico methods are used throughout the pharmaceutical industry and have rapidly advancing application in the areas of drug metabolism, pharmacokinetics, and drug transport. The ability to screen large numbers of compounds for multiple parameters with little to no experimental data will soon become the reality of drug discovery.

### A.  In Vitro Methods

The primary in vitro systems employed in drug metabolism involve liver tissue preparations from either animal or human origin. The liver is the predominant

**Table 1**  Organs and Organelles Often Associated with Metabolic Biotransformations

| Capacity | Organ | Organelle |
|---|---|---|
| High | Liver | Endoplasmic reticulum (microsomes[a]) and cytosol[a] |
| Moderate | Lung, kidney, intestine | Mitochondria |
| Low | Brain, adrenals, skin, testis, placenta, olfactory | Peroxisomes and lysosomes |

[a] Considered subcellular fractions.

site of xenobiotic biotransformation; however, organs such as the brain, intestine, and kidney can perform some limited biotransformation reactions. Depending on the metabolic issue to be addressed, there are a wide variety of in vitro preparations ranging from the individual enzymes to the perfused organ. Each in vitro model has its own set of advantages and disadvantages as they range from simple to more complex systems: individual enzymes, subcellular fractions, cellular systems, liver slices and the whole organ, respectively. In vitro systems are very useful to evaluate subtle or dramatic changes within the enzyme environment. Adding or removing various enzymes or cofactors or changes in pH or temperature can often help in elucidating which enzymes are involved in a particular enzymatic event or chain of events. Table 2 indicates the various in vitro systems, their complexity with regard to metabolic events, utility in drug discovery, and some of the applications of each in discovery. All of the applications listed in Table 2 involve the use of the mass spectrometer either for metabolite characterization or quantitation of parent drug or metabolites.

## 1. Single-Enzyme Systems

Single-enzyme systems, typically obtained from cDNA expression systems, are an effective technique for the study of individual enzyme reactions and their products. The incubation of enzyme, buffer, necessary cofactors, and drug produces extremely clean samples that are amenable to either simple cleanup procedures or direct injection into the LC/MS instrument (Chap. 6). These isolated enzymes can be stored at $-80°C$ for extended periods of time, making them readily available whenever necessary. In addition, they are reasonably easy to work with either by hand or in a robotic environment where high-throughput incubation, sample preparation, and analysis are possible.

**Table 2** Various In Vitro Systems and Their Application to Drug Metabolism

| System | Complexity[a] | Utility in drug discovery[a] | Application |
|---|---|---|---|
| Single enzyme | + | +++ | MS, DDI |
| Microsomes | ++ | +++ | MS, DDI, XSP, PK |
| S9 fraction | ++ | ++ | MS, DDI, XSP |
| Hepatocytes | +++ | +++ | MS, DDI, XSP, PK |
| Liver slices | +++ | + | MS, DDI, XSP |
| Perfused liver | ++++ | + | MS, PK |

[a] +, low; ++, moderate; +++, high; and ++++, very high.
Abbreviations: MS, metabolic stability; DDI, drug–drug interaction screening; XSP, cross-species comparison; and PK, determination of pharmacokinetic parameters.

## 2. Subcellular Fractions

As early as the 1940s, liver homogenates became one of the first in vitro techniques to be employed. Since that time, the isolation of subcellular fractions from liver homogenate have become more popular and useful. Liver homogenate centrifuged at 9000$g$ results in a supernatant (S9 fraction) rich in drug-metabolizing enzymes and a pellet containing predominately nuclei, lysosomes, peroxisomes, and mitochondria. Centrifuging the S9 fraction at 100,000$g$ results in a supernatant (cytosol) and pellet (microsomes) each containing different types or forms of drug-metabolizing enzymes. Figure 2 illustrates the preparation of these subcellular fractions.

Microsomes are the most commonly used subcellular fraction and contain the majority of oxidative drug-metabolizing enzymes, such as the cytochromes P450 (CYP) and flavin monooxygenases (FMO). Microsomes are microvesicles of endoplasmic reticulum (approximately 100 nm in diameter) containing large amounts of membrane-bound enzymes. In addition to the oxidative enzymes, microsomes contain glucuronosyltransferases, epoxide hydrolases, alcohol/aldehyde dehydrogenases, esterases, amidases, and methyltransferases. Similar to individual enzymes, microsomes can be stored at −80°C and are readily available and commonly used with robotic systems for high-throughput metabolic studies. Microsomes are one of the predominant in vitro systems employed to evaluate metabolic stability (a prediction of how fast or slow a drug will be metabolized

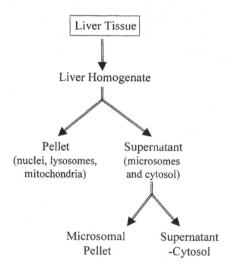

**Figure 2** Origin of subcellular fractions prepared by differential centrifugation of liver tissue: homogenate, S9 fraction, microsomes, and cytosol.

in vivo) due to the fact that the majority of rapidly metabolized compounds are via the oxidative enzymes found in microsomes. The cytosolic fraction contains the majority of conjugative drug metabolizing enzymes, such as sulfotransferases, *N*-acetylases, amino acid transferases, and glutathione *S*-transferases. Due to the lack of oxidative enzymes, the cytosol is not commonly used in drug discovery, except in those specific cases where an enzyme present in cytosol needs to be studied. In addition, more specialized enzymes found in other organelles, such as those involved in β-oxidation (mitochondria or peroxisomes), are not frequently used in drug discovery.

## 3. Hepatocytes

Hepatocytes (liver cells) are a complete drug-metabolizing unit, with an intact cell membrane, cellular environment (organelles, temperature, and pH), drug-metabolizing enzymes, and cofactors. These liver cells contain the majority of drug-metabolizing enzymes in their natural orientation (architecture) and concentration as well as cofactors at their normal concentrations. Similar to the in vivo situation, drugs must cross a cell membrane and compete with other exogenous or endogenous substrates for the multiple drug-metabolizing systems. The advantage of a functional cell membrane is that drugs and metabolites must cross this barrier either by active or passive transport. Transporters may be involved in increasing or decreasing the concentration of drug and/or metabolites inside or outside of the cell, thereby dramatically altering the concentration. These characteristics are in contrast to the previously mentioned cell-free in vitro systems (expressed enzymes, cytosol, and microsomes) that do not contain a cell membrane nor normal concentrations of cofactors and employ unusual buffering systems.

Hepatocytes have many applications in both drug discovery and development. In drug development, they are employed in every aspect of drug evaluation. For example, cross-species comparisons of metabolic profiles; drug–drug interactions (inhibition or induction); reaction phenotyping (determining the enzymes involved in the biotransformation of a drug); and, most importantly, the first evaluation of metabolites that can be formed in vivo in humans. In addition, the estimation of in vivo pharmacokinetic parameters has been shown with animal and human hepatocytes with reasonable success [6,7]. In drug discovery, the applications are somewhat more limited, merely due to the cost and availability of human hepatocytes. However, hepatocytes are used to evaluate the metabolic stability of compounds, much like the use of microsomes for the same purpose, with the advantage of a broader complement of drug-metabolizing enzymes. Caldwell et al., who examined eight β-adrenolytic drugs in rat, monkey, and human hepatocytes, showed this application by high-throughput methods utilizing 96-well plates and LC/MS [8].

Animal hepatocytes, in particular rat hepatocytes, are readily accessible and are used throughout both drug discovery and development. Unfortunately, human hepatocytes have not been as accessible or available when needed, therefore they were predominately employed in drug development. Newer advancements in the cryopreservation of hepatocytes have allowed many researchers to freeze and store hepatocytes for routine screening in drug discovery [9]. The ability to store hepatocytes from multiple different species has a distinct advantage in drug discovery over the laborious nature of isolating cells and the random access to individual human livers.

## 4. Liver Slices

Liver slices have historically been employed to study multiple biochemical processes; however, they fell out of use due to the inability to consistently produce viable slices of uniform thickness. In the 1980s, commercial tissue slicers became available that circumvented all of the limitations previously encountered in preparing liver slices. In the 1980s and 1990s, the use of liver slices resurfaced as an important tool to study metabolism, pharmacokinetics, toxicology, and other biochemical functions. Liver slices afford the researcher with an in vitro system of higher architecture, e.g., containing all of the liver cell types with the cell–cell junctions intact as well as bile canuliculi. These interconnections allow the slice to maintain a higher level of function through cellular communication and compound/cofactor trafficking. Like the hepatocytes, the slices also contain all of the Phase I (oxidative) and Phase II (conjugative) enzymes necessary to obtain a complete metabolic profile. Today, liver slices have begun to slowly fall out of use in drug metabolism due predominately to issues associated with drug transport into and out of the slices and the increased use of hepatocytes to study similar reactions [7]. However, slices remain a popular tool to study processes that require multiple organelles or regulatory machinery within the cell, such as toxicology and biochemistry.

## 5. Organ Perfusion

Organ perfusion is the ultimate in vitro (in situ) model system employed to study many aspects of metabolism or pharmacokinetics of not only the liver, but also the intestine, brain, and kidney. The major advantage of the organ perfusion system is the complete preservation of organ architecture, such as vasculature, elimination pathways (bile canuliculi or renal tubules), cellular contacts, and a zonal distribution of enzymes within the organ. In contrast to all of the other techniques described previously, organ perfusion is typically a single experiment with a single test article, whereas the other in vitro systems will typically use the subcellular fraction or hepatocytes from a single liver to perform hundreds of experiments. Therefore, organ perfusions are typically employed to answer specific questions and are not routinely used to screen discovery compounds for ADME properties.

Depending on the in vitro technique employed, only certain types of metabolic reactions are possible or require the addition of specific cofactors for reactions to occur. For example, microsomes do not have the ability to form sulfate metabolites due to the lack of both the sulfotransferase enzyme and the necessary cofactor. Table 3 illustrates the metabolic events that are possible within each in vitro system, assuming the necessary cofactors are added in the appropriate concentrations. The location of these enzymes will be important to remember when choosing an in vitro technique to use and which products (metabolites) to expect from each system.

## B. In Vivo Methods

In vivo studies have generally been much more labor and time intensive compared to in vitro experiments; however, with the advent of cassette dosing (see Chap. 11) and higher throughput methods of sample preparation and analysis, in vivo experiments are essential in drug discovery. Unlike the in vitro systems described above, in vivo experiments are much more descriptive in overall metabolic events and drug disposition due to the complexity and arrangement of the biological systems incorporated into living organisms. In vivo studies are much more appropriate to examine how drugs are absorbed, distributed, metabolized, and eliminated and how these processes are interconnected to take in drugs and eliminate them from the body. Dosing animals or humans with new chemical

**Table 3** Metabolic Reactions Found in Various In Vitro Systems

| In vitro system | Metabolic reactions |
|---|---|
| Microsomes | CYP, FMO, GT, EH, AD, ES/AM, MT |
| Cytosol | ST, GST, EH AD, ES/AM, DHD, COMT, NA, AA |
| Hepatocytes | CYP, FMO, GT, EH, AD, ES/AM, MT, ST, GST, DHD, COMT, NA, AA, β-O |
| Liver slices | CYP, FMO, GT, EH, AD, ES/AM, MT, ST, GST, DHD, COMT, NA, AA, β-O |
| Liver perfusion | Bile, CYP, FMO, GT, EH, AD, ES/AM, MT, ST, GST, DHD, COMT, NA, AA, β-O |

Abbreviations: CYP, cytochrome P450; FMO, flavin monooxygenase; GT, glucuronyl transferase; EH, epoxide hydrolase; AD, alcohol/aldehyde dehydrogenase; ES/AM, esterase/amidase; MT, methyl transferase; ST, sulfotransferase; GST, glutathione S-transferase; DHD, dihydrodiol dehydrogenase; COMT, catechol O-methyltransferase; NA, N-actetyl transferase; AA, amino acid acyltransferase; β-O, β-oxidation.

entities (NCE) and sampling from blood, urine, bile, feces, or a combination of these matrices can lead to a great deal of information on the rate at which a drug (metabolite) is eliminated (e.g., pharmacokinetics), whether the drug gets to the target organ, by which excretory route the drug is eliminated, and if the drug is metabolized to any great extent. With regard to metabolism, most of the investigations deal with routes of excretion, extent of metabolism, and the amount of drug (metabolites) found circulating in plasma or excreted as well as the identification of metabolites found in the various matrices. Today, all of these parameters involve the use of mass spectrometry in some form, either through quantitation or characterization of parent drug and metabolites.

## C. Mass Spectrometry

The need for rapid, sensitive methodologies for the identification and quantitation of drugs and metabolites to accelerate drug discovery and development has given atmospheric pressure ionization liquid chromatography mass spectrometry (API LC/MS) its central position in the drug metabolism laboratory. There are many strategies and LC/MS techniques that are available which provide varying sensitivities and levels of structural information. This section attempts to orient the reader with respect to the most commonly used MS techniques. The instruments described below range from those that can provide an approximate molecular weight for a drug or metabolite to those that can provide absolute structural identification or high-resolution exact mass determination.

Mass spectrometry work for drug metabolism can be divided into two groups: quantitative and qualitative (or structural elucidation). The former attempts to quantify the amount of a metabolite formed during a time interval (producing a time/metabolite concentration curve). The latter attempts to identify the chemical structure of a metabolite. These efforts are substantially different and require different sets of tools and techniques. As described in Chapter 3, triple-quadrupole mass spectrometers are a premier tool for quantitative work, while better full-scan data can be obtained from ion traps and time-of-flight (TOF) instruments. Time-of-flight instruments are advantageous because they allow for exact mass determinations (to the third or fourth decimal place) of unknowns. Ion traps are useful because they can produce fragmentation and structural information through a facile implementation of MS/MS. Quadrupole time-of-flight instruments offer a combination of these features, including the capability of doing MS/MS and the higher resolution of TOF.

Although APCI can accommodate higher flow rates (up to approximately 2 mL/min), the source is usually operated at higher temperatures than an electrospray ionization source, which can lead to thermal degradation of the sample. This is especially problematic with the detection of fragile glucuronides, N-oxides, and sulfates. Either ion source can accommodate a wide range of HPLC mobile-phase components, so long as nonvolatile salts are avoided (Chap. 5). In developing mass spectrometric methods to accommodate large numbers of structurally diverse molecules it is useful to utilize the most generic LC conditions at the source of the mass spectrometer to allow the formation of ions for mass spectral characterization.

## III. METABOLIC REACTIONS

Metabolic reactions are categorized into two types of biotransformation, Phase I or Phase II reactions. Phase I reactions are considered functionalization reactions that introduce or uncover functional groups with increased polarity or nucleophilicity. Phase I reactions generally involve oxidation, reduction, and hydrolytic reactions. In contrast, Phase II reactions are considered conjugative reactions whereby endogenous cofactors are activated and covalently bound to the parent drug or metabolites, increasing their ability to be excreted. The following discussion includes multiple examples of drugs and organic compounds that undergo Phase I and/or Phase II reactions. Additional information on these and other metabolic reactions can be found in several excellent articles and book chapters [3–5,10,11]. It should be noted that metabolites are generally formed after multiple reactions that work in concert to produce the final metabolite; therefore one must expect to use several combinations of these reactions to yield the final product. Table 4 lists some of the most common oxidative reactions with a few examples of mass changes expected from these biotransformations for both Phase I and II reactions.

### A. Phase I or Functionalization Reactions

### 1. Oxidation Reactions

Of the Phase I reactions, oxidative biotransformations are by far the most common. These reactions are carried out by several oxidative enzyme systems, the most predominant of which is the CYP superfamily of enzymes. Additional oxidative enzymes include FMO, xanthine oxidase, aldehyde oxidase, alcohol and aldehyde dehydrogenases monoamine oxidases, and various peroxidases. Determining the enzyme(s) employed to biotransform any particular substrate will depend on the substrates chemical and physical characteristics as well as functional substituents. This chapter does not describe in detail the mechanism of these various enzymes; however, it does illustrate the product(s) (i.e., metabolites) produced by each reaction.

Many of the oxidative reactions involve the incorporation of an oxygen atom into the product, replacement of another atom with oxygen, or a change in

**Table 4**  Various Forms of
Biotransformation Reactions and the
Associated Mass Changes

| Metabolic reaction | Change in mass[a] |
|---|---|
| Carbon hydroxylation | +16 |
| Epoxidation | +16 |
| N-Hydroxylation | +16 |
| N-Oxidation | +16 |
| Phosphorus oxidation | +16/17[b] |
| Sulfur oxidation | +16/32[c] |
| Oxidative desulfuration | −32 |
| Glucuronidation | −176 |
| Sulfation | +80 |
| Methylation | +14 |
| Acetylation | +42 |
| Amino acid conjugation | |
|   Glycine | +57 |
|   Glutamine | +145 |
|   Taurine | +107 |

[a] +, addition of mass; −, loss of mass.
[b] Phosphorus oxidation and hydroxylation, respectively.
[c] Sulfoxide and sulfone, respectively.

the oxidative state of the substrate. Aliphatic or aromatic oxidations; epoxidation; and N-, S-, P-oxidations are examples of oxygen atom incorporation (Figs. 3–6). Examples of carbon atom oxidation (aliphatic, aromatic, olefinic) are illustrated in Fig. 3. The saturated fatty acid lauric acid (dodecanoic acid), used in soaps and cosmetics, undergoes aliphatic oxidation to both the 11- and 12-hydroxy metabolites (ω-1 and ω-positions, respectively) as the primary metabolites [12]. Amphetamine [a central nervous system (CNS) stimulant] undergoes both p-hydroxylation (aromatic hydroxylation) and oxidative deamination (described below) as illustrated in Fig. 3 [13,14]. Epoxides will generally react with epoxide hydrolase to form the corresponding dihydrodiol; however, there are some cases where the epoxide is stable. In some instances, the epoxide can react with other nucleophiles, such as glutathione, or, in the extreme cases, other macromolecules, resulting in toxicity. The commonly prescribed anticonvulsant carbamazepine is unique in that it forms a stable epoxide metabolite due to stabilization of the two adjacent aromatic rings; however, the epoxide does undergo further metabolism to the *trans*-dihydrodiol metabolite [15,16].

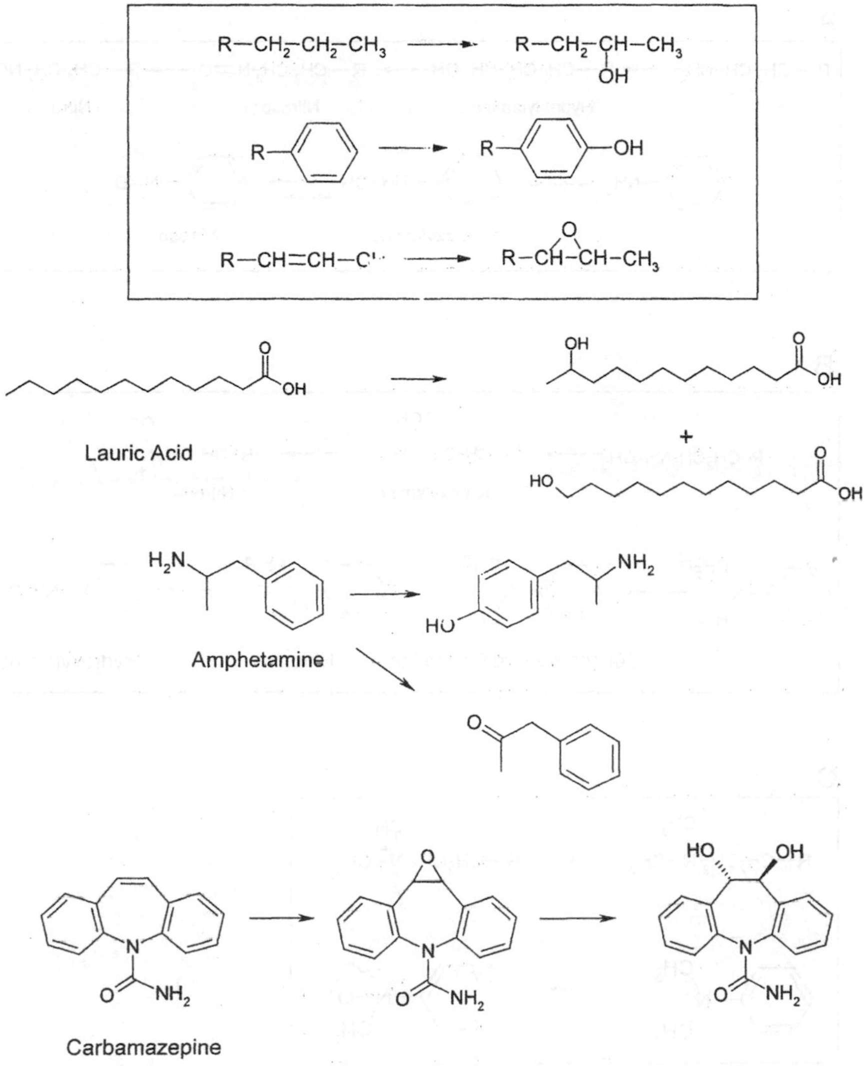

**Figure 3** Examples of carbon atom oxidation: aliphatic (lauric acid), aromatic (amphetamine), and olefinic (carbamazepine).

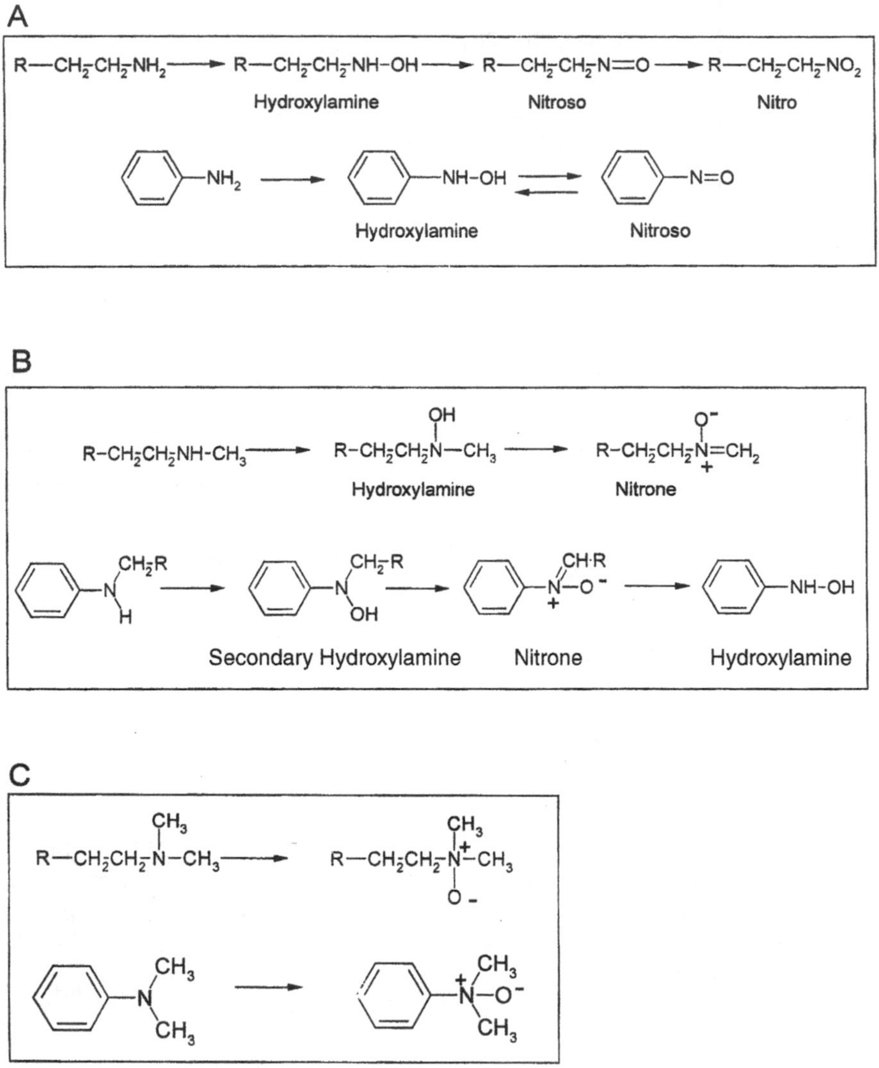

**A**

R—CH$_2$CH$_2$NH$_2$ $\longrightarrow$ R—CH$_2$CH$_2$NH–OH $\longrightarrow$ R—CH$_2$CH$_2$N=O $\longrightarrow$ R—CH$_2$CH$_2$NO$_2$

Hydroxylamine                Nitroso                Nitro

Ph—NH$_2$ $\longrightarrow$ Ph—NH–OH $\rightleftarrows$ Ph—N=O

Hydroxylamine                Nitroso

**B**

R–CH$_2$CH$_2$NH–CH$_3$ $\longrightarrow$ R–CH$_2$CH$_2$N(OH)–CH$_3$ $\longrightarrow$ R–CH$_2$CH$_2$N$^+$(O$^-$)=CH$_2$

Hydroxylamine                Nitrone

Ph–N(H)–CH$_2$R $\longrightarrow$ Ph–N(OH)–CH$_2$R $\longrightarrow$ Ph–N$^+$(O$^-$)=CH·R $\longrightarrow$ Ph–NH–OH

Secondary Hydroxylamine          Nitrone                Hydroxylamine

**C**

R—CH$_2$CH$_2$N(CH$_3$)–CH$_3$ $\longrightarrow$ R—CH$_2$CH$_2$N$^+$(CH$_3$)(O$^-$)–CH$_3$

Ph–N(CH$_3$)–CH$_3$ $\longrightarrow$ Ph–N$^+$(CH$_3$)(O$^-$)–CH$_3$

**Figure 4** Representative illustrations of primary amine oxidation (A), secondary amine oxidation (B), tertiary amine oxidation (C), and xenobiotics that undergo N-hydroxylation or oxidation reactions (D).

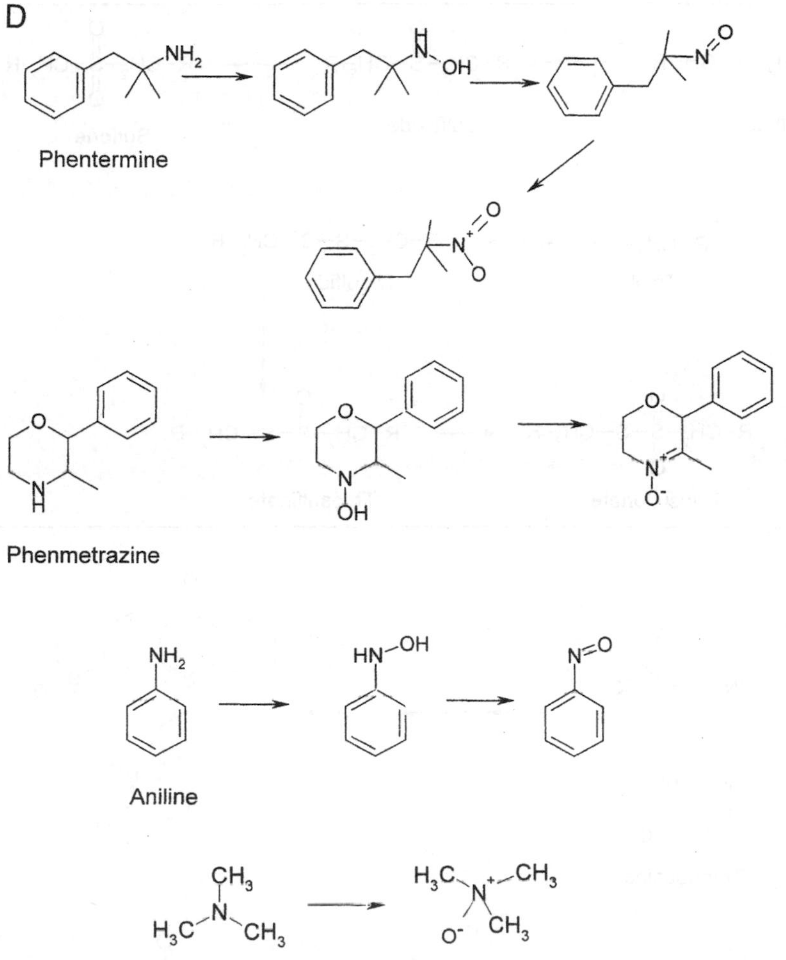

D

Phentermine

Phenmetrazine

Aniline

**Figure 4** Continued

Representative and actual examples of nitrogen, sulfur, and phosphorus oxidations are illustrated in Figs. 4A–4D, 5, and 6, respectively. Figures 4A–4C are representative reactions, while Fig. 4D illustrates actual examples of both aliphatic and aromatic nitrogen oxidation reactions. The oxidation of primary amines and in some cases tertiary amines is somewhat common, whereas the oxidation of secondary amines is found less frequently. More common are the oxidative dealkylation reactions of amines that are discussed below. Phentermine is an example of a primary amine that undergoes oxidation to the hydroxylamine,

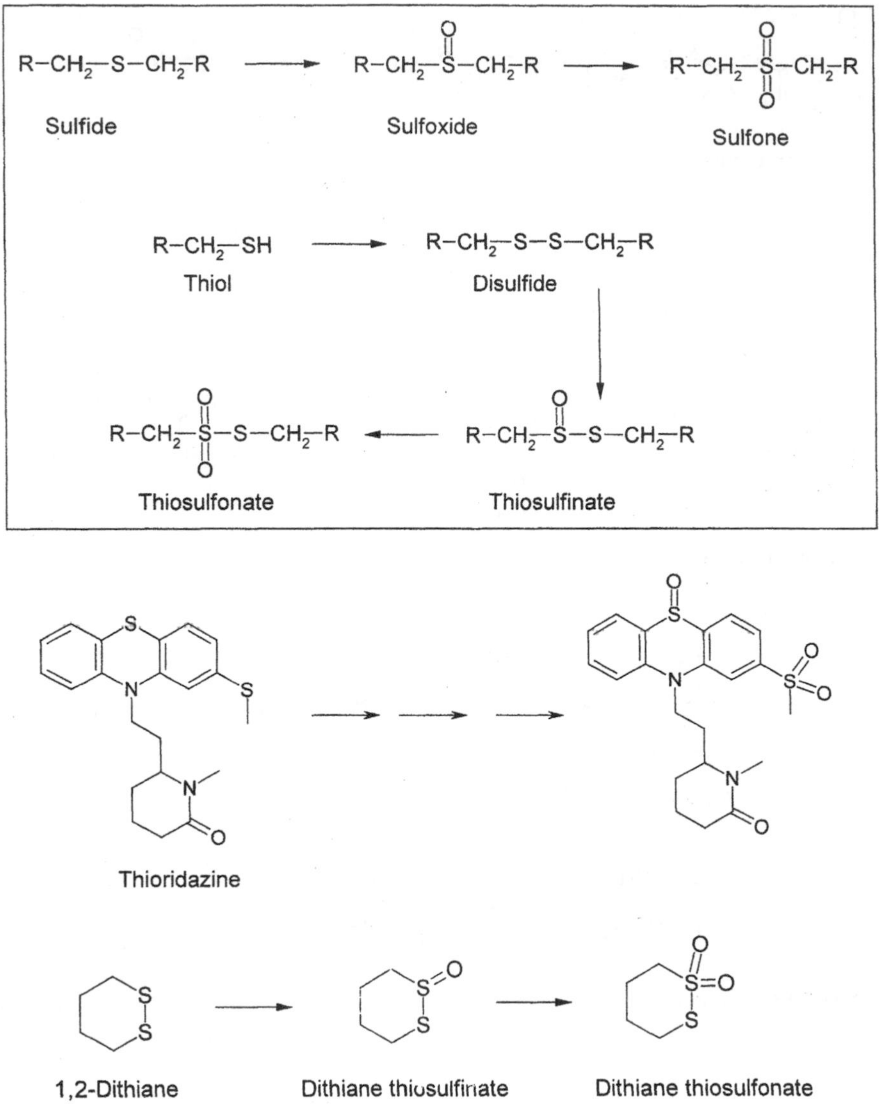

**Figure 5**  Various forms of sulfur oxidation: sulfoxide/sulfone and disulfide formation.

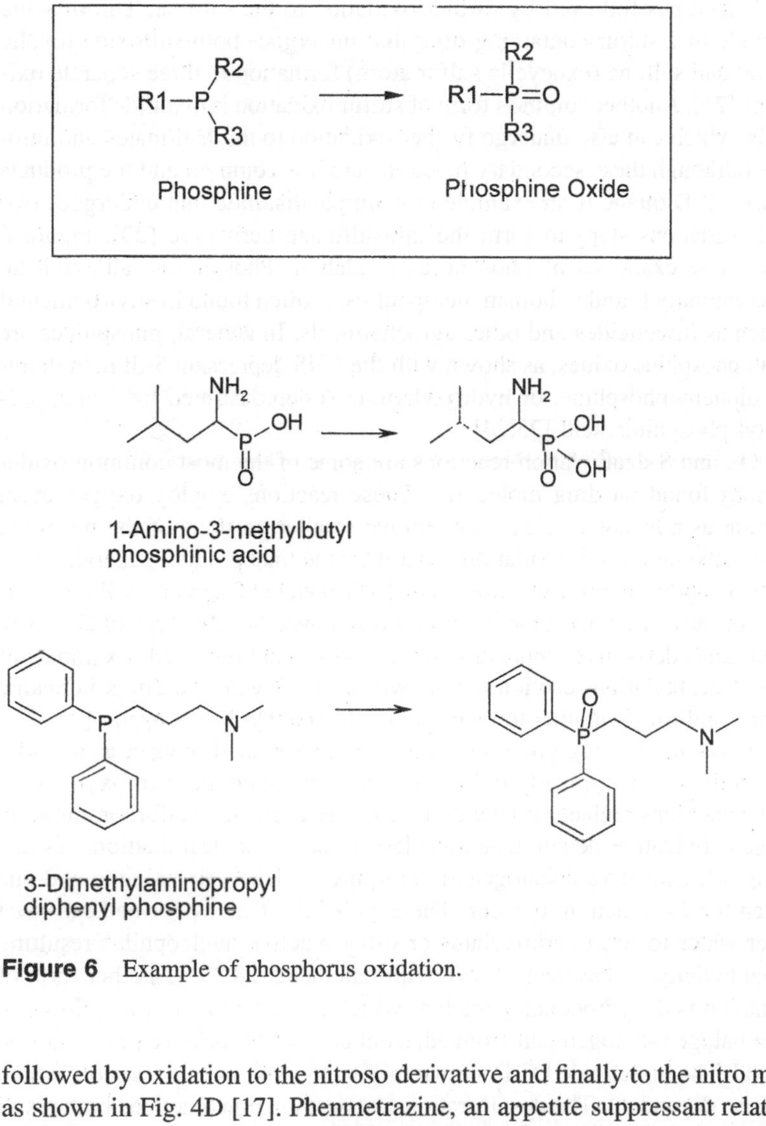

1-Amino-3-methylbutyl
phosphinic acid

3-Dimethylaminopropyl
diphenyl phosphine

**Figure 6** Example of phosphorus oxidation.

followed by oxidation to the nitroso derivative and finally to the nitro metabolite, as shown in Fig. 4D [17]. Phenmetrazine, an appetite suppressant related to amphetamine, is a secondary aliphatic amine that is oxidized to the secondary hydroxylamine and further to the nitrone metabolite [18]. Simple examples, such as aniline and the endogenous substrate trimethylamine, also undergo nitrogen oxidation to nitroso and *N*-oxide metabolites, respectively [19,20].

Oxidations of sulfur and phosphorus occur readily when present in xenobiotics. Figure 5 shows several examples of sulfur-containing compounds that undergo oxidation. The most common form of sulfur oxidation is formation of a sufoxide, sometimes followed by further oxidation to the sulfone. Thioridazine is an example of a sulfur-containing drug that undergoes both sulfoxide (cyclic sulfur atom) and sulfone (exocyclic sulfur atom) formation in three separate oxidation steps [21]. Another common form of sulfur oxidation is disulfide formation from thiols, which can also undergo further oxidation to thiolsulfinates and thiosulfonates (although these secondary reactions are less common and the products less stable). 1,2-Dithiane is an example of a simple disulfide that undergoes two sequential oxidations steps to form the thiosulfonate derivative [22]. Figure 6 illustrates a few examples of phosphorus oxidation. Phosphorus, although an atom not commonly found in human therapeutics, is often found in environmental agents, such as insecticides and other agrochemicals. In general, phosphines are oxidized to phosphine oxides, as shown with the CNS depressant 3-dimethylaminopropyl diphenylphosphine, or hydroxylation, as demonstrated by 1-amino-3-methylbutyl phosphinic acid [23,24].

N-, O-, and S-dealkylation reactions are some of the most common oxidative reactions found on drug molecules. These reactions employ oxygen atom incorporation as a means to cleave or remove smaller portions of the molecule (N-, O-, S-dealkylation or β-oxidation); therefore the final product generally does not contain an additional oxygen atom. The hydroxylated fragment will be formaldehyde (formic acid) when a methyl group is removed or some type of aldehyde (carboxylic acid) derivative, depending on the substituent removed. Examples of each type of dealkylation reaction are shown in Fig. 7 with the drugs lidocaine and codeine and the fungicide tetrachloro-1,4-bis(methylthio)benzene [25–27]. Note again how the leaving group contains the incorporated oxygen as an aldehyde metabolite that is typically oxidized to the corresponding carboxylic acid.

Some reactions replace another atom, such as a halogen, sulfur, or nitrogen, with oxygen (oxidative dehalogenation, desulfuration, or deamination, respectively) (Fig. 8). Oxidative dehalogenation requires at least one halogen and one α-hydrogen for this reaction to occur. The acyl halides that are formed can react with either water to form carboxylates or other reactive nucleophiles resulting in potential toxicity, as has been shown with halothane [28,29]. Another form of dehalogenation is dehydrodehalogenation, which is the removal of a hydrogen–halogen or halogen–halogen pair from adjacent carbons to form a carbon–carbon double bond. The insecticide DTT, shown in Fig. 18, typifies this reaction, which is glutathione dependent. The desulfuration and deamination reactions both result in loss of either sulfur or nitrogen and direct replacement with oxygen. The oxidative deamination of amphetamine was shown in Fig. 3. Thiopental (Pentothal) is an intravenous anesthetic drug that undergoes oxidative desulfuration ($C=S$) to form another common intravenous anesthetic drug pentobarbital (i.e., active metabolite) [30]. The insecticide parathione also undergoes oxidative desulfuration ($P=S$) to form a phosphate ester [31].

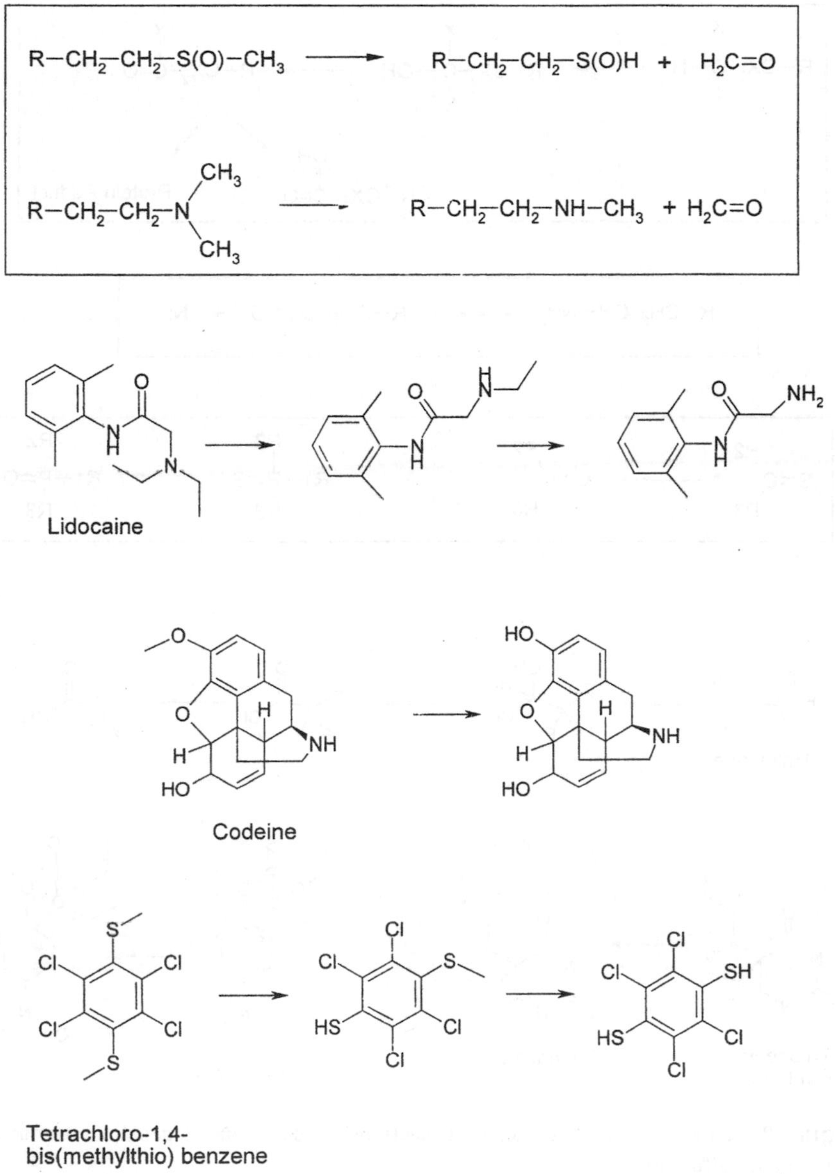

$$R-CH_2-CH_2-S(O)-CH_3 \longrightarrow R-CH_2-CH_2-S(O)H \; + \; H_2C=O$$

$$R-CH_2-CH_2-N \overset{CH_3}{\underset{CH_3}{\big\langle}} \longrightarrow R-CH_2-CH_2-NH-CH_3 \; + \; H_2C=O$$

Lidocaine

Codeine

Tetrachloro-1,4-
bis(methylthio) benzene

**Figure 7** Examples of hetereoatom (N—, O—, and S—) dealkylation reactions.

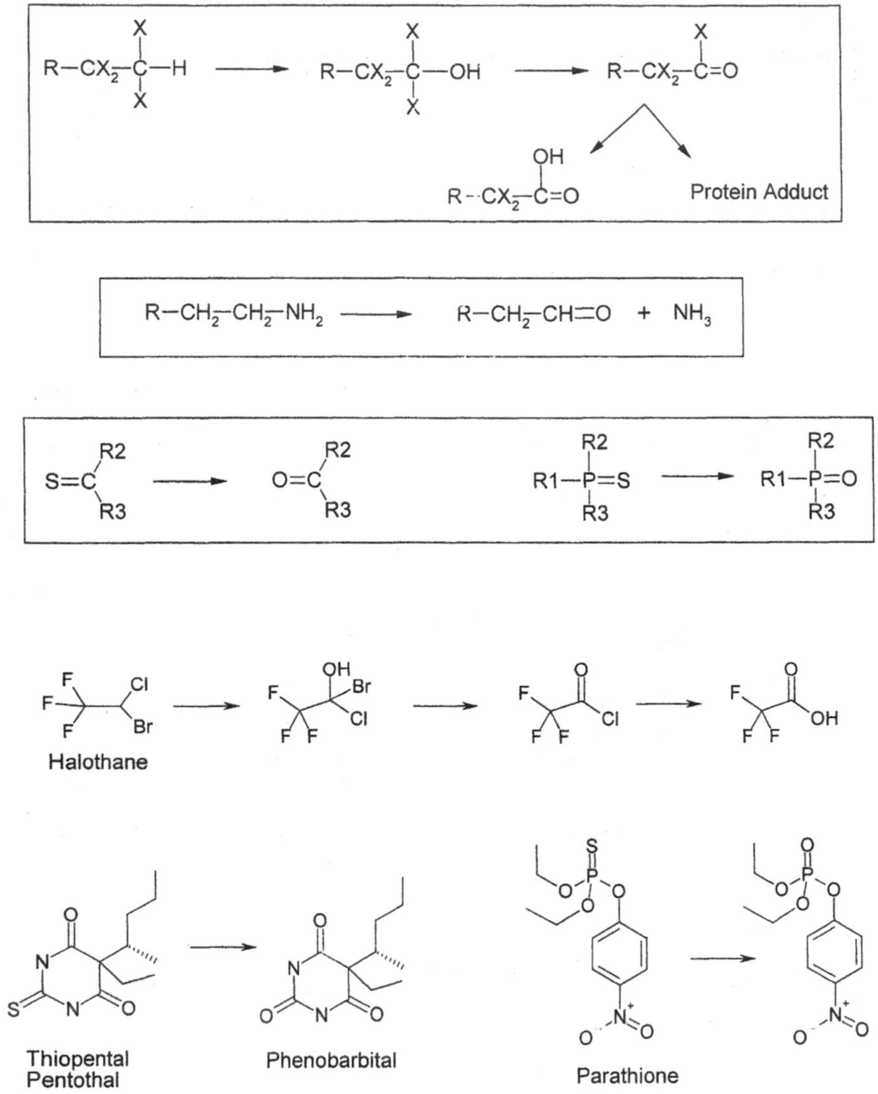

**Figure 8**  Various forms of replacement reactions: oxidative dehalogenation, deamination, and desulfuration.

Still, there are some reactions that merely change the oxidative state of the substrate, such as alcohol and aldehyde dehydrogenase catalyzed conversion of substrate to aldehydes and carboxylic acids, respectively. Figure 9A shows the terminal hydroxylation of an alkyl group, followed by oxidation to the aldehyde and subsequently to the carboxylic acid. This reaction is also observed with $\omega$-hydroxy lauric acid (Fig. 3). Interestingly, these terminal carboxylic acid group can be subjected to another form of oxidation (or hydrolysis) known as $\beta$-oxidation. $\beta$-Oxidation involves one or more sequential cycles of oxidative metabolism (hydrolysis) on an alkyl side chain or long-chain fatty acid. Starting from an alkyl side chain, oxidation occurs at the terminal ($\omega$-position), as opposed to the thermodynamically more stable $\omega$-1, position (Fig. 9A). Oxidation of the alcohol to the carboxylic acid gives the substrate for $\beta$-oxidation. Each step in the oxidation process begins with the formation of an acyl coenzyme A derivative followed by oxidation of the $\beta$-carbon. Subsequent steps result in the release of two carbon units as acetyl-CoA, leaving behind a fatty acid CoA derivative. After the alkyl chain cannot be shortened any further, the final product is released as a terminal carboxylate. In general, $\beta$-oxidation results in the release of even-numbered carbon units (two at a time); however, there are some instances where odd numbers of carbon units have been removed. Figure 9B illustrates an example of carbon oxidation and $\beta$-oxidation of the cholesterol lowing agent CI-976 [32].

## 2.  Hydrolytic/Cleavage Reactions

There are several different types of hydrolytic reactions and enzymes: carboxylesterases, epoxide hydrolases, drug conjugate cleavage enzymes (glucuronidase, sulfatase, or phosphatase), and peptidases. The process of $\beta$-oxidation described previously could also be considered a cleavage reaction. These reactions generally produce a metabolite that is more polar and susceptible to conjugation by Phase II enzymes.

The carboxylesterases are enzymes that commonly hydrolyze carboxylic acid esters as well as amide, thio-, and phosphoric acid esters [33]. These enzymes are found in many tissues and fluids, such as liver, kidney, intestine, and plasma. There are several other functional groups that are susceptible to hydrolytic cleavage, such as the $\beta$-lactam ring of penicillin derivatives (cyclic amide) [34], carbamates, hydrazides, and urea-type compounds [35] (Fig. 10A). In general, these cleavage reactions result in the splitting of the molecule (except for cyclic compounds) into two separate components: a carboxylic acid and the reduced form of the heteroatom (nitrogen, oxygen, or sulfur in the remaining portion). The name carboxylesterase is outdated and not truly appropriate, as it encompasses much more than carboxylic acid esters; however, this is the term one encounters in much of the literature. Figures 10A and 10B illustrate some general examples

**A**

**CI-976**

**B**

and actual metabolic reactions involving carboxyesterase enzymes, respectively [35–40].

Another important group of hydrolytic enzymes are the epoxide hydrolases, also known as epoxide hydrase or epoxide hydratase, most commonly found in liver tissue. Epoxide hydrolases catalyze the hydration of arene oxides and aliphatic epoxides to their corresponding *trans*-dihydrodiols or diols, respectively, by activating a water molecule to attack one of the carbons of the arene oxide or epoxide [41]. Although one of the major metabolites of carbamazepine is the stable epoxide, this metabolite also undergoes hydrolysis to form the *trans*-diol metabolite (Fig. 3). Likewise, the anticonvulsants phenytoin and mephenytoin form arene oxides, which then form *trans*-dihydrodiol that undergo further oxidation to form the catechols by the enzyme dihydrodiol dehydrogenase (Fig. 11) [42,43].

Also considered hydrolytic reactions are the cleavage of glucuronide, sulfate, and phosphate conjugates by β-glucuronidase, sulfatase, and phosphatase enzymes, respectively. These enzymes are found predominately in the intestinal microflora of animals and humans and to some extent in the mammalian host. These enzymes can cleave newly formed conjugates back to the parent drug or Phase I metabolite.

The peptidases are not as common in the metabolism of xenobiotics; however, with the advent of protein expression and molecular biology, we are starting to see an increased prevalence of proteins as drugs. The peptidases are categorized as aminopeptidases, carboxypeptidases, or endopeptidases and catalyze the cleavage of the amide linkage between two amino acids at either the amino terminus, carboxy terminus, or internally, respectively.

## 3. Reduction Reactions

Given the oxidative cellular environment of higher animals, the reduction of certain functional groups is not as common as oxidative, hydrolytic, or conjugative reactions; however, the reduction of drugs and metabolites has been shown to occur. It should be noted that intestinal microorganisms have a greater propensity to reduce xenobiotics, therefore bacteria found in the intestine can actually form a significant amount of the reduced metabolites eliminated in the feces of mammals. Reduction of ketones, nitro and azo compounds, $N$- and $S$-oxides, disulfides, quinones, and alkenes ($\alpha,\beta$-unsaturated carbonyls) as well as reductive dehalogenation have all been noted in the literature. Some of the most common reactions

**Figure 9** β-Oxidation reaction beginning with initial alkane hydroxylation (A) and the hydroxylation and β-oxidation of CI-976 (B).

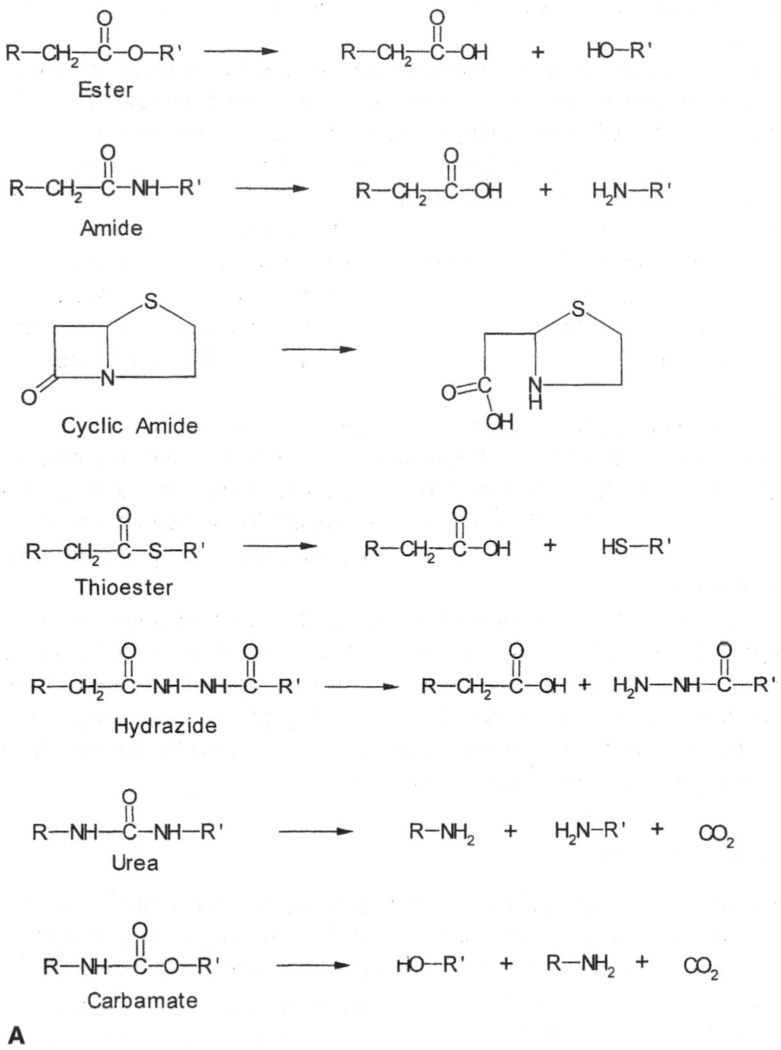

**Figure 10** General examples (A) and actual examples (B) of hydrolytic/cleavage reactions by carboxyesterase enzymes.

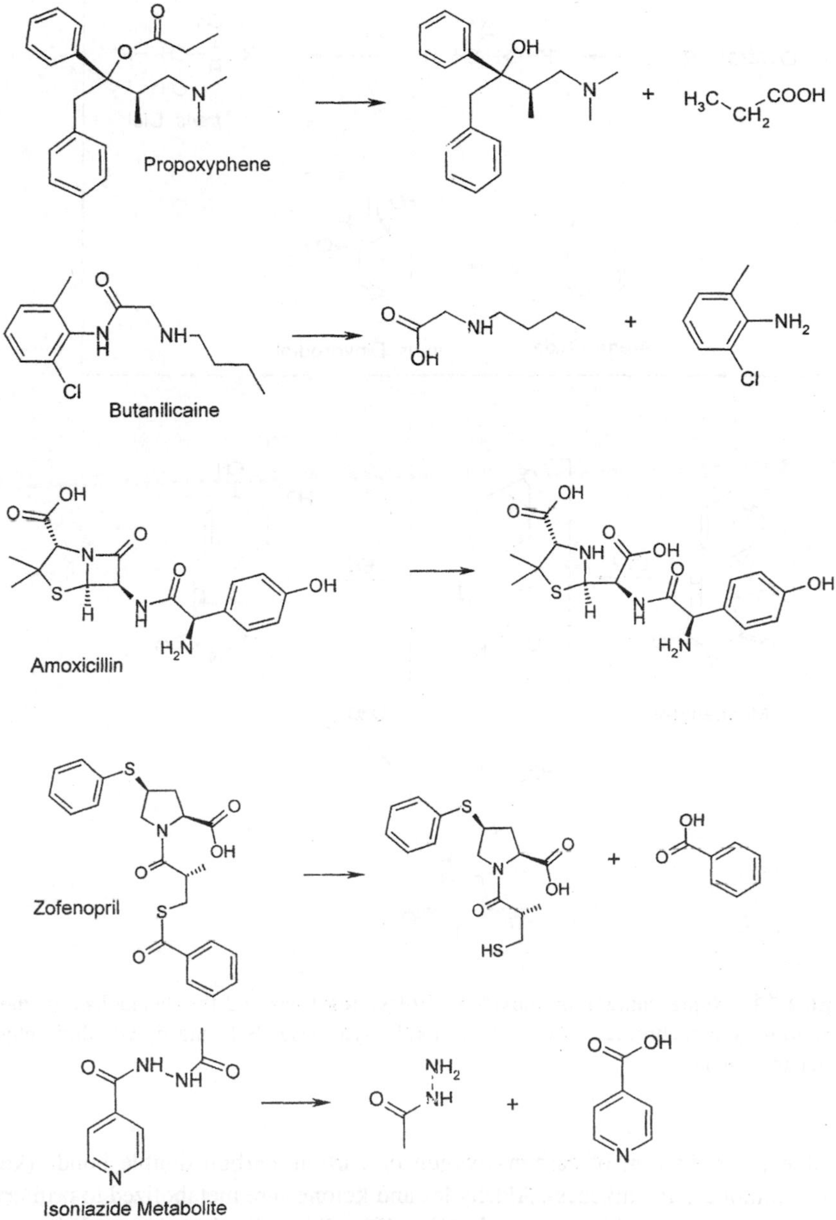

Propoxyphene

Butanilicaine

Amoxicillin

Zofenopril

Isoniazide Metabolite

**B**

**Figure 10** Continued

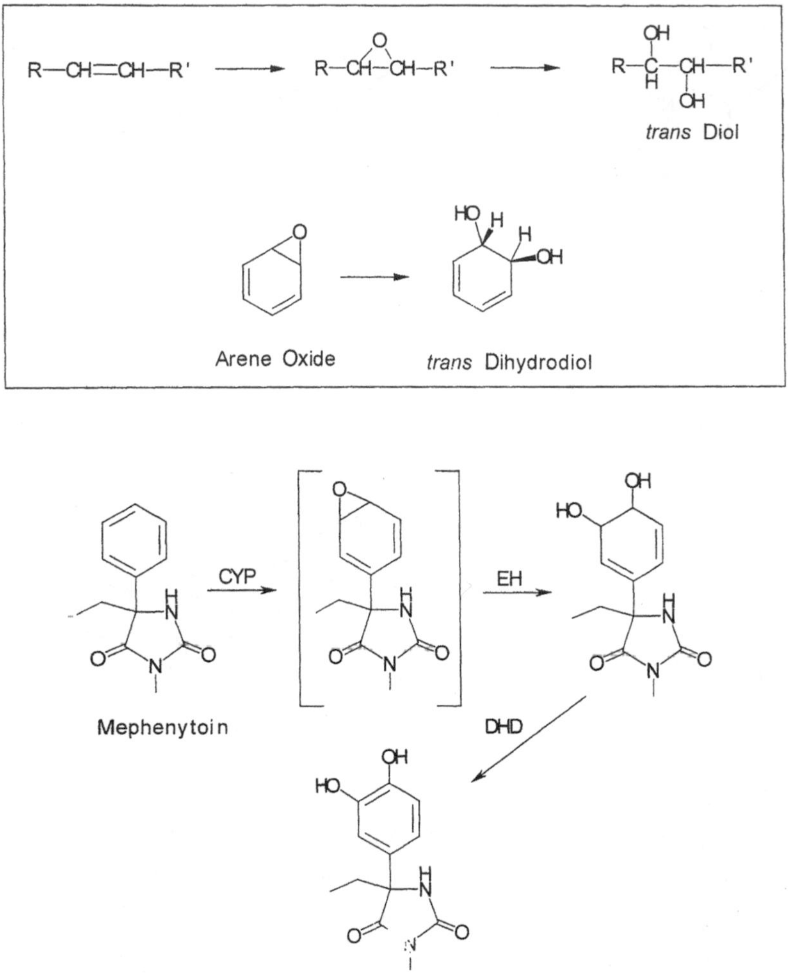

**Figure 11** Representation of epoxide hydrolase reactions and the metabolism of mephenytoin by cytochrome P450 (CYP), epoxide hydrolase (EH) and dihydrodiol dehydrogenase (DHD).

involve the reduction of carbon–oxygen or carbon–carbon double bonds (ketones, quinones, and alkenes). Aldehydes and ketones are metabolized to primary
and secondary alcohols, respectively (Fig. 12). Chloral hydrate is a well-known
example of a compound that initially forms an aldehyde which is then reduced
to the primary alcohol [44]. The reduction of ketones, such as metyrapone, often
leads to the generation of a chiral center and two secondary alcohol enantiomers,

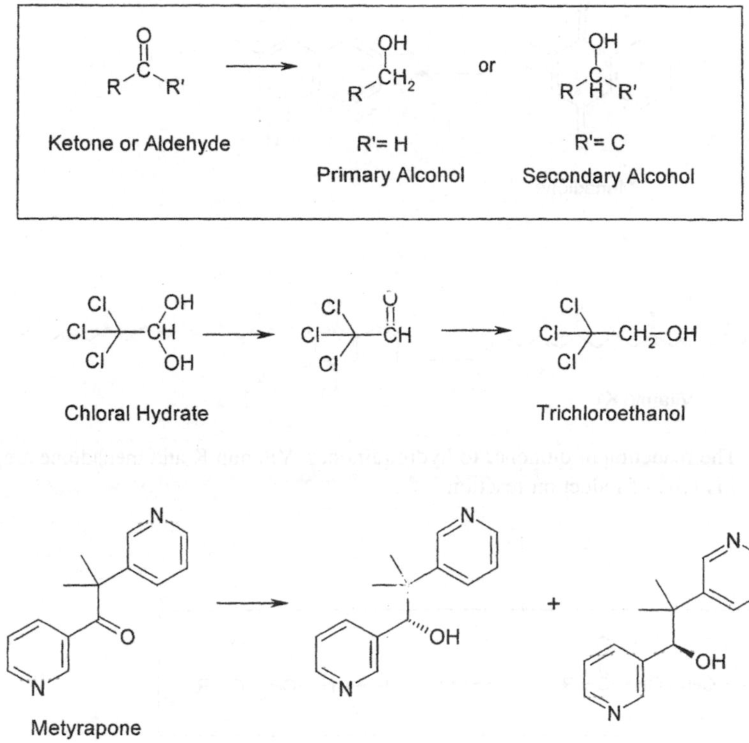

Chloral Hydrate Trichloroethanol

Metyrapone

**Figure 12** The reduction of aldehydes and ketones to primary and secondary alcohols, respectively.

as illustrated in Fig. 12. Typically, one enantiomer is preferrentially formed over the other; however, both can generally be found [45]. Quinones, such as those found in menadione and vitamin K, can be reduced to the corresponding hydroquinones by an enzyme called NADPH quinone reductase (or DT-diaphorase) (Fig. 13) [46,47]. Due to the ease of oxidation and reduction of hydroquinones and quinones, they can lead to a cyclic generation of reactive oxygen species, such as superoxide anion, hydrogen peroxide, and hydroxyl radical, potentially causing toxicity.

The reductions of alkenes are a special case and are found associated with α,β-unsaturated carbonyls and not with isolated double bonds. Both norethindrone and 5-fluorouracil have been shown to undergo carbon–carbon double-bond reduction [48–49], as shown in Fig. 14.

Generalized reductions of nitrogen-containing groups are illustrated in Fig. 15. These include the reduction of aromatic nitro and azo compounds as well as

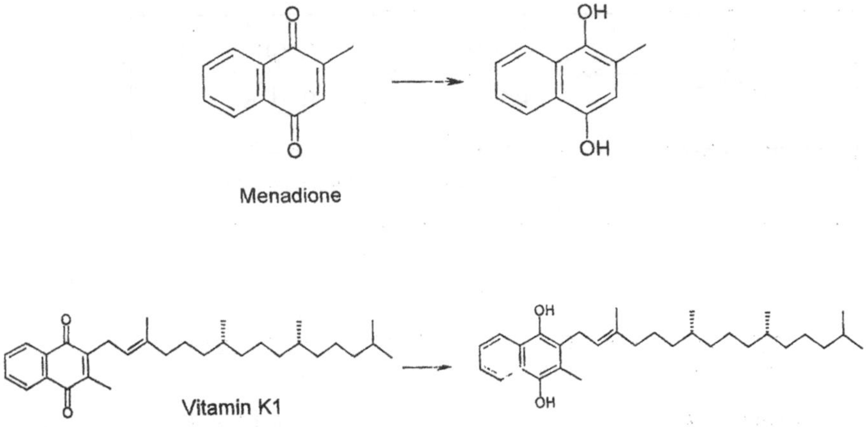

Menadione

Vitamin K1

**Figure 13** The reduction of quinones to hydroquinones. Vitamin K and menadione are examples of this type of reduction reaction.

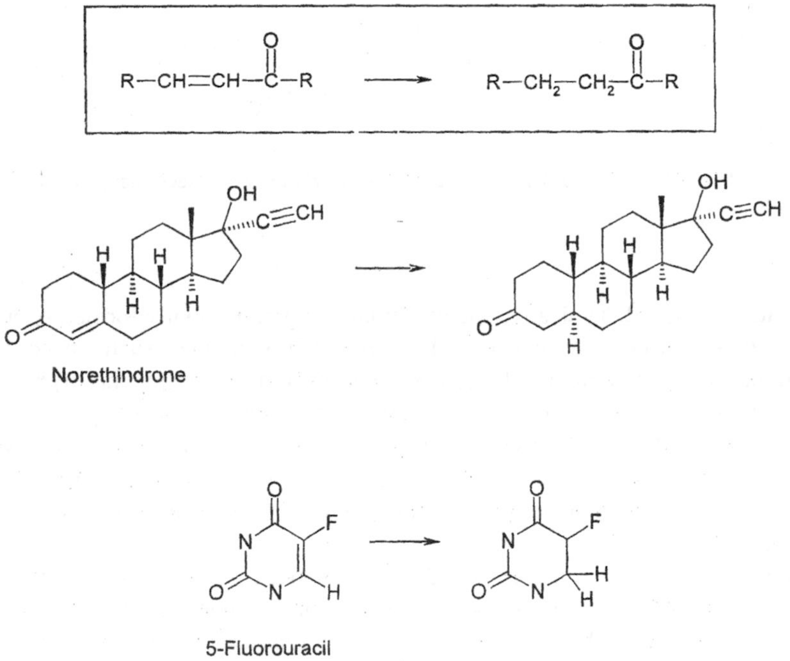

Norethindrone

5-Fluorouracil

**Figure 14** The reduction of alkenes (α,β-unsaturated carbonyls) as illustrated with norethindrone and 5-fluorouracil.

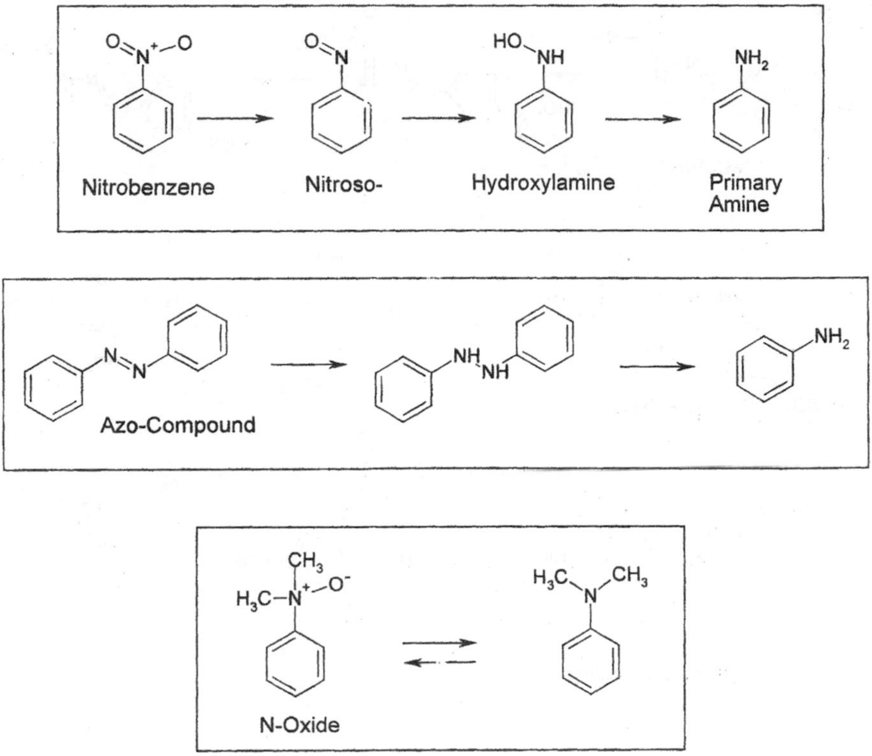

**Figure 15** Representative examples of nitrogen reductions (nitro-, azo-, and *N*-oxides).

the reduction of *N*-oxides to their corresponding primary and tertiary amines, respectively. The reduction of aromatic nitro compounds begins with a two-electron reduction to form the nitroso derivative, followed by two additional two-electron reductions to form the hydroxyl amine and primary amine, respectively [50,51]. Compounds do not necessarily need to proceed all the way to the primary amine; some compounds such as nitrofuran derivatives generally terminate at the hydroxylamine step. Nitrobenzene and the antimicrobial nitrofural are examples of compounds that undergo nitro reduction (Fig. 15 and 16A) [50–53].

The reduction of aromatic azo compounds proceeds through a hydrazino intermediate that is cleaved to give two primary amines (Fig. 15). The best cited illustration of a drug-related azo reduction is found with the prodrug prontosil, which undergoes azo reduction to form the active sulfonamide antibiotic sulfanilamide (Fig. 16B) [54,55]. The urinary analgesic phenazopyridine is another compound that undergoes azo reduction to form both 2,3,6-triaminopyridine and ani-

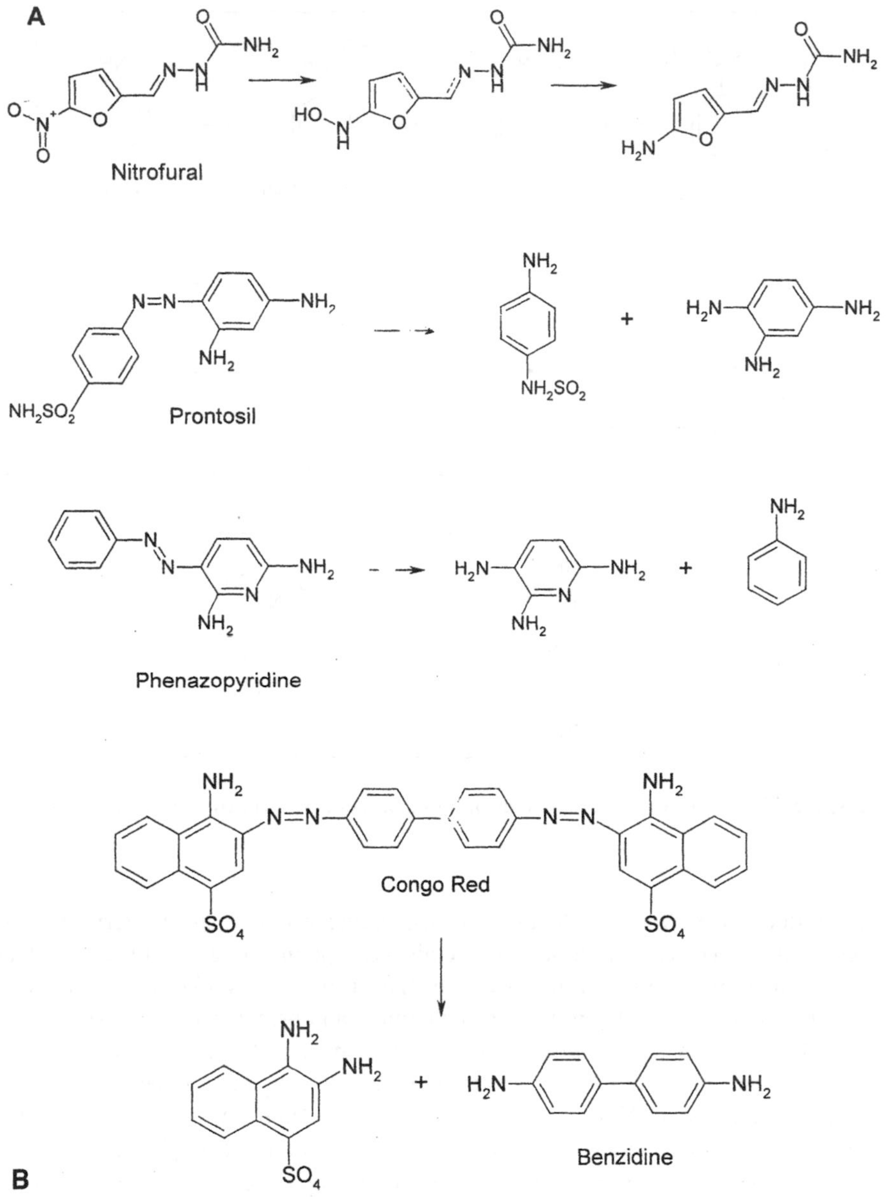

**Figure 16** Nitro reduction of nitrofual (A), azo reductions of prontosil, phenazopyridine and Congo red (B), and the reduction of chlopromazine and nicotine *N*-oxides (C).

line (Fig. 16B) [56]. Another class of azo-containing compounds that readily undergo reduction either by mammals or intestinal bacteria are the azo dyes, such as Congo red, which is metabolized to the carcinogen benzidine (Fig. 16B) [57].

The reduction of N-oxides in most cases represents a minor pathway of metabolism; however, this reaction can lead to the regeneration of a significant amount of parent drug that was initially formed by N-oxidation [see Fig. 15 for an example of this reversible reaction (N,N-dimethylanaline N-oxide)]. Therefore, if the original N-oxide is not eliminated from the body reasonably quickly and reduction has a chance to occur, the plasma concentration and the pharmacological effect of the drug may be extended. Chlorpromazine N-oxide and nicotine 1'-N-oxide are examples of N-oxides that are reduced to their respective tertiary amines [58,59], (Fig. 16C). Reductions of sulfur-containing xenobiotics are similar to those found with nitrogenous substrates. Disulfides and sulfoxides can both be reduced to the corresponding sulfides. Figure 17 illustrates several examples of sulfur-containing compounds that have been found to undergo sulfur reduction. The garlic constituent diallyl disulfide and disulfiram (Antabuse) are both reduced to their respective sulfides [60,61]. Also, the vasodilator flosequinan, which contains a sulfoxide substituent, is reduced to its sulfide metabolite (note: flosequinan is also oxidized to the sulfone metabolite) [62].

The last reductive pathway examined in this chapter is reductive dehalogenation, the replacement of a halogen with hydrogen. This is a common reaction

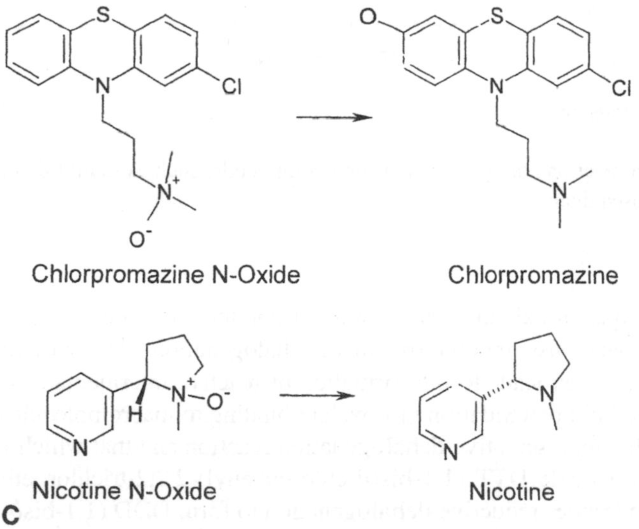

Chlorpromazine N-Oxide       Chlorpromazine

Nicotine N-Oxide       Nicotine

C

**Figure 16** Continued

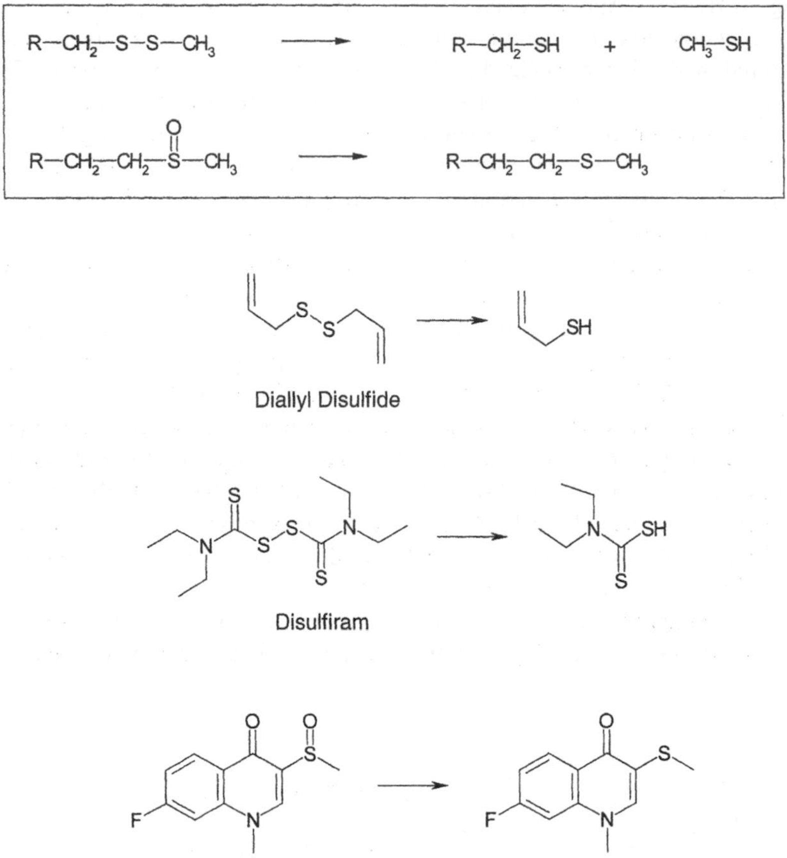

$$R-CH_2-S-S-CH_3 \longrightarrow R-CH_2-SH \quad + \quad CH_3-SH$$

$$R-CH_2-CH_2-\overset{O}{\underset{\|}{S}}-CH_3 \longrightarrow R-CH_2-CH_2-S-CH_3$$

Diallyl Disulfide

Disulfiram

Flosequinan

**Figure 17** Reduction reactions of sulfur-containing compounds, such as diallyl disulfide, disulfiram, and flosequinan.

with many different types of haloalkanes, such as carbon tetrachloride and halothane (previously shown to also undergo oxidative dehalogenation). This particular type of metabolism often leads to the formation of reactive intermediates or metabolites that cause lipid peroxidation or covalent binding to macromolecules. Figure 18 shows both a representative dehalogenation reaction and that which is observed with the insecticide DTT (1,1-bis(4-chlorophenyl)-2,2,2-trichloroethane) [63,64]. DDT undergoes reductive dehalogenation to form DDD (1,1-bis(4-chlorophenyl)-2,2-dichloroethane) as well as DDE (1,1-bis(4-chlorophenyl)-2,2-

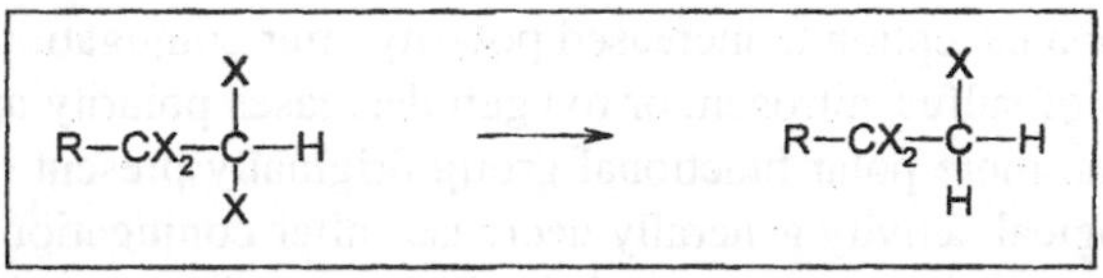

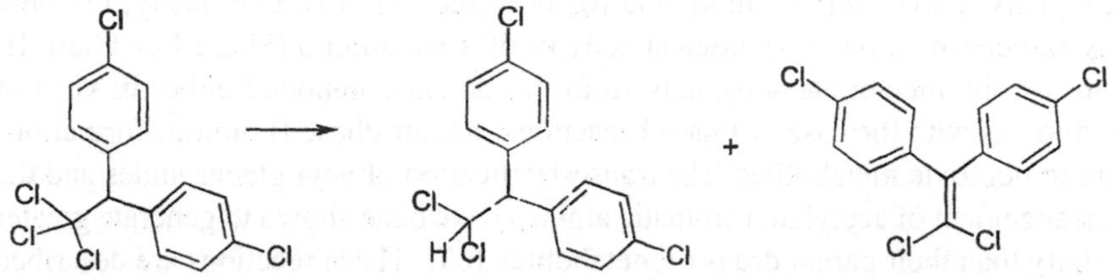

Figure 18 The reductive dehalogenation of the insecticide DDT.

dichloroethene), which has been shown to be mediated by a glutathione-dependent mechanism (discussed above under Sec. III.A.1).

## B. Phase II or Conjugative Reactions

Phase II or conjugative reactions include glucuronidation, sulfation, methylation, acetylation, and glutathione and amino acid conjugation. These conjugative reactions are typically preceded by one of the above-mentioned Phase I reactions; however, direct conjugation of a xenobiotic can occur if the appropriate functional group is present. Most Phase II reactions occur on heteroatoms, such as sulfur, nitrogen, oxygen, and in some instances direct conjugation can occur on carbon atoms should the conditions be favorable. For example, glutathione conjugation can occur on carbon atoms, and for highly acidic carbons, glucuronidation has been shown to occur. Some common functional groups where Phase II reactions occur would include primary amines, carboxylic acids, hydroxyl and thiol groups, epoxides (arene oxides), or highly reactive (electrophilic or nucleophilic) sites. All of these reactions are catalyzed by enzymes and require various cofactors, with the exception of glutathione conjugation of highly electrophilic sites that can occur independently of transferase enzyme.

In general, conjugation reactions increase hydrophilicity and mass, promoting the elimination of metabolites in bile or urine, and decrease pharmacological activity. A common exception to increased polarity after conjugation is methylation. Methylation of sulfur, nitrogen, or oxygen decreases polarity and excretion by covering up the more polar functional group originally present on the molecule. Pharmacological activity generally decreases after conjugation due to poor cellular penetration of the conjugate and the increased clearance that generally occurs with the more polar and larger metabolites. It should be noted that there are several exceptions to this general rule; for example, morphine 6-*O*-glucuronide is an equally active form of morphine [65,66] (see Fig. 19). Ultimately, the only way to determine pharmacological activity of a metabolite (Phase I or Phase II) is to test the metabolite separately from the parent compound either in vitro or in vivo. As with the case of Phase I reactions, not all Phase II biotransformations lead to nontoxic metabolites. The transesterification of acyl glucuronides and the rearrangement of acetylated aromatic amines have been shown to generate greater toxicity than their parent drugs or metabolites [67]. These reactions are described in greater detail under their respective sections.

The following discussion describes the types of reactions that are possible for the most common forms of conjugation. Whereas glucuronidation and sulfation are two of the most common forms of conjugation, greater emphasis is placed on these reactions.

## 1. Glucuronidation

Glucuronidation is the most prevalent form of conjugation due to the number of functional groups that can be glucuronidated, catalytic activity of the enzyme, and high concentration of cofactor (UDPGA). The aglycone (parent drug or metabolite prior to conjugation) is conjugated through an enzyme-mediated process that employs UDP-glucuronosyl transferases (microsomal enzymes) and UDPGA (uridine-5′-diphospho-α-D-glucuronic acid) (Fig. 20) [68]. The cofactor UDPGA contains a carboxylic acid (p$K_a$ 3–4) and multiple hydroxy groups, which, when combined, add significant polarity to any substrate. It should be noted that the configuration of the C1 carbon of UDPGA prior to the conjugation reaction is alpha and is converted to the β-configuration upon attack by the aglycone, therefore the products are known as β-D-glucuronides [69].

Glucuronidation occurs by nucleophilic attack of an electron-rich heteroatom, typically oxygen, nitrogen, sulfur, and sometimes carbon. Table 5 describes the common functional groups that participate in glucuronidation reactions and Figs. 19–21 illustrate several examples of glucuronidation reactions (valproic acid, aniline, and methimazole, [70–73]). Figure 21 illustrates two of the more unusual glucuronidation reactions: formation of quaternary ammonium-linked glucuronides and carbon-linked glucuronides. Quaternary-linked glucuronides have been isolated from the urine of humans and other primates. Historically, the formation of these glucuronides has been reported to be negligible or nonexis-

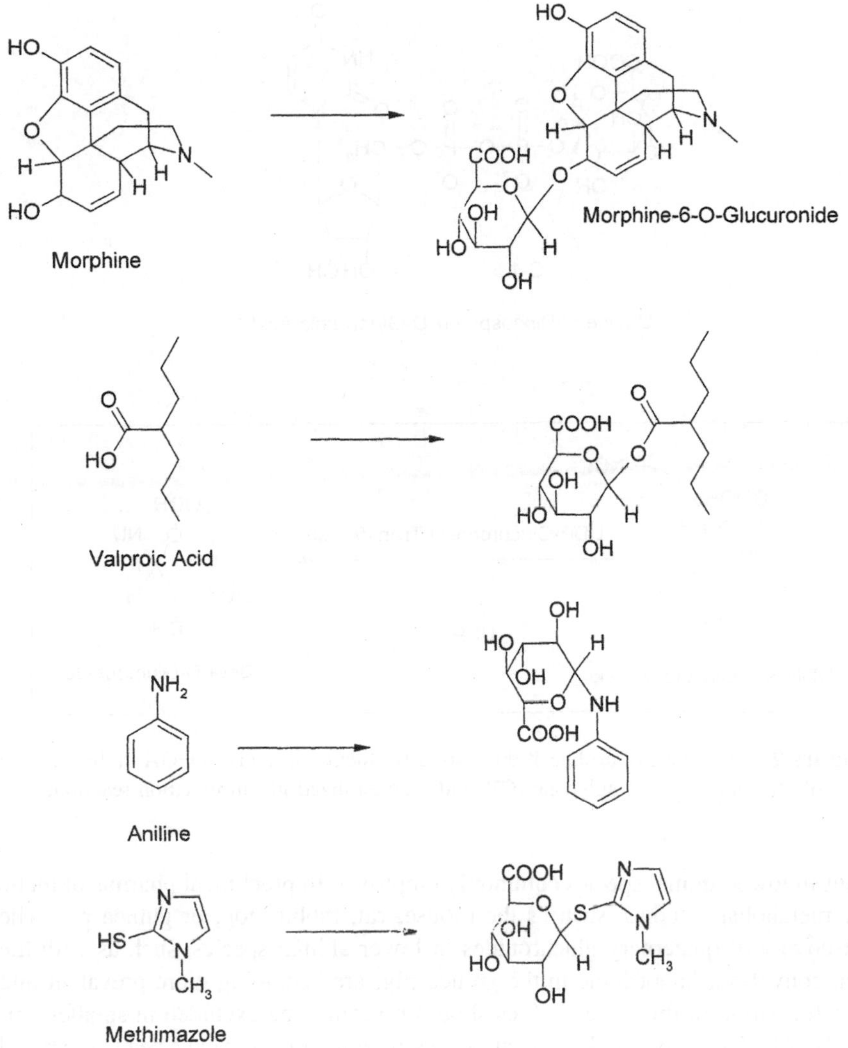

**Figure 19** Examples of glucuronidation reactions.

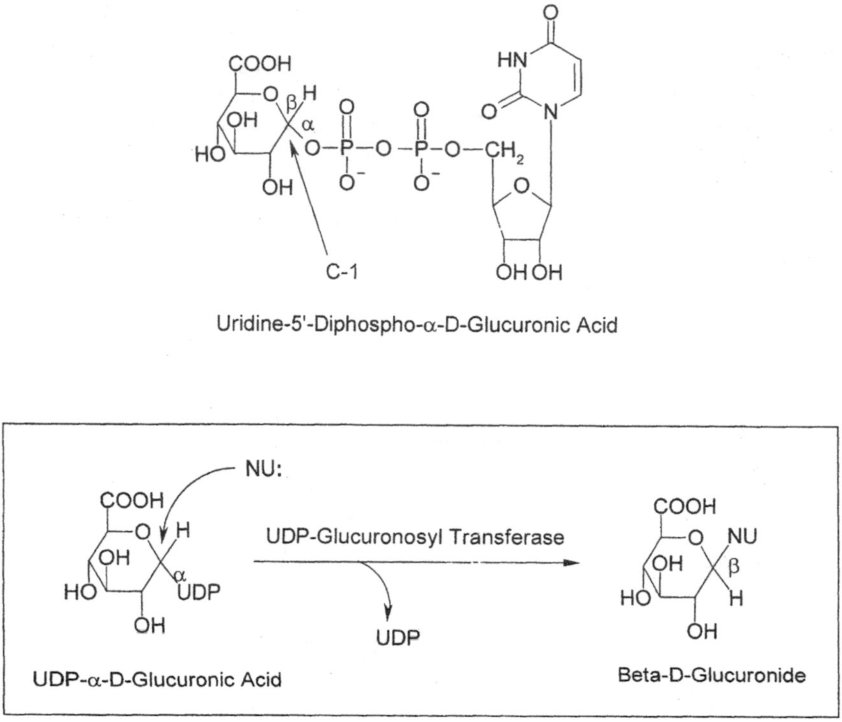

Uridine-5'-Diphospho-α-D-Glucuronic Acid

**Figure 20** Structure of uridine 5'-diphospho-D-glucuronic acid (UDPGA) indicating the site of attachment to the aglycone (C1) and a generalized glucuronidation reaction.

tent in lower animal species commonly employed in preclinical pharmacokinetic or metabolism studies, such as the mouse, rat, rabbit, dog, or guinea pig. The discovery of quaternary glucuronides in lower animal species, such as with the anticonvulsant lamotrigine in the guinea pig, are becoming more prevalent and the formation of these metabolites should no longer be excluded in smaller animals [74]. The carbon-linked glucuronides are quite rare and have been associated with 1,3-dicarbonyl compounds that contain highly electron-rich (highly acidic) carbons. Phenylbutazone and sulfinpyrazone are two commonly cited examples of carbon-linked glucuronides [75,76].

Ester or acyl glucuronides are potentially reactive metabolites that have been shown to react in several different ways under physiological conditions [77]. The first reaction common to acyl glucuronides is their susceptibility to hydrolysis in vivo, resulting in reformation of the aglycone. This hydrolysis, either by chemical or enzymatic means (β-glucuronidase), can lead to confusion on the total extent of glucuronidation as the aglycone goes through several rounds of

**Table 5** Sites of Glucuronidation

| Compound | Functional groups | Examples |
|---|---|---|
| *O*-Glucuronides | | |
| | Alcohols | Chloramphenicol |
| | Phenols | *p*-Hydroxy phenytoin |
| | Carboxylic acids | Valproic acid |
| | Enols | 4-Hydroxy coumarin |
| | *N*-Hydroxylamines (amides) | *N*-Hydroxyacetylaminofluorene |
| *N*-Glucuronides | | |
| | Aromatic amines | Aniline |
| | Tertiary amines | Lamotrigine |
| | Sulfonamides | Sulfisoxazole |
| | Amides/carbamates | Meprobamate |
| *S*-Glucuronides | | |
| | Aromatic thiols | Methimazole |
| | Aliphatic thiols | 2-Mercaptobenzothiazole |
| *C*-Glucuronides | | |
| | 1,3-Dicarbonyls | Sulfinpyrazone |

conjugation and deconjugation [78]. In addition, deconjugation (reformation of the aglycone) can elicit a pharmacological response. Acyl glucuronides can also undergo intra- and intermolecular rearrangements through acyl migration (RXN-1) or transesterification (RXN-2), respectively (Fig. 22). Acyl glucuronides undergo an intramolecular rearrangement by migration of the aglycone from the C1 position to the C2 hydroxyl group and further to the C3 and C4 positions. These reactions are reversible except for the reformation of the original C1 acyl glucuronide [79]. The rearrangement products have the same molecular weight as the original glucuronide, but differ in their chemicophysical properties of cleavage by chemical or enzymatic treatment. The transesterification of acyl glucuronides occurs when macromolecules covalently bind to the aglycone, releasing the glucuronic acid moiety and forming a drug–protein adduct (hapten). These drug–protein adducts have been associated with various forms of toxicity and can take much longer to be cleared from the body, as the protein must be turned over to ultimately eliminate the drug [80,81].

A minor conjugation pathway that also involves a sugar molecule is glucosidation, where a glucose molecule is transferred to the parent drug or metabolite [82]. As with glucuronidation, glucosidation is an enzyme-mediated reaction involving nucleophilic attack of the aglycone on the C1 carbon of UDP-α-D-glucose. Functional groups that have been associated with this form of conjugation include aromatic amines, carboxylic acids, alcohols, and thiols [82]. Glucosida-

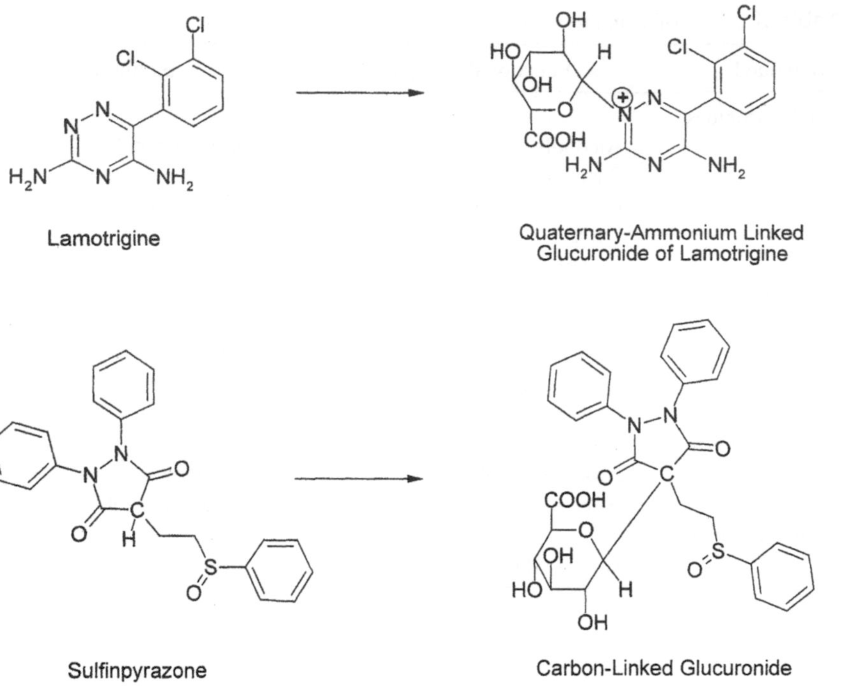

Lamotrigine

Quaternary-Ammonium Linked
Glucuronide of Lamotrigine

Sulfinpyrazone

Carbon-Linked Glucuronide

**Figure 21**  Examples of unusual glucuronidation reactions: lamotrigine (quaternary ammonium glucuronide) and sulfinpyrazone (carbon-linked glucuronide).

tion, once thought to be exclusive to plants and insects, has also been shown to occur in mammals. Several xenobiotics undergo this form of conjugation, such as phenobarbital, amobarbital, sulfamethazine, and sulfamethoxozole, albeit generally as a minor pathway [83–87]. Figure 23 illustrates the β-D-glucoside of phenobarbital that has been found in human urine.

## 2.  Sulfation

Sulfation is another common form of conjugation predominately found with phenolic compounds; however, sulfate esters can also be formed with alcohols, arylamines, and N-hydroxy compounds. Sulfation involves the transfer of $SO_3^-$ from 3′-phosphoadenosine-5′-phosphosulfate (PAPS) to one of the above-mentioned functional groups by an enzyme-catalyzed reaction involving the cytosolic enzymes (sulfotransferases), as illustrated in Fig. 24 [11]. It is common to find phenolic compounds that are metabolized by both sulfation and glucuronidation, as they are often competing pathways. However, sulfation has a significant limita-

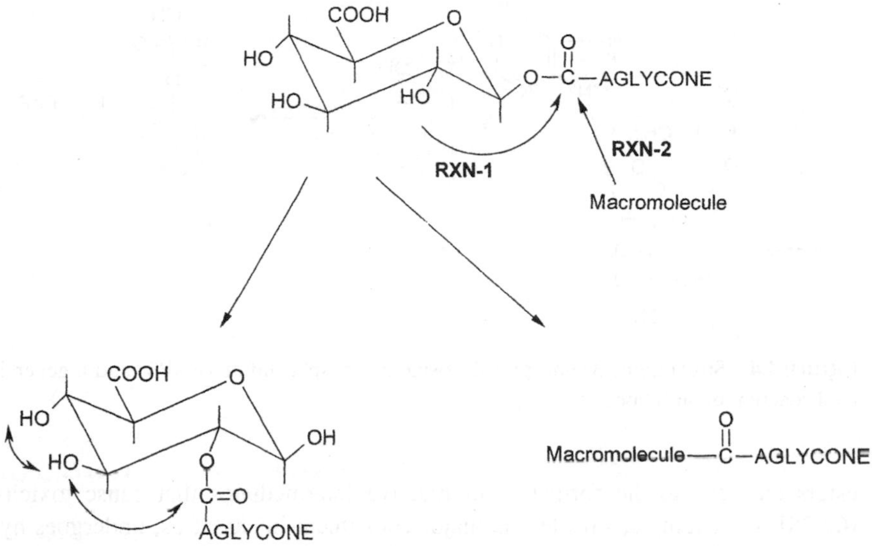

**Figure 22** Acyl glucuronide rearrangements: acyl migration (RXN-1) and transesterification (RXN-2).

tion in the total extent of conjugation that can occur with any xenobiotic, that being the amount of cofactor, PAPS, available for the reaction, whereas glucuronidation has a much larger pool of UDPGA. Therefore, sulfation is considered a saturable form of conjugation and can lead to decreased clearance of some compounds given in high doses where sulfation is the major pathway for elimination. In comparison to the glucuronide conjugates, sulfate esters can be cleaved in vivo by sulfatases (either from the host or microorganisms) and some sulfate

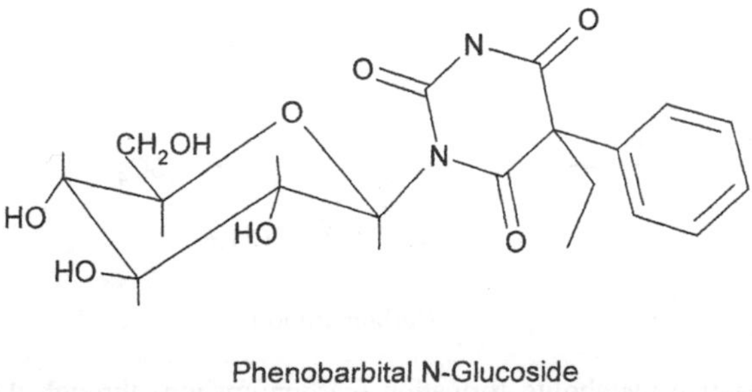

Phenobarbital N-Glucoside

**Figure 23** Glucosidation: β-D-glucoside of phenobarbital.

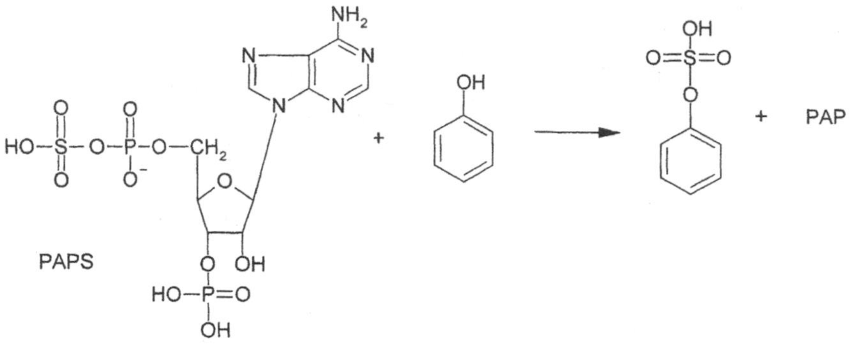

**Figure 24**  Structure of 3′-phosphoadenosine-5′-phosphosulfate (PAPS) and a generalized reaction of sulfation.

esters can lead to the formation of reactive intermediates that cause toxicity [67,78]. For example, safrole, the major constituent in sassafras, undergoes hydroxylation followed by sulfation (Fig. 25). The sulfate ($SO_4^-$), being a good leaving group, can be eliminated by rearrangement of the terminal double bond of safrole, leading to the generation of a highly electrophilic carbonium ion [88]. Sulfate conjugates have been associated with the formation of both carbonium and nitrenium ions that, by their electrophilic nature, bind to DNA and can be tumorigenic [89].

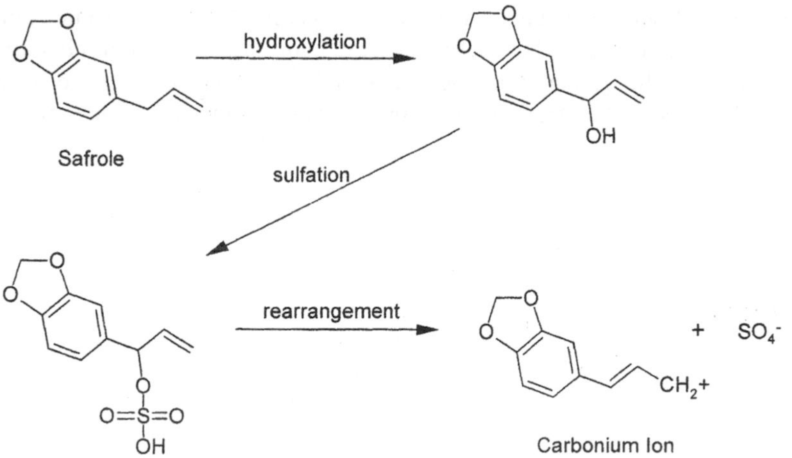

**Figure 25**  Illustration of reactive metabolite formation (carbonium ion) through the oxidation, sulfation, and rearrangement of safrole.

## 3.  Glutathione Conjugation

Glutathione conjugation is a significant pathway of biotransformation, especially for xenobiotics that inherently or through metabolism have the ability to be reactive [90–92]. Glutathione conjugation is often referred to as a detoxification pathway because it masks reactive electrophiles that could bind to other nucleophilic centers, such as those found on proteins and nucleic acids. Glutathione is a tripeptide ($\gamma$-glutamylcysteinylglycine) (Fig. 26) found in high concentrations (10 mM) in the liver [11]. This form of conjugation is catalyzed by a family of glutathione $S$-transferases found predominately in the cytosolic fraction of liver cells; however, there are some minor forms found in microsomal preparations. Glutathione conjugation is a reaction between the nucleophilic sulfur atom of cysteine (found in glutathione) and an electrophilic site on the xenobiotic, thereby forming a thiol ester (glutathione conjugate). Depending on the electrophilicity of the reactive xenobiotic, glutathione conjugation can be nonenzymatic; therefore strong electrophiles can form thiol esters with glutathione in the absence of glutathione $S$-transferase. Conjugation can occur with parent drug or a reactive metabolite generally by one of two mechanisms: (1) direct addition of glutathione at an electrophilic center or (2) addition of glutathione by displacement of an electrophilic group.

Once formed, glutathione conjugates have two routes of elimination: bile or urine. Glutathione conjugates eliminated in bile ultimately will be found in

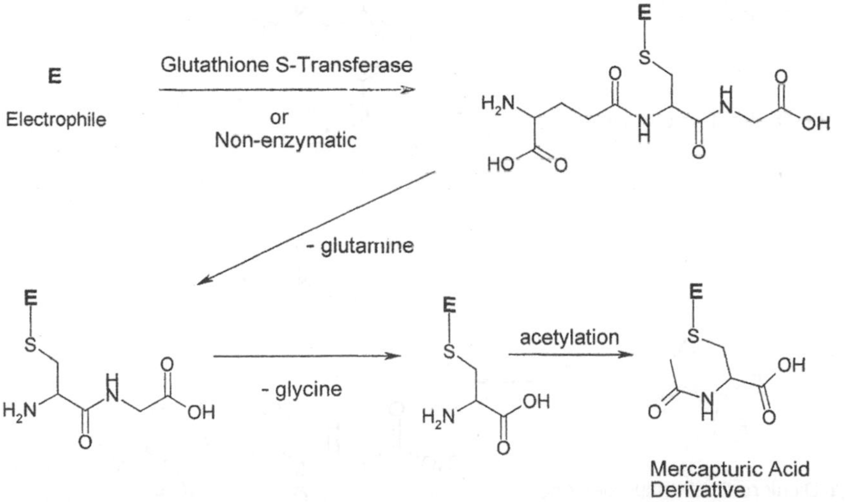

**Figure 26**   Representation of a glutathione conjugation reaction and formation of mercapturic acid.

the feces, predominately as intact glutathione conjugates [90]. Conjugates that are eliminated via urine are further metabolized in the kidney to mercapturic acids. As glutathione conjugates enter the kidney, they are metabolized by amino acid cleavage enzymes that sequentially remove the glutamic acid and glycine amino acids, resulting in a cysteine conjugate (Fig. 26). The cysteine conjugate is then N-acetylated to form the mercapturic acid, which is then eliminated in urine.

As mentioned previously, glutathione conjugation takes place with highly electrophilic substrates and occurs by either direct addition or displacement. Figures 27 and 28 illustrate typical examples of electrophilic substrates known to form glutathione conjugates. These substrates often contain halogens (or other good leaving groups); α,β-unsaturated double bonds with suitable electron withdrawing groups, such as ketones; and aromatic or aliphatic epoxides (often formed during metabolism). Figure 27 typifies how an aromatic compound (2,4-dinitro-1-chlorobenzene) and an aliphatic compound (1,1-dichloro-2,2,2-

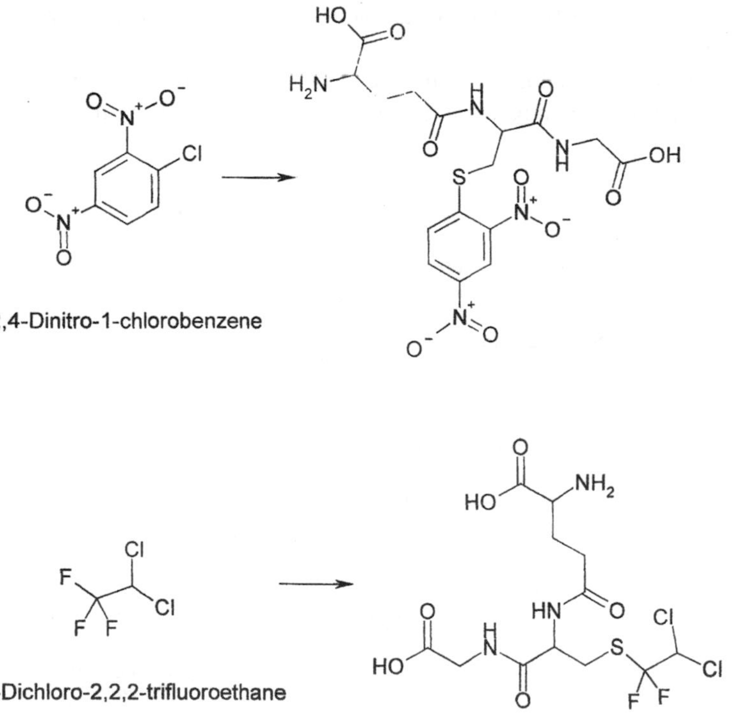

2,4-Dinitro-1-chlorobenzene

1,1-Dichloro-2,2,2-trifluoroethane

**Figure 27** Glutathione conjugation reactions of halogenated compounds, 2,4-ditro-1-chlorobenzene and 1,1-dichloro-2,2,2-trifluoroethane.

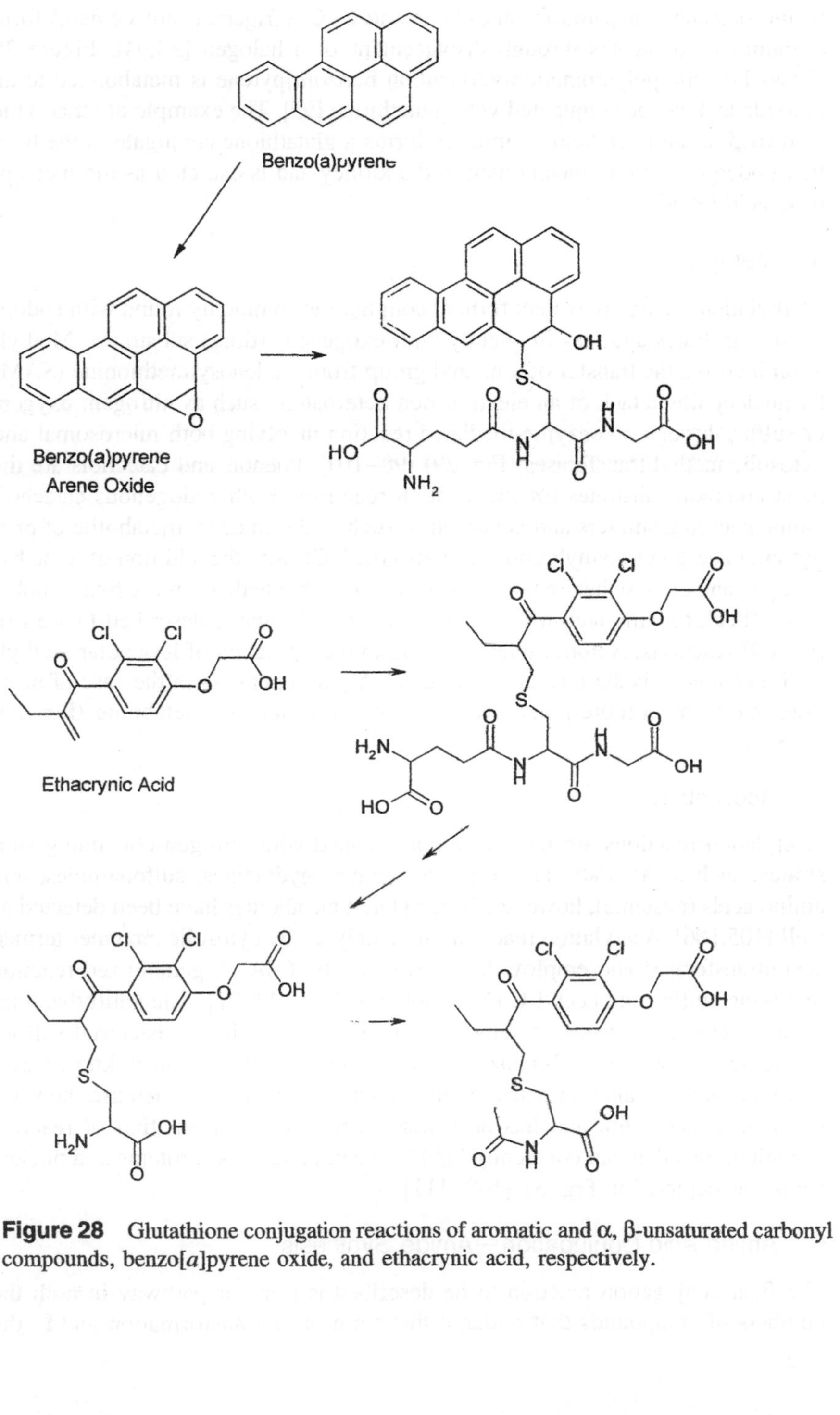

**Figure 28** Glutathione conjugation reactions of aromatic and α, β-unsaturated carbonyl compounds, benzo[*a*]pyrene oxide, and ethacrynic acid, respectively.

trifluoroethane, a hepatotoxic anesthetic and CFC refrigerant replacement) form glutathione conjugates through displacement of a halogen [93,94]. Figure 28 shows how the polyaromatic hydrocarbon benzo[a]pyrene is metabolized to an epoxide that can be conjugated with glutathione [95]. The example of ethacrynic acid ($\alpha,\beta$-unsaturated ketone) initially forms a glutathione conjugate in the liver that undergoes further metabolism in the kidney and is excreted as the mercapturic acid [96,97].

## 4.  Methylation

Methylation is a less prevalent form of conjugation commonly found with endogenous substrates and less frequently with exogenous (drug) substrates. Methylation involves the transfer of a methyl group from $S$-adenosylmethionine (SAM) by nucleophilic attack of an electron-rich heteroatom, such as nitrogen, oxygen, or sulfur, through an enzyme-mediated reaction involving both microsomal and cytosolic methyl transferases (Fig. 29) [98–101]. Phenols and catechols are the most common substrates for methylation reactions. Both endogenous catecholamine neurotransmitters and xenobiotics, such as the catechol metabolite of phenytoin, undergo O-methylation reactions [102]. Clearly, the addition of a methyl group to any of these heteroatoms results in a metabolite that is more hydrophobic and difficult to eliminate, which is in contrast to the above-described Phase I or Phase II reactions. A noticeable exception to the formation of less polar methylated metabolites is the formation of quaternary nitrogens, as in the case of nicotine, that forms a more polar, readily excreted quaternary metabolite (Fig. 29) [103,104].

## 5.  Acetylation

Acetylation reactions are most commonly found with nitrogen-containing substrates, such as aromatic and aliphatic amines, hydrazines, sulfonamides, and amino acids (cysteine); however, O-acetylated metabolites have been detected as well [105,106]. Acetylation reactions are catalyzed by cytosolic enzymes termed acetyltransferases and employ the cofactor acetyl-CoA. A generalized reaction involving aniline and acetyl-CoA is shown in Fig. 30 [107]. The antiturbercular agent isoniazid undergoes extensive acetylation as do the antibacterial sulfonamide agents, such as sulfadiazine [108]. Although acetylation is known as a detoxification/elimination pathway, this reaction does little to increase the solubility of a metabolite and has been associated with the formation of reactive nitrenium ions that can covalently bind to macromolecules (proteins and nucleic acids), as depicted in Fig. 31 [109–111].

## 6.  Amino Acid Conjugation—Amide Synthesis

The final conjugation reaction to be described is a minor pathway in both the numbers of compounds that undergo this form of biotransformation and in the

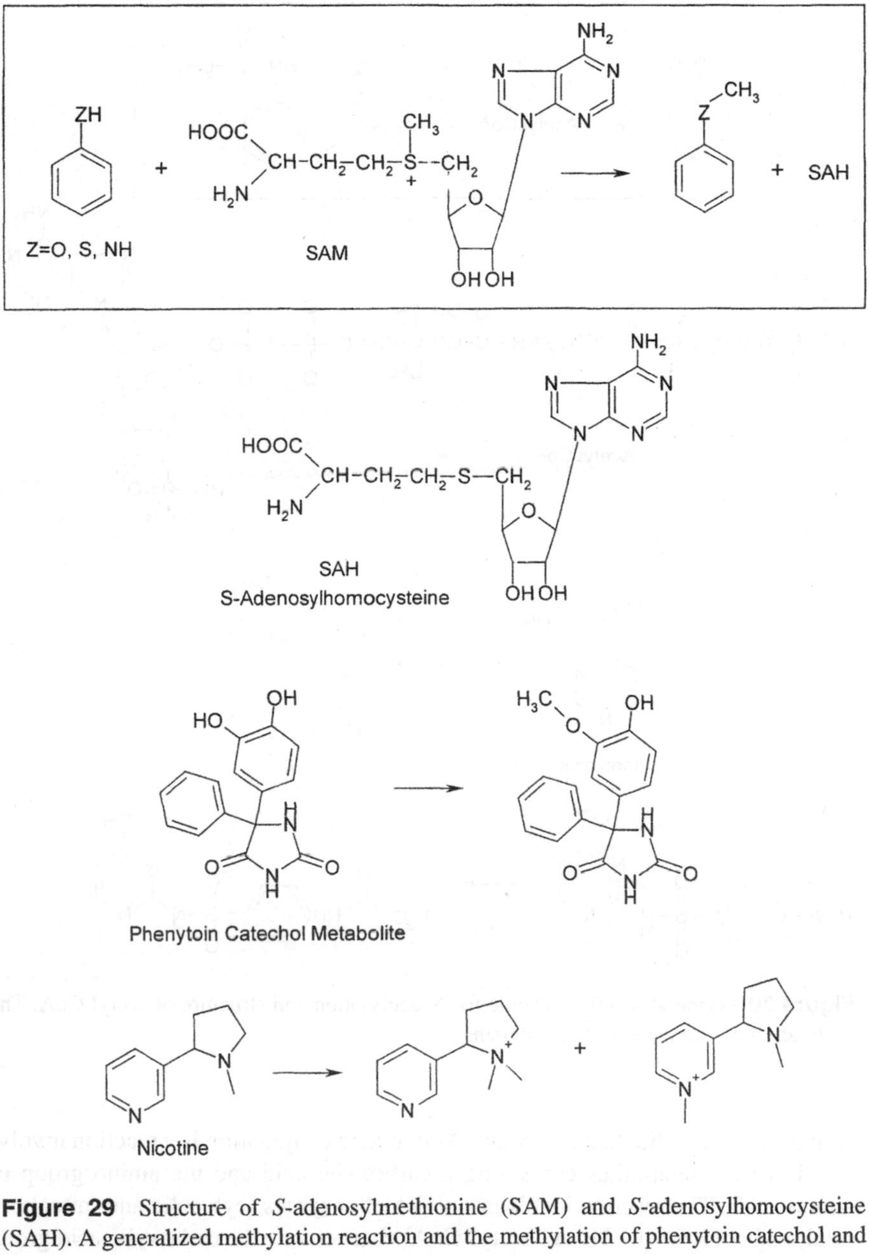

**Figure 29** Structure of *S*-adenosylmethionine (SAM) and *S*-adenosylhomocysteine (SAH). A generalized methylation reaction and the methylation of phenytoin catechol and nicotine.

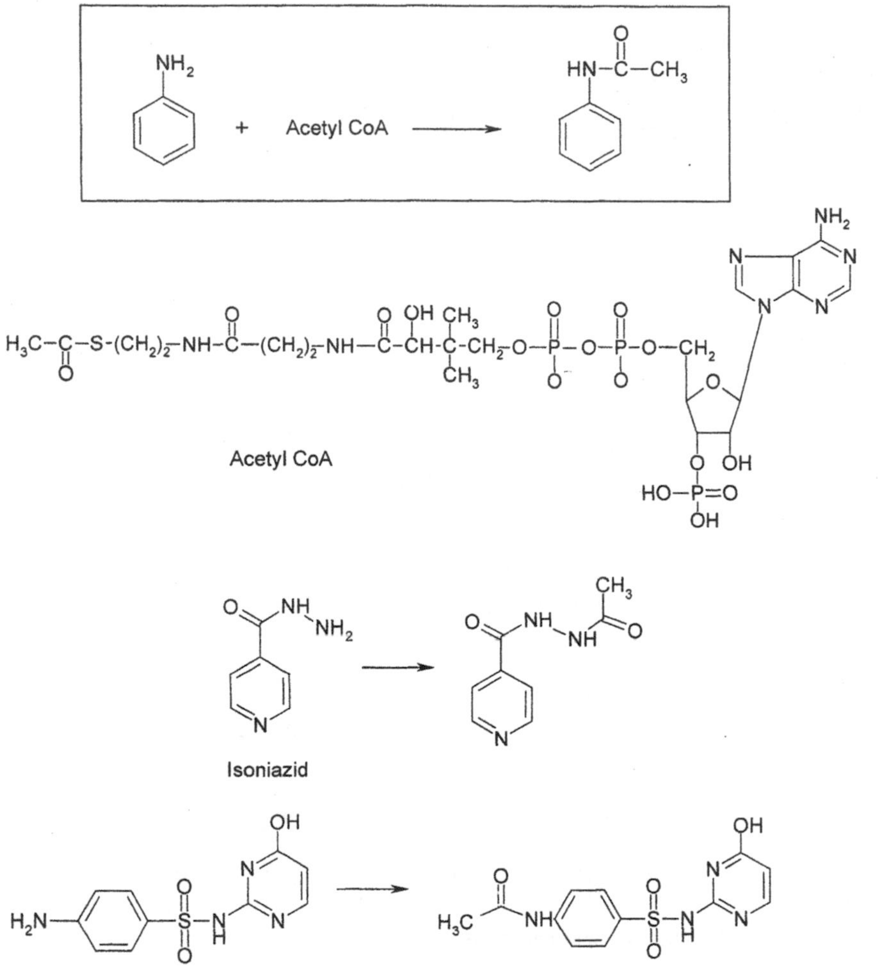

**Figure 30**   General reaction scheme for N-acetylation and structure of acetyl CoA. The acetylation of isoniazid and sulfadiazine.

amount of metabolite that is formed. Amino acid conjugation is a reaction involving drugs or metabolites containing a carboxylic acid and the amino group of amino acids. The reaction involves a series of enzymes (cytosolic and mitochondrial) and cofactors (ATP and acetyl-CoA) that activate the carboxylic acid group and result in the formation of an amide bond. Figure 32 is a simplified representation of this reaction between glycine and benzoic acid [112]. The most common amino acids involved in this type of conjugation include glycine, glutamine, tau-

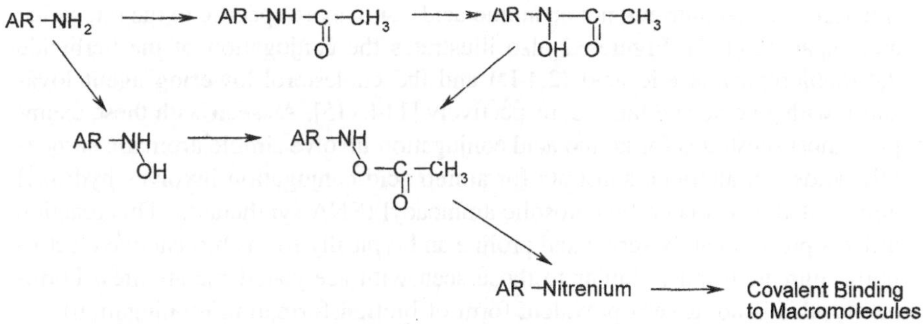

**Figure 31** Illustration of reactive metabolite formation (nitrenium ion) by re-arrangement of an N-acetylated derivative (formed by either initial N-hydroxylation or direct N-acetylation) that has the ability to covalently bind to endogenous macromolecules such as proteins and nucleic acids.

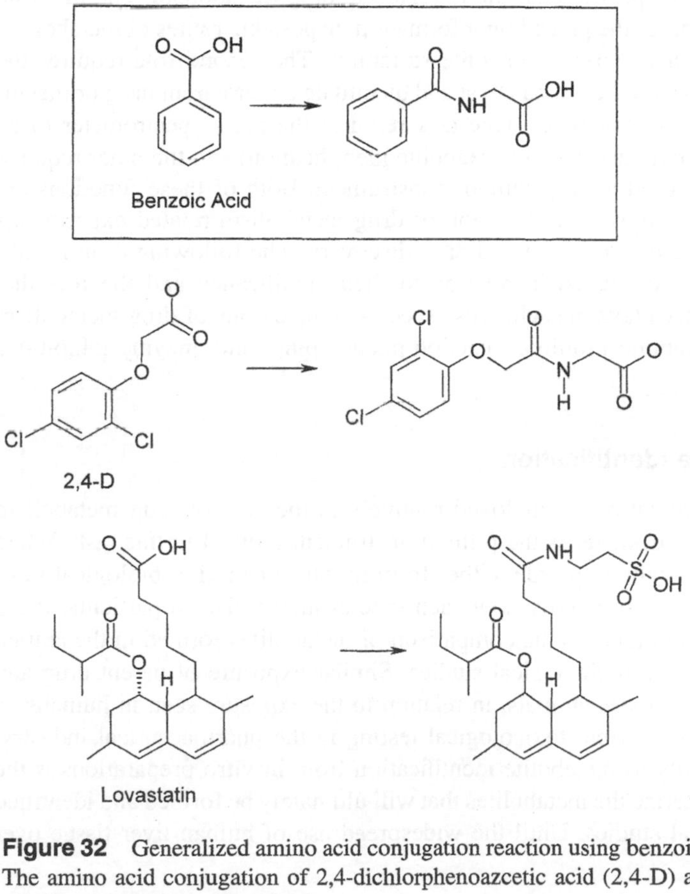

**Figure 32** Generalized amino acid conjugation reaction using benzoic acid as substrate. The amino acid conjugation of 2,4-dichlorphenoazcetic acid (2,4-D) and lovastatin with glycine and taurine, respectively.

rine (an essential nutrient, not an amino acid), and, more specific to the rat, serine and aspartate [113]. Figure 32 also illustrates the conjugation of the herbicide 2,4-dichlorophenoacetic acid (2,4-D) and the cholesterol-lowering agent lovastatin with glycine and taurine, respectively [114,115]. As seen with these examples, most substrates for amino acid conjugation involve simple aromatic carboxylic acids. An alternate substrate for amino acid conjugation involves hydroxyl amines and is mediated by cytosolic aminoacyl tRNA synthetases. This reaction utilizes predominately serine and proline and typically forms bioreactive electrophilic nitrenium ions, similar to those seen with acetylated metabolites. Fortunately, this is not a very prevalent form of biotransformation (conjugation).

## IV.  DRUG DISCOVERY AND DRUG METABOLISM

With the generation of samples from both in vitro and in vivo models, the mass spectrometrist must play a dual role in regard to drug metabolism. The first role would be to use all of the preceding information on possible routes of metabolism and identify or characterize metabolite structures. The second role requires the mass spectrometrist to be a bioanalyst and quantitate either parent drug or metabolites from biological samples. One task requires the mass spectrometer to be used as a qualitative instrument (metabolite identification) and the other requires the device to be used as a quantitative instrument. Both of these functions are presented in the following description of drug metabolism-related experiments employed in drug development and drug discovery. The following drug metabolism studies are discussed in relation to their significance and the role that mass spectrometry plays in each: cross-species comparisons of drug metabolism and in vivo metabolic profiling, reaction phenotyping, and enzyme inhibition/ induction.

### A.  Metabolite Identification

Metabolite identification is employed routinely in the area of drug metabolism and the mass spectrometer is the primary instrument utilized in this task. When comparing the metabolic profile, either from in vitro or in vivo biological samples, it is important to identify how each species metabolizes a particular drug. Of prominent importance is the comparison of metabolites formed in the animal species employed in toxicological studies. Similar exposure of parent drug and metabolites across these animals, in relation to the exposure seen in humans, is the foundation for current toxicological testing in the pharmaceutical industry. The greatest utility to metabolite identification from in vitro preparations is the ability to characterize the metabolites that will ultimately be formed and identified in human clinical studies. Until the widespread use of human liver tissue over the past decade, the identification of human metabolites and the comparison to metabolites from animal species could not be determined until clinical development of the compound. Therefore, the human metabolism of a compound can now be determined in drug discovery.

The data resulting from a typical procedure for tentatively identifying the metabolites of a discovery-phase drug candidate are described below and in Figs. 33–36. Although these data were collected on an electrospray LC/MS ion trap and display the signal-to-noise advantages of that instrument, the processes described here are generic and independent of the platform used. In more traditional metabolite structure elucidation work for drug development, the use of a radiolabeled compound and a radiometric flow detector can be valuable in quickly locating major metabolite peak retention times in a chromatographic run. Unfortunately, radiolabeled drug is typically not available in early discovery, therefore metabolite peak detection is occasionally done with a spectrophotometric LC detector (UV or fluorescence). For compounds with inadequate chromophores, such as the example presented below, locating metabolite peaks is performed by visual inspection or through the use of specialized software tools. Because the ionization efficiency of different functional moieties varies greatly, it is difficult to accurately estimate the relative proportions of metabolites from mass spectral response and this type of information is not usually provided.

After a drug candidate is incubated in the presence of hepatocytes, microsomes, or other liver fractions, samples are taken at various incubation times and compared to a preincubation sample (blank). Typically an aliquot of acetonitrile is added to this sample to quench enzyme activity and to assist in the removal of proteins prior to chromatographic injection. Figure 33A shows a time-zero, total-ion chromatogram for a drug incubated in rat liver microsomes. A positive-ion, extracted-ion chromatogram of the parent drug ($m/z$ 303; Fig. 33B) isolates the signal for this compound. A gradient liquid chromatography separation helps separate the parent drug from the various metabolites. The gradient employed for this separation is described in the figure caption. Selected ion chromatograms are then constructed for suspected metabolites. In this example, chromatograms for ring hydroxides ($m/z$ 319; Fig. 33C) and desmethyl compounds ($m/z$ 289; Fig. 33D) were examined and found to contain no peaks, verifying that these metabolites are not present in the preincubation sample. Because this sample was obtained from microsomes, Phase II metabolites did not form and their corresponding selected-ion chromatograms were not considered.

Metabolites to be identified are present in the incubated sample. In this example, Fig. 34 represents a microsomal incubation aliquot taken at 40 min. The total ion chromatogram (Fig. 34A) is somewhat difficult to interpret because of the low intensities of metabolites relative to background. Therefore, selected ion chromatograms are constructed based on a general understanding of metabolism principles described earlier and through use of specialized software capable

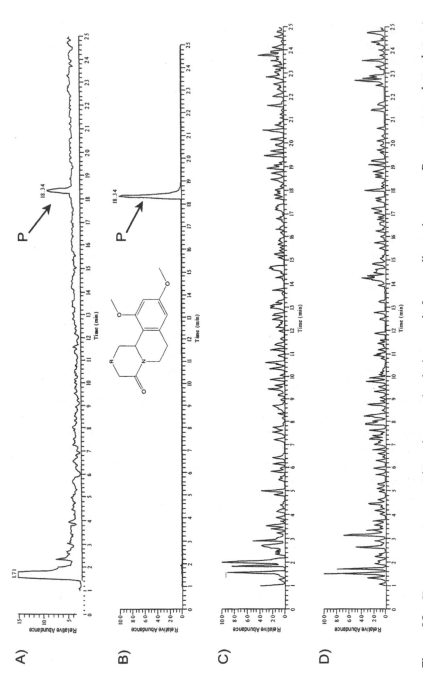

**Figure 33** Chromatograms representing a time-zero incubation sample from rat liver microsomes. P represents unchanged parent drug with molecular weight 302 Da. (A) Total ion chromatogram from 100 to 900 Da showing P. (B) Selected ion chromatogram at 303 Da, showing M + 1 peak for P. (C) Selected ion chromatogram at 319 Da. (D) Selected ion chromatogram at 289 Da. Gradient conditions: 5 min at 10% acetonitrile: 90% 25 mM ammonium acetate (pH 4.0), linear gradient over 20 min to 90% acetonitrile. Flow rate-1.0 ml/min on a 150 × 4.6-mm Phenomenex Polaris C$_{18}$ column.

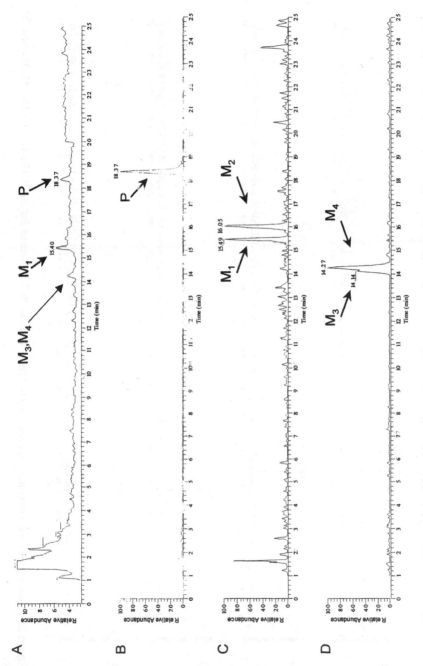

**Figure 34** Chromatograms representing a 40-min incubation sample from rat liver microsomes. $M_1$ and $M_2$ represent metabolites of molecular weight 318 Da. $M_3$ and $M_4$ represent metabolites of molecular weight 288 Da. (A) Total ion chromatogram from 100 to 900 Da, showing P, $M_1$, $M_3$, and $M_4$. (B) Selected ion chromatogram at 303 Da, showing M + 1 peak for P. (C) Selected ion chromatogram at 319 Da showing $M_1$ and $M_2$ peaks. (D) Selected ion chromatogram at 289 Da, showing $M_3$ and $M_4$ peaks.

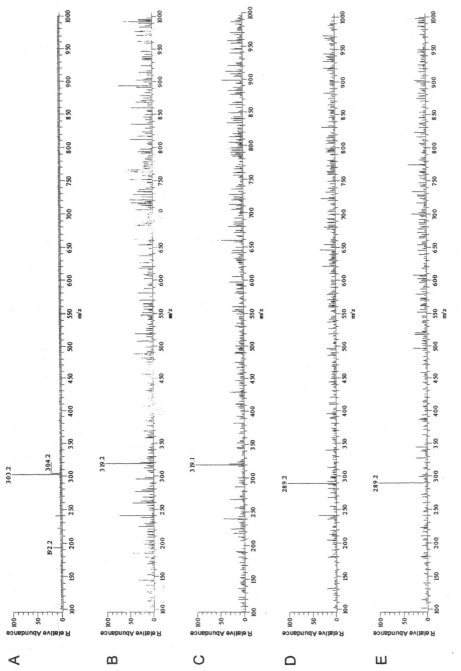

**Figure 35** Electrospray ionization mass spectra extracted from total ion chromatograms for (A) P, (B) $M_1$, (C) $M_2$, (D) $M_3$, and (E) $M_4$.

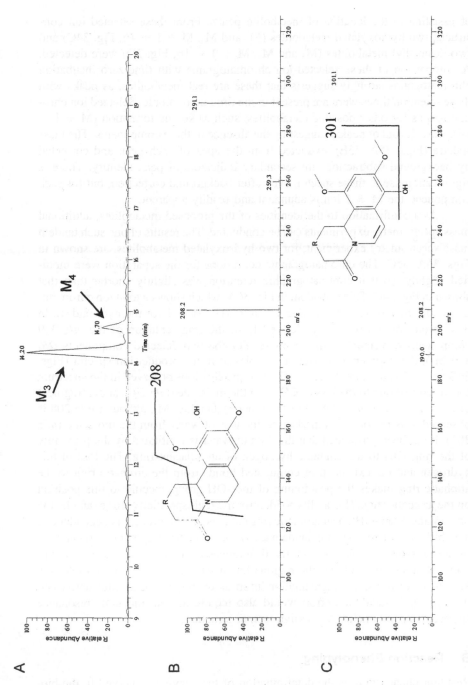

**Figure 36** (A) Reconstructed product ion chromatogram for $m/z$ 319. Product ion spectra for (B) $M_1$ at retention time 14.20 min and (C) $M_2$ at 14.70 min.

of assisting in the location of metabolite peaks. From these selected ion constructs, two hydroxylated metabolites ($M_1$ and $M_2$, M + 1 + 16; Fig. 34C) and two desmethyl metabolites ($M_3$ and $M_4$, M + 1 − 16; Fig. 2D) were detected. A comparison of these selected-ion chromatograms with time-zero incubation chromatograms strongly suggests that these are real metabolites, as peaks with these retention times were not present in the time-zero sample. Selected ion chromatograms for other possible metabolites, such as ketone formation (M + 1 + 14), were devoid of peaks, suggesting the absence of these components. The mass spectra (Figs. 35A–35E), extracted from the apex of each peak and corrected by background subtraction, are secondary indicators of peak identity. There is significant noise in these spectra, even after background correction, but for each component, the M + 1 ion is abundant and readily apparent.

As a confirmation to the identities of the proposed metabolites, additional mass spectrometry experiments can be conducted. The results of one such tandem mass spectrometry experiment, for two hydroxylated metabolites, are shown in Figs. 36A–36C. The chromatographic conditions for the separation were modified slightly, so that chromatographic retention was slightly shorter than that shown in Fig. 34C. This is evident in Fig. 36A, which shows a total ion chromatogram constructed from the products of $m/z$ 319 (M + 1 for the hydroxides). In this chromatogram only, ions derived from the fragmentation of the $m/z$ 319 components contribute to the response. For the two detected components ($M_1$ and $M_2$), the fragmentation patterns displayed in the product ion spectra (Figs. 36B and 36C) are quite different. The two product ions observed in the spectrum for $M_1$ correspond to the loss of CO from the molecule ($m/z$ 291) and a fragment of the ring system ($m/z$ 208). For the other metabolite, $M_2$, although $m/z$ 208 is observed, it is minor in comparison to the loss of water from the molecule ($m/z$ 301). It has been postulated that this loss of water is facilitated by the proximity of the ring -OH to the aliphatic hydrogen of an adjacent ring. The loss of this hydrogen and the extension of conjugated bonds from the aromatic ring to the aliphatic ring makes the positioning of the -OH group specific to this position on the aromatic ring. This ability to identify the regiospecificity (for example, the positioning of an -OH on an aromatic ring) of a metabolic process is dependent on the ability to obtain revealing tandem mass spectrometric fragmentation patterns. In the example described above, the differences in fragmentation strongly supports the assignment of specific regions for metabolic changes to the molecule. For other situations, the fragmentation information may not be as informative and the structural elucidation effort would also require nuclear magnetic resonance spectrometry to conclusively assign isomer structures.

## B.  Reaction Phenotyping

Reaction phenotyping is the determination of the enzymes involved in the biotransformation of a xenobiotic. In the pharmaceutical industry, that implies predominately the oxidative enzymes, cytochromes P450. Reaction phenotyping can elucidate potential drug interactions by: knowing the number of enzymes involved in the metabolism of a compound, knowing what coadministered compound can effect the metabolic elimination of a compound, and knowing if the drug is metabolized by an enzyme that is polymorphically distributed throughout the population. When a compound is metabolized by a single pathway, any effect (inhibition or induction) of that pathway will have a significant effect on the metabolic elimination of the compound. For example, the metabolic elimination of a compound only metabolized by CYP3A4 will be greatly diminished by coadministration of the CYP3A4 inhibitor, ketoconazole. Where as, if the same compound were equally metabolized by CYP3A4, CYP2C9, and CYP1A2, the effect on metabolic elimination by ketoconazole would be less. Similarly, coadministration of inhibitors or substrates that are metabolized by the same CYP enzyme can have an effect on the elimination of the compound. For example, a compound that is predominantly metabolized by CYP2C19 when coadministered with S-mephenytoin (another CYP2C19 substrate) can create an interaction between the two compounds. The most characterized example of polymorphic drug metabolism is CYP2D6, an enzyme which is not present (deficient) in 5–10% of the Caucasian population [116]. Other polymorphic CYP enzymes include CYP2C9 and CYP2C19 [117]. The lack of any of these enzymes in patients results in a dramatic reduction in the clearance of a drug that requires these enzymes for elimination. For example, a drug metabolized only by CYP2D6 will have significant consequences in poor metabolizers of CYP2D6; however, a compound that is metabolized by CYP2D6 and CYP3A4 will be of less significance in poor metabolizers. In the course of reaction phenotyping, mass spectrometry can be used to characterize the metabolites formed, thereby assigning which enzyme(s) is involved in each metabolic step. Therefore, the researcher can predict which metabolites and routes of metabolism will be affected by various inducers, inhibitors, and polymorphic phenotypes.

## C. Quantitation of Parent Drug or Metabolites

Many instances of parent drug quantitation are illustrated throughout this book. In some cases, the quantitation of metabolites is also warranted. Quantitation of metabolite is necessary when the metabolite is either toxic or pharmacologically active or when the concentrations of metabolite reach or exceed the parent drug concentration in plasma. The latter case can be important because "total drug exposure" is more an exposure to metabolite than parent drug and most safety and efficacy parameters are in relation to plasma exposure.

A concern in the development of an NCE from discovery programs is the potential to inhibit or induce the enzyme activity of major cytochrome P450 enzymes found in the human liver. Changes, either increases or decreases, in the enzyme activity of drug-metabolizing enzymes can have a dramatic effect on the metabolism and disposition of coadministered drugs. The biochemical assays that measure increases or decreases in enzyme activity have already been well characterized (previously by LC/UV or fluorescence detection). The utility of mass spectrometry in this case is to more efficiently analyze large numbers of samples in a relatively short period of time. For example, the typical CYP3A4 assay that measures the formation of 6β-hydroxy testosterone from testosterone would take approximately 1 week using LC/UV instrumentation; however, with LC/MS, this same assay can be completed in a single overnight run. In addition, the lower limit of quantitation will be improved with the sensitivity and selectivity of the mass spectrometer compared to the UV detector. The reduced time needed to generate this type of data is essential in any drug discovery program where speed of analysis is vital.

## V.  SUMMARY

The speed at which we collect data (information), interpret data, and reach decisions will ultimately determine how successful each drug discovery program will be in moving compounds with good drug profiles (ADME, safety, and efficacy) forward to development. In order to remain competitive and successful, each drug metabolism organization must continue to improve their current practices and methodologies in the area of metabolism: for example, higher throughput assays that are more predicable; data-handling software and databases that input, export, and analyze data more efficiently; as well as more predictive software tools that require minimal to no actual laboratory work to make predictions of human ADME properties, i.e., in silico.

    The flexibility of liquid-handling systems and robotic workstations today allow for their increased utilization in the area of routine in vitro incubations and sample preparations. In addition, the employment of multiwell plates to perform incubations and multiwell plate readers (UV, fluorescence, and radiometric) has increased the capacity to study multiple compounds simultaneously. Some of the early assays converted to higher throughput screening using robotics were the metabolic stability and drug inhibition analysis of discovery compounds [118,119]. Obviously the data generated from these higher throughput systems becomes overwhelming to the individual researcher and requires the use of data-handling systems and software. More versatile and simpler means of placing data into databases and in the export and analysis of data become a critical factor in all high-throughput systems. Other types of databases used in the ADME area are knowledge-based or predictive databases that contain large numbers of metabolic reactions from the literature or a set of metabolism rules that predict metabolic reactions, respectively [120]. These types of databases are useful learning tools for uncommon types of metabolic reactions; however, they do not as yet have the ability to accurately predict a priori the metabolic fate of an unknown NCE with a high-enough degree of certainty. Examples of such knowledge databases include MetabolExpert from ComGenex, Meteor from Lhasa, and Metabolite from MDL Information Systems. Additional software advances have occurred in the area of metabolite identification by mass spectrometry. In standard drug development, a great deal of time and effort is placed on the examination of mass spectra by metabolite identification experts to characterize different forms or sites of biotransformation. The time needed for an individual researcher to diligently pore over this information is not within the time frame in which discovery information needs to be generated. Therefore, mass spectrometry investigators and manufacturers of mass spectrometers have developed software systems, such as Metabolynx from Micromass, to rapidly identify drug metabolites [121,122]. Again, these software packages are useful for straightforward metabolic reactions but are, however, not sophisticated enough to identify all metabolites. Their application in discovery is warranted due to the need to profile metabolites in a rapid fashion; however, they should not be employed when more detailed analysis of metabolic profiles are necessary, such as in drug development.

An entirely different type of software advance is in the area of modeling drug metabolizing enzymes to predict whether a discovery compound will be a substrate for a particular enzyme; for example, the polymorphic enzyme CYP2D6 [123,124]. In addition, software companies, such as Camitro Corp (see Chap 1), are developing software to predict the site(s) and rates of oxidative metabolism by modeling the major cytochrome P450 enzymes. Both of these innovative approaches allow thousands of compounds to be screened in a virtual environment with minimal use of laboratory experiments to make decisions.

All of these current and future advances in instrumentation and predictive software, when applied appropriately, will result in screening larger numbers of compounds, making discovery decisions more rapidly, developing better compounds, and delivering a safer and more efficacious drug product to the patient.

## ACKNOWLEDGMENTS

The authors thank Ms. Karen Metzler for her assistance in typing the references as well as Ms. Kristen Siebenaler and Laura Egnash for their review of the text. A special acknowledgment to Dr. David Rossi for his additional comments and editorial review.

## REFERENCES

1. RT Williams. Detoxification Mechanisms: The Metabolism of Drugs and Allied Organic Compounds. New York: Wiley.

2. RT Williams. 1959, Detoxification Mechanisms: The Metabolism and Detoxification of Drugs, Toxic Substances and Other Organic Compounds, New York: Wiley.
3. A. Parkinson. Biotransformation of xenobiotics. In: C Klaassen, ed. Casarett and Doull's Toxicology: The Basic Science of Poisons. 5th ed. New York: McGraw–Hill, 1996, pp 113–186.
4. L Low, N Castagnoli Jr. Metabolic changes of drugs and related organic compounds. In: RF Doerge, ed. Wilson and Grisvold's Textbook of Organic Medicinal Chemistry and Pharmaceutical Chemistry, 8th ed., Philadelphia: Lippincott, 1982.
5. T Woolf, ed. Handbook of Drug Metabolism. New York: Marcel Dekker, 1999.
6. MK Baylis, JA Bell, WN Jenner, GR Park, K. Wilson. Utility of hepatocytes to model species differences in the metabolism of loxtidine and to predict pharmacokinetic parameters in rat, dog, and man. Xenobiotica 29:253–267, 1999.
7. DJ Carlile, N Hakooz, JB Houston. Kinetics of drug metabolism in rat liver slices. IV. Comparison of ethoxycoumarin clearance by liver slices, isolated hepatocytes, and hepatic microsomes from rats pretreated with known modifiers of cytochrome P-450 activity. Drug Metab Dispos 24:526–532, 1999.
8. GW Caldwell, JA Masucci, E Chacon. High throughput liquid chromatography-mass spectrometry assessment of the metabolic activity of commercially available hepatocytes from 96-well plates. Com Chem High Throughput Screen 2:39–51, 1999.
9. AP Li, C Lu, JA Brent, C Pham, A Fackett, CE Ruegg, PM Silber. Cryopreserved human hepatocytes: characterization of drug-metabolizing enzyme activities and applications in higher throughput screening assays for hepatotoxicity, metabolic stability, and drug–drug interaction potential. Chem–Biol Interact 121:17–35, 1999.
10. B Testa. The Metabolism of Drugs and Other Xenobiotics: Biochemistry of Redox Reactions. London: Academic Press, 1995, pp 41–458.
11. GJ Mulder. Conjugation Reactions in Drug Metabolism: An Integrated Approach. London: Taylor & Francis, 1990, pp 51–364.
12. BS Cummings, RC Zanger, RF Novak, LH Lash. Cellular distribution of cytochromes P-450 in the rat kidney. Drug Metab Dispos 27:542–548, 1999.
13. A Hutchaleelaha, J Sukbuntherng, HH Chow, M Mayersohn. Disposition kinetics of D- and L-amphetamine following intravenous administration of racemic amphetamine to rats. Drug Metab Dispos 22:406–411, 1994.
14. RT Coutts, GR Jones, RE Townsend. Species differences in the in vitro metabolic reduction of amphetamine metabolite, 1-phenyl-2-propanone. J Pharm Pharmacol 30:415–418, 1978.
15. S Eto, N Tanaka, H Noda, A Noda. Chiral separation of 10,11-dihyro-10,11-trans-dihyrdoxycarbamazepine, a metabolite of carbamazepine with two asymmetric carbons, in human serum. J Chromatogr B Biomed Appl 677:325–330, 1996.
16. A Frigerio, R Fanelli, P Biandrate, G Passerini, PL Morselli, S Garaattini. Mass spectrometric characterization of carbamazepine-10,11-epoxide, a carbamazepine metabolite isolated from human urine. J Pharm Sci. 61:1144–1147, 1972.
17. MA Mori, H Uemura, M Kobayashi, T Miyahara, H Kozuka. Metabolism of phentermine and its derivatives in the male Wistar rat. Xenobiotica 23(6):709–717, 1993.

18. RB Franklin, LG Dring, RT Williams. The metabolism of phenmetrazine in man and laboratory animals. Drug Metab Dispos 5(3):223–233, 1977.
19. BJ Blaauboer, CWM Van Holsteijn. Formation and disposition of N-hydroxylated metabolites of aniline and nitrobenzene by isolated rat hepatocytes. Xenobiotica 13(5):295–302, 1983.
20. DH Lang, CK Yeung, RM Peter, C Ibarra, R Gasser, K Itagaki, RM Philpot, AE Rettie. Isoform specificity of trimethylamine N-oxygenation by human flavin-containing monooxygenase (FMO) and P450 enzymes: selective catalysis by FMO3. Biochem Pharmacol. 56(8):1005–1012, 1998.
21. G Lin, EM Hawes, G McKay, ED Korchinski, KK Midha. Metabolism of piperidine-type phenothiazine antipsychotic agents. IV. Thiorizidine in dog, man, and rat. Xenobiotica 23(10):1059–1074, 1993.
22. D Fukshima, YH Kim, T Iyanagi, S Oae. Enzymatic oxidation of disulfides and thiolsulfinates by both rabbit liver microsomes and a reconsituted system with purified cytochrome P-450. J Biochem 83(4):1019–1027, 1978.
23. VV Schumyantseva, SV Meshkov, YUD Ivanov, OV Alexandrova, VYU Uvarov, AI Archakov. Interaction of organophosphorus analogues of amino acids with P450. Xenobiotica 25(3):219–227, 1995.
24. RA Wiley, LA Sternson, HA Sasame, JR Gillette. Enzymatic oxidation of diphenylmethylphosphodine and 3-dimethylaminopropyldiphenylphosphine by rat liver microsomes. Biochem Pharm 21:3235–3247, 1972.
25. Y Masubuchi, J Araki, S Narimatsa, I Suzuki. Metabolic activation of lidocaine and covalent binding to rat liver microsomal protein. Biochem Pharmacol 43(12): 2551, 1992.
26. H He, SD Shay, Y Caraco, M Wood, AJJ Wood. Simultaneous determination of codeine and its seven metabolites in plasma and urine by high-performance liquid chromatography with ultraviolet and electrochemical detection. J Chromatogr B 708(1/2):185–193, 1998.
27. G Renner, PT Nguyen. Biotransformation of derivatives of the fungicides pentachloronitrobenzene and hexachlorohemzene in mammals. Xenobiotica 14(9): 693–704, 1984.
28. DK Spracklin, DC Hankins, JM Fisher, KE Thummel, ED Kharasch. Cytochrome P450 2E1 is the principle catalyst of human oxidative halothane metabolism in vitro. J Pharmacol Exp Ther 281(1):400–411, 1997.
29. RC Lind, AJ Gandolfi. Late dimethyl sulfoxide administration provides a protective action against chemically induced injury in both the liver and the kidney. Toxicol Appl Pharmacol 142(1):201–207.
30. JL Huang, LE Mather, CC Duke. High-performance liquid chromatographic determination of thiopentone enantiomers in sheep plasma. J Chromatogr B Biomed Appl 673(2):245–250, 1995.
31. H Kexel, E Schmelz, HL Schmidt. Oxygen transfer in micosomal oxidative delulfuration. In: V Ullrich, A Hildebrant, J Roots, R Eastabrook, A Conney, eds. Microsomes and Drug Oxidations. New York: Pergamon Press, 1977, pp 269–274.
32. MW Sinz, AE Black, SM Bjorge, A Holmes, BK Trivedi, T Woolfe. In vitro and in vivo disposition of 2,2-dimethyl-n-(2,4,6-trimethoxyphenyl) dodecanamide (CI-976). Drug Metab Dispos 25(1):123–130, 1997.

33. T Satoh, M Hosokawa. The mammalian carboxylases: from molecules to functions. Annu Rev Pharm Toxicol 38:257–288, 1998.

34. K Miyazakai, O Ogino, H Sato, M Nakano, T Arita. Flurometric determination of amoxicillin. Chem Pharm Bull 25(2):253–258, 1977.

35. C Pantarotto, L Cappellini, A De Pascale, A Frigerio. Epoxide-diol pathway in the metabolism of 5H-dibenzo[b,f]-azepine (iminostilbene). J Chromatogr 134:307–314, 1977.

36. RE McMahon, HR Sullivan, SL Due, FJ Marshall. The metabolite pattern of D-propoxyphene in man. The use of heavy isotopes in drug disposition studies. Life Sci 12(II):463–473, 1973.

37. M Hosokawa, T Endo, M Fujisawa, S Hara, N Iwata, Y Sato, T Satoh. Interindividual variation in carboxylesterase levels in human liver microsomes. 23(10):1022–1027, 1995.

38. RA Morrison, KJ Kripalani, AM Marino, AV Dean, BH Migdalof, SH Weinstein, NB Jain, MS Bathala, SM Singhivi. Intestinal absorption of captopril and two thio-ester analogs in rats and dogs. Biopharm Drug Dispos 18:25–39, 1997.

39. BH Thomas, LT Wong, W Zeitz, G Solomonraj. Isoniazid metabolism in the rabbit, and the effect of rifampin pretreatment. Res Commun Chem Pathol Pharmacol 33:235–247, 1981.

40. TC Sarich, SP Adams, G Petricca, JM Wright. Inhibition of isoniazid-induced hepatotoxicity in rabbits by pretreatment with an amidase inhibitor. J Pharm Exp Ther 289:695–702, 1999.

41. DJ Harvey, L Glazener, DB Johnson, CM Butler, MG Horning. Comparative metabolism of four allylic barbituates and hexobarbital by the rat and guinea pig. Drug Metab Dispos 5:527–546, 1977.

42. RK Lynn, JE Bauer, RG Smith, D Griffin, ML Deinzer, N Gerber. Identification of glucuronide metabolites of mephenytoin in human urine and in bile from the isolated perfused rat liver. Drug Metab Dispos 6(4):494–501, 1978.

43. RK Lynn, JE Bauer, WP Gordon, RG Smith, D Griffin, RM Thompson, R Jenkins, N Gerber. Characterization of mephenytoin metabolites in human urine by gas chromatography and mass spectrometry. Drug Metab Dispos 7(3):138–144, 1979.

44. JP Kassam, BK Tang, D Kadar, W Kalow. In vitro studies of human liver alcohol dehydrogenase variants using a variety of substrates. Drug Metab Dispos 17(5):567–572, 1989.

45. UCT Oppermannn, G Nagel, I Belai, JE Bueld, S Genti-Raimondi, J Koolman, KJ Netter, E Maser. Carbonyl reduction of an anti-insect agent imidazole analogue of metyrapone in soil bacteria, invertebrate and vertebrate species. Chem–Biol Interact 114:211–224, 1998.

46. M Nakamura, T Hayashi. One- and two-electron reduction of quinones by rat liver subcellular fractions. J Biochem 115:1141–1147, 1994.

47. AD MacNicoll, AK Nadian, MG Townsend. Inhibition by warfarin of liver microsomal vitamin K-reductase in warfarin-resistant and susceptible rats. Biochem Pharmacol 33(8):1331–1336, 1984.

48. DJT Porter, JA Harrington, MR Almond, GT Lowen, T Spector. (5)-5-fluoro-5,6-dihydrouracil: kinetics of oxidation by dihydropyrimidine by dihydropyrimidine aminohydrolase. Biochem Pharmacol 48(4):775–779, 1994.

49. H He, A Chandrasekran, M Zang, N Molinaro, K Chan, J Scatina, SF Sisenwine. In vitro metabolism of norethindrone by rat liver microsomes. Pharm Res 13(9): S400, 1996.

50. AA Levin, JG Dent. Comparison of the metabolism of nitrobenzene by hepatic microsomes and cecal microflora from Fischer-344 rats in vitro and the relative importance of each in vivo. Drug Metab Dispos 10(5):450–454, 1982.

51. S Hamill, DY Cooper. The role of cytochrome P-450 in the dual pathways of N-demethylation of N, N'-dimethylaniline by hepatic microsomes. Xenobiotica 14 (1/2):139–149, 1984.

52. RC Bender, HE Paul. Metabolism of the nitrofurans II. Incubation of furacin with mammalian tissues. J Biol Chem 191:217–222, 1951.

53. DR McCalla, A Reuvers, C Kaiser. Activation of nitrofuraznoe in animal tissues. Biochem Pharmacol 20:3532–3537, 1971.

54. J Trefouel, J Trefouel, F Nitti, D Bovet. Activite du p-aminophenylsulfamide sur les infections streptococciques. Comptes Rend Seanc Soc Biol 120:756–762, 1935.

55. AT Fuller. Is p-aminobenzenesulphonamide the active agent in prontosil therapy? Lancet 1:194–198, 1937.

56. WJ Johnson, A Chartrand. The metabolism and excretion of phenazopyridine-hydrochloride in animals and man. Toxicol Appl Pharmacol 37:371–376, 1976.

57. WG Levine. Metaboilsm of azo dyes: implication for detoxification and activation. Drug Metab Dispos 23:253–309, 1937.

58. TJ Jaworski, EM Hawes, G McKay, KK Midha. The metabolism of chloropromazine N-oxide in man and dog. Xenobiotica 20(1):107–115, 1990.

59. MJ Duane. Disposition kinetics and metabolism of nicotine-1'-N-oxide in rabbits. Drug Metab Dispos 19(3):667–672, 1991.

60. C Egen-Schwind, R Eckard, FH Kemper. Metabolism of garlic constituents in the isolated perfused rat liver. Planta Med 58:301–305, 1992.

61. P Hu, L Jin, T Baillie. Studies on the metabolic activation of disulfiram in rat: evidence for electrophilic S-oxygenated metabolites as inhibitors of aldehyde dehydrogenase and precursors of urinary N-acetylcysteine conjugates. J Pharmacol Exp Ther 281(2):611–617, 1997.

62. E Kashiyama, T Todaka, M Odomi, Y Tanokura, DB Johnson, T Yokoi, T Kamataki, T Shimizu. Stereoselective pharmacokinetics and interconversions of flosequinan enantiomers containing chiral sulphoxide in rat. Xenobiotica 24(4):369–377, 1994.

63. SC Fawcett, LJ King, PJ Bunyan, PI Stanley. The metabolism of [14C]-DDT, [14C]-DDD, [14C]-DDE, and [14C]-DDMU in rats and Japanese quail. Xenobiotica 17:525–538, 1987.

64. MJ Kelner, JC McLenithan, MW Anders. Thiol stimulation of the cytochrome P-450-dependent reduction of 1,1,1-trichloro-2,2-bis(p-chlorophenyl)ethane (DDT) to 1,1-dichloro-2,2-bis(p-chlororphenyl)ethane (DDD). Drug Metab Dispos 35(11): 1805–1807, 1986.

65. J Hasselstroem, JO Svensson, J Saewe, Z Wiesenfeld-Halin, Q-Y Yue, X-J Xu. Disposition and analgesic effects of systemic morphine, morphine-6-glucuronide and normorphine in rat. Pharmacol Toxicol (Copenhagen) 79:40–46, 1996.

66. BL Coffman, CD King, GR Rios, TR Tephly. The glucuronidation of opiods, other xenobiotics, and androgens by human UGT2B7Y(268) and UGT2B7H(268). Drug Metab Dispos 26(1):73–77, 1998.
67. PE Hanna, RB Banks. Arylhydroxylamines and arylhydroxamic acids: conjugation reactions. In: Bioactivation of Foreign Compounds. London: Academy Press, 1985, pp 375–402.
68. B Burchell, MWH Coughtrie. UDP-glucoronsyltransferases. Pharm Ther 43:261–289, 1989.
69. KH Dudley. Commentary: stereochemical formulas of β-D-glucuronides. Drug Metab Dispos 13(5):524–526, 1985.
70. RG Dickinson, WD Hooper, MJ Eadie. pH Dependent rearrangement of the biosynthetic ester glucuronide of valproic acid to beta-glucuronidase-resistant forms. Drug Metab Dispos 12(2):247–252, 1984.
71. AM Taburet, P Aymard. Valproate gluconidation by rat liver microsomes: interaction with parahydroxyphenobarbital Biochem Pharm 32(24):3859–3861, 1983.
72. W Lilienblum, KW Bock. N-Glucuronide formation of carcinogenic aromatic amines in rat and human liver microsomes. Biochem Pharmacol 33(13):2041–2046, 1984.
73. DS Sitar, DP Thornhill. Methimazole: absorption, metabolism, and excretion in the albino rat. J Pharmacol Exp Ther 184:432–439, 1973.
74. RP Remmel, MW Sinz. A Quaternary ammonium glucuronide is the major metabolite of lamotrigine in guinea pig in vitro and in vivo studies. Drug Metab Dispos 19(3):630–636, 1991.
75. W Dieterle, JW Faigle, F. Fruh, H Mory, W Theoblod, KO Alt, WJ Richter. Metabolism of phenylbutazone in man. Arzneim Forsch 26:572–577, 1976.
76. W Dieterle, JW Faigle, H Mory, W Theoblod, WJ Richter. Biotransformation and pharmacokinetics of sulfinpyrazone. Eur J Clin Pharmacol 9:135–145, 1975.
77. H Spahn-Langguth, LZ Benet. Acyl glucuronides revisited: is the glucuronidation process a toxification as well as a detoxification mechanism? Drug Metab Rev 24(1):5–48, 1992.
78. AB Roy. Review: evaluation of sulfate esters and glucuronides by enzymatic means. Analyt Biochem 165:1–12, 1987.
79. KA McGurk, RP Remmel, VP Hosagrahara, D Tosh, B Burchell. Reactivity of mefenamic acid 1-0-acyl glucuronide with proteins in vitro and ex vivo. Drug Metab Dispos 24(8):842–849, 1996.
80. PJ Hayball. Formation and reactivity of acyl glucuronides: the influence of chirality. Chirality 7:1–9, 1995.
81. H Spahn-Langguth, M Dahms, A Hermening. Acyl glucuronides: covalent bonding and its potential relevance. In: Biological Reactive Intermediates V. New York: Plenum, 1996, pp 313–327.
82. BK Tang. Drug glucosidation. Pharmac Ther 46:53–56, 1990.
83. DE Duggan, JJ Baldwin, BH Arison, RE Rhodes. N-glucoside formation as a detoxification mechanism in mammals J Pharmacol Exp Ther 190(3):563–569, 1974.
84. T Gessner, A Jacknowitz, CA Vollmer. Studies of mammalian glucoside conjugation. Biochem J 132:249–258, 1973.

85. BK Tang, W Kalow. Amobarbital metabolism in man: N-glucoside formation. Res Comm Chem Pathol Pharmacol 21:45–53, 1978.

86. LE Richards, HJ Pieniaszek, S Shatzmiller, GO Page, KF Blom, JM Read, AF Davidson, PN Confalone. Human moricizine metabolism. I. Isolation and identification of metabolites in human urine. Xenobitotica 27:217–229, 1997.

87. BK Tang, W Kalow, AA Grey. Metabolic fate of phenobarbital in man: N-glucoside formation. Drug Metab Dispos 7(5):315–318, 1979.

88. EW Boberg, EC Miller, JA Miller, A Poland, A Liem. Strong evidence from studies with brachymorphic mice and pentachlorophenol that 1′-sulfooxysafrole is the major ultimate electrophilic and carcinogenic metabolite of 1′-hydroxysafrole in mouse liver. Cancer Res 43:5163–5173, 1983.

89. EC Miller, JA Miller, EW Boberg, KB Delclos, CC Lai, TR Fennell, RW Wiseman, A Liem. Sulfuric acid esters as ultimate electrophilic and carcinogenic metabolites of some alkenylbenzenes and aromatic amines in mouse liver. Carcinog Compr Surv 10:93–107, 1985.

90. GJ Mulder, JM Te Koppele, COGM Schipper, WHM Snel, KS Pang, M Polhuijs. Stereoselectivity of glutathione conjugation in vivo, in the perfused liver and in isolated hepatocytes. Drug Metab Rev 23:311–330, 1991.

91. CB Pikett, AYH Lu. Glutathione S-transferase: gene structure, regulation, and biological function. Annu Rev Biochem 58:743–764, 1989.

92. RN Armstrong. Glutathione S-transferase: structure and mechanism of an archetypical detoxification enzyme. Adv Enzymol Relat Areas Mol Biol 69:1–44, 1994.

93. P Steinberg, T Fischer, S Kiulies, K Biefang, KL Platt, F Oesch, T Boettger, C Bulitta, P Kempf, J Hengstler. Drug metabolizing capacity of cryopreserved human rat, and mouse liver parenchymal cells in suspension. Drug Metab Dispos 27:1415–1422, 1999.

94. G Urban, W Dekant. Metabolism of 1,1-dichloro-2,2,2-trifluoroethane in rats. Xenobiotica 24:881–892, 1994.

95. A Ramesh, F Inyang, DB Hood, A Archibong, AM Nyanda, ME Knucles. Oral bioavailability and metabolic profile of benzo(a)pyrene [B(a)p] in F-344 rats. Proceedings of the International Society for the Study of Xenobiotics, Nashville, 1999, p. 236.

96. RG Tirona, KS Pang. Bimolecular glutathione conjugation kinetics of ethacrynic acid in rat liver: in vitro and perfusion studies. J Pharmacol Exp Ther. 290:1230–1241, 1999.

97. RG Tirona, E Tan, G Meier, KS Pang. Uptake and glutathione conjugation of ethacrynic acid and efflux of the glutathione adduct by periportal and perivenous rat hepatocytes. J Pharmacol Exp Ther 291:1210–1219, 1999.

98. DR Thakker, CR Creveling. O-Methylation. In: GJ Mulder, ed. Conjugation Reactions in Drug Metabolism: An Integrated Approach. London: Taylor & Francis, 1990, pp 193–232.

99. SS Ansher, WB Jakoby. N-Methyltransferases. In: GJ Mulder, ed. Conjugation Reactions in Drug Metabolism: An Integrated Approach. London: Taylor & Francis, 1990, pp 233–250.

100. JL Stevens, JE Bakke. S-Methylation. In: GJ Mulder, ed. Conjugation Reactions

in Drug Metabolism: An Integrated Approach. London: Taylor & Francis, 1990, pp 251–272.

101. GM Pacifici, S Santerini, L Giuliani, A Rane. Thiol methyltransferase in humans: development and tissue distribution. Dev Pharmacol Ther 17:8–15, 1991.

102. RE Billings. Mechanisms of catechol formation from aromatic compounds in isolated rat hepatocytes. Drug Metab Dispos 13(3):287–290, 1985.

103. PA Crooks. N-methylation of nicotine enantiomers by human liver cytosol. J Pharm Pharmacol 40:153–154, 1988.

104. G Schepers, K Rustmeier, R-A Walk, U Hackenberg. Metabolism of S-nicotine in noninduced and Aroclor-induced rats. Eur J Drug Metab Pharmacokinet 18(2):187–197, 1993.

105. DAP Evans. N-Acetyltransferase. In: W Kalow, ed. Pharmacogenetics of Drug Metaboilsm. New York: Pergamon Press, 1984, pp 95–178.

106. WW Weber, GN Levy, DW Hein. Acetylation. In: GJ Mulder, ed. Conjugation Reactions in Drug Metabolism: An Integrated Approach. London: Taylor & Francis, 1990, pp 163–192.

107. L Estrada-Rodgers, GN Levy, WW Weber. Substrate selectivity of mouse N-acetyltransferas 1,2 and 3 expressed in COS-1 cells. Drug Metab Dispos 26(5):502–505, 1998.

108. JL Woolley Jr., CW Siegel. Metabolism and disposition by the rat of 35S-sulfadiazine alone and in the presence of trimethoprim. Drug Metab Dispos 7:94–99, 1979.

109. DK Monteith, G Michalopoulos, SC Strom. Conjugation of chemical carcinogens by primary cultures of human hepatocytes. Xenobiotica 20(8):753–763, 1990.

110. CL Ritter, D Malejka-Giganti. Activation of the carcinogens N-hydroxy-N-fluorenylbenzamide and N-hydroxy-N-2-fluorenylacetamide via deacylations and acetyl transfers by rat peritoneal serosa and liver. Carcinogenesis 15(2):163–170, 1994.

111. Y Feng, W Jiang, DW Hein. 2-Aminofluorene-DNA adduct levels in tumor-target and nontarget organs of rapid and slow acetylator Syrian hamsters congenic at the NAT2 locus. Toxicol Appl Pharmacol 141(1):248–255, 1996.

112. T Kanazu, T Yamaguchi. Comparison of in vitro carnitine and glycine conjugation with branched-side chain and cyclic side chain carboxylic acids in rats. Drug Metab Dispos 25(2):149–153, 1997.

113. AJ Hutt, J Caldwell. Amino Acid Conjugation. In: GJ Mulder, ed. Conjugation Reactions in Drug Metabolism: An Integrated Approach. London: Taylor & Francis, 1990, pp 273–306.

114. RJ Griffin, VB Godfrey, YC Kim, LT Burka. Sex dependent differences in the disposition of 2,4-dichorophenoxyacetic acid in Sprague–Dawley rats, B6C3F1 mice and Syrian hamsters. Drug Metab Dispos 25:1065–1071, 1997.

115. RA Halpin, EH Ulm, AE Till, PH Kari, KP Vayas, DB Hunninghake, DE Duggan. Biotransformation of Lovastatin. V. Species differences in in vivo metabolite profiles of mouse, rat, dog, and human. Drug Metab Dispos 21(6):1003–1011, 1993.

116. UA Meyer, RC Skoda, UM Zanger M Heim, F Broly. The genetic polymorphism of debrisoquine/sparteine metabolism—molecular mechanisms. In: W Kalow, ed. Pharmacogenetics of Drug Metabolism. New York: Pergamon Press, 1984, pp 609–624.

117. GR Wilkinson, FP Guengerich, RA Branch. Genetic polymorphism of S-mephenytoin hydroxylation. In: W Kalow, ed. Pharmacogenetics of Drug Metabolism. New York: Pergamon Press, 1984, pp 657–688.

118. J-M Linget, P du Vignaud. Automation of metabolic stability studies in microsomes, cytosol and plasma using a 215 Gilson liquid handler. J Pharm Biom Anal 19:893–901, 1999.

119. I Chu, L. Faveau, T. Soares, CC Lin, AA Nomeir. Validation of higher-throughput high performance liquid chromatography/atmospheric pressure chemical ionization tandem mass spectrometry assays to conduct cytochrome P450s CYP2D6 and CYP3A4 enzyme inhibition studies in human live microsomes. Rapid Commun Mass Spectrom 14:207–214, 2000.

120. PW Erhardt. Drug metabolism data: past, present, and future considerations. In Drug Metabolism, Databases and High Throughput Testing During Drug Design and Development. London: Blackwell, 1999, pp 2–15.

121. CL Fernandez- Metzler, KG Owens, TA Ballie, RC King. Rapid liquid chromatography with tandem mass spectrometry-based screening procedures for studies on the biotransformation of drug candidates. Drug Met Dispos 27:32–40, 1999.

122. X Yu, D Cui, MR Davis. Identification of in vitro metabolites of Indinavir by "intelligent automated LC-MS/MS" (INTAMS) utilizing triple quadrupole tandem mass spectrometry. J Am Soc Mass Spectrom 10:175–183, 1999.

123. MJ de-Groot, MJ Ackland, VA Horne, AA Alex, BC Jones. A novel approach to predicting P450 mediated drug metabolism. CYP2D6 catalyzed N-dealkylation reactions and qualitative metabolite predictions using a combined protein and pharmacophore model for CYP2D6. J Med Chem 42:4062–4070, 1999.

124. MJ de-Groot, MJ Ackland, VA Horne, AA Alex, BC Jones. Novel approach to predicting P450 mediated drug metabolism: development of a combined protein and pharmacophore model for CYP2D6. J Med Chem 42:1515–1524, 1999.

# 10

## Mass Spectrometry-Assisted Applications of Stable-Labeled Compounds in Pharmacokinetic and Drug Metabolism Studies

**Krishna Ramalingam Iyer**
*Bombay College of Pharmacy, Mumbai, India*

> The man who has no imagination has no wings.
> —*Muhammad Ali*

## I. INTRODUCTION

Atoms that have the same atomic number ($Z$) but different atomic mass numbers ($A$) are termed isotopes [1]. Isotopes are classified into atoms that are unstable and decay by emission of radiation to stable products (radioisotopes) and those that have never been observed to decay (stable isotopes). Most of the elements associated with organic compounds possess naturally occurring stable isotopic variants. Stable isotopes frequently encountered in the course of pharmacokinetic and drug-metabolism studies and their natural abundances, relative to the most abundant mass, are listed in Table 1 [2,3].

All organic molecules present in nature or synthesized by standard methods (using reagents derived from natural sources) are inevitably composed of stable-isotope variants, due to the natural presence of isotopes of constituent atoms. These molecular isotopic variants have one or more constituent isotopic atoms present in the molecule to the extent of their natural abundance. Reagents enriched in the content of a heavier isotope variant (more than 99% atom percentage) are also available for artificial/intentional labeling of a molecule. Generally,

**Table 1**  Relative Abundances of Some of the Commonly Used Stable Isotopes

| Element | Most abundant mass (100%) | Stable isotope | Relative natural abundance (%) |
|---|---|---|---|
| Hydrogen | $^1H$ | $^2H$ | 0.016 |
| Carbon | $^{12}C$ | $^{13}C$ | 1.11 |
| Nitrogen | $^{14}N$ | $^{15}N$ | 0.38 |
| Oxygen | $^{16}O$ | $^{17}O$ | 0.04 |
| | | $^{18}O$ | 0.20 |
| Sulfur | $^{32}S$ | $^{33}S$ | 0.78 |
| | | $^{34}S$ | 4.40 |
| Chlorine | $^{35}Cl$ | $^{37}Cl$ | 32.5 |
| Bromine | $^{79}Br$ | $^{81}Br$ | 98.0 |
| Silicon | $^{28}Si$ | $^{29}Si$ | 5.10 |
| | | $^{30}Si$ | 3.4 |

the enhancement in molecular weight due to the presence of stable isotopes does not significantly alter the physicochemical properties, and in most cases, these compounds behave similarly to their unlabeled counterparts. More importantly, stable labeled compounds and their unlabeled analogs cannot be distinguished by common detection systems, like UV, fluorescence, or colorimetry [4,5]. Historically, it was the absence of isotope-specific detection methods that hindered the utility of stable-labeled compounds in pharmaceutical research [5–7]. Stable-isotope-labeled compounds were therefore deemed "silent tracers," in contrast to radiolabeled compounds that could be easily detected and traced due to the emission of radioactivity [5]. Advances in chromatography and mass spectrometry, more specifically high-performance liquid chromatography (HPLC) coupled to selected ion monitoring or multiple reaction monitoring, have been instrumental in the increased utility of stable-labeled compounds in several elegant studies involving pharmacokinetics and drug metabolism [4–13]. In addition, the pronounced influence from the areas of metabolism and pharmacokinetics in the drug discovery process have allowed these researches to incorporate techniques once exclusive to drug development to this new fast-paced arena.

Stable isotopes occurring naturally in a molecule are detected by presence of satellite peaks in a mass spectrum. These satellite peaks appear at $m/z$ values higher than that calculated from standard tables for parent or fragment ions constituted of the naturally more abundant mass. The shift in $m/z$ values and intensity of peaks depend on the number and natural abundance of isotopic variants in the molecule and are fixed within limits of experimental error [2,3]. When stable-labeled compounds are synthesized, the mass spectra of the parent or fragment ion show peaks shifted to $m/z$ values equal to $(M + x)$ depending on the nature and number of stable labels introduced (M + 1 or M + 2 if one or two deuterium atoms are introduced, respectively). The relative ion intensity with respect to the unlabeled counterpart is dependent on the %atom incorporation of the stable label, after correction for contribution of the naturally more abundant mass to the (M + x) channel [14]. Thus, mass spectrometry allows both qualitative and quantitative determination of the presence of stable-labeled compounds.

Nuclear magnetic resonance spectroscopy (NMR) is an alternate method for detection of stable isotopes [4,5,15]. Both $^{13}$C and $^{15}$N have a nuclear spin of 1/2 and possess magnetic moments. Thus, magnetic resonances characteristic of the environment of these atoms can be generated and recorded. Although $^{15}$N NMR yield signals with poor signal-to-noise ratios, $^{13}$C NMR with proton decoupling yields sharp peaks whose chemical shift values depend on the environment of the carbon atoms. In contrast, $^{2}$H has no magnetic moment and is NMR silent. However, indirect methods involving comparison of NMR spectra of the labeled and unlabeled analogs can be utilized for determination of the presence/location of $^{2}$H in a molecule. Thus, NMR allows determination of both the initial location of the stable label and change in chemical shift indicative of metabolism at the labeled site. Although NMR techniques are less useful for quantitative purposes, they are complementary to mass spectrometry in that they allow unequivocal determination of the position of the stable label in a molecule, which may not always be possible by mass spectrometry [5–7,15–17].

With this brief background on stable isotopes and the principles behind their detection, this chapter attempts to provide a window to some unique applications of stable isotopes in pharmacokinetic and drug metabolism studies. These and additional applications have been subjects of some excellent reviews that provide more elaborate discussions of the use of stable isotopes in research [4–13,18]. The emphasis, for obvious reasons, is placed on applications that utilize mass spectrometry as the analytical method. The applications are given below, followed by a brief discussion of each application and considerations involved therein.

## II. APPLICATIONS OF STABLE-LABELED COMPOUNDS

### A. Internal Standards for Quantitative Analysis

As discussed in Chapters 5 and 6, stable isotopes find a wide application in quantitative mass spectral assays using direct stable-isotope dilution methods (wherein stable-labeled compounds are quantitated using the unlabeled analog as the internal standard) or reverse stable-isotope dilution procedures (wherein unlabeled compounds are quantitated using the stable-labeled analog as the internal standard). Whereas labeled and unlabeled species are presumed to have identical physicochemical properties, they progress through the sample work-up, chromatography, and inonization steps in an identical manner. Their differential determination occurs only in the ultimate step of the assay and is based on the unique ability of the mass spectrometer to provide an added dimension of separation, whereby the labeled and unlabeled species are separated and subsequently detected due to differences in their mass-to-charge values. Thus, this methodology maximizes the inherent selectivity/specificity and sensitivity that characterize mass spectrometry [5,19]. In a typical assay based on reverse isotope dilution, a standard curve (ratio of ion intensities of analyte to internal standard vs. concentration of analyte) is generated using differing concentrations of analyte while keeping the concentration of the internal standard constant. This is followed by determination of the ratio of ion intensities of analyte to stable-labeled internal standard in the test sample. Finally, the concentration of analyte in a test sample is obtained by interpolation of ratio of ion intensities to the concentration axis.

There are several considerations in the use of stable-labeled compounds for quantitative purposes when using mass spectrometry as part of the assay methodology [4,5,10,19]. The stable label must be incorporated in a chemically and/or metabolically stable position and should be retained in the ions selected for monitoring. In this regard, the preferred isotopes are those of carbon, nitrogen, and oxygen because deuterium is notorious for undergoing both nonenzymatic and enzymatic exchange reactions that result in loss of the label. Thus, the propensity for loss of deuterium label must be evaluated during assay development. Ultimately, deuterium labeling is often easier to incorporate into molecules and this is a distinct advantage for the shorter time frames encountered in drug discovery. Further, the stable-labeled analog must preferably differ in molecular weight by 2–5 Da so that its measured ion intensity is devoid of any significant satellite contributions from the natural isotopic abundance of its unlabeled counterpart. This is critical, especially in case of compounds containing chlorine, bromine, and sulfur, whose isotopes are present in high natural abundance. However, introduction of larger numbers of stable labels, especially deuterium, may result in significant differences in physicochemical properties. These differences in physicochemical properties can result in fractionation of the labeled and unlabeled compounds during sample work-up or chromatography [5]. It is not uncommon to find small but significant differences in retention times between peaks of unlabeled and labeled compounds while using high-resolution columns that are available at present. Likewise, to enhance the lower limit of quantitation, the internal standard labeling procedure should yield an analog with high isotopic purity (99% or more) so that the contribution of any unlabeled impurity to the analyte ion channel is minimal.

Within the framework of the above considerations (which apply to other applications of stable isotopes as well) stable-labeled internal standards, arguably "ideal" internal standards [5,19,20], have been used for routine determinations of several compounds and metabolites [21–31].

## B.  Determination of Absolute and Relative Bioavailability

The conventional method for bioavailability determination involves administration of a reference dose (by iv administration for obtaining absolute bioavailability and by any other route/dosage form for relative bioavailability determinations) and the test dose/dosage form, in a randomized crossover design, with a suitable washout period between the treatments. This is followed by determination and comparison of relevant pharmacokinetic parameters (generally, dose-normalized AUC or fraction excreted unchanged) of the test versus reference dose/dosage form for calculation of absolute or relative bioavailability. The underlying assumption in this study design is that pharmacokinetic parameters of drug, in the same subject, are identical on different dosing occasions. However, this may not always be true, especially in case of drugs that have high hepatic extraction ratios and consequently exhibit high first-pass metabolism [32–34]. Stable-labeled analogs offer an elegant solution to this problem [5,35]. One of two doses/dosage forms that is part of the bioavailability study is prepared using a stable-labeled analog. Because mass spectrometry can easily distinguish between the two forms, the two doses/dosage forms can be administered simultaneously, biological samples withdrawn and analyzed, and concentrations of labeled and unlabeled compounds determined in a single analytical run.

Important considerations in the use of stable-labeled compounds for bioavailability studies, in addition to those mentioned in the preceding section for use as internal standards, are that pharmacokinetic parameters of the unlabeled and labeled forms are identical and that administration of two doses at the same time does not result in nonlinear kinetics (via saturation of elimination or excretion pathways) [4–6,12]. The first issue may be addressed by dosing a mixture of unlabeled and labeled compounds of known composition and confirming that the ratio of ion intensities of peaks related to labeled and unlabeled drug is maintained in all samples obtained in course of the experiment. Possible nonlinear pharmacokinetics can be investigated in a pilot study to determine pharmacokinetic parameters at the expected higher doses. Thus, use of stable-labeled drugs for bioavailability determinations (1) obviates the consideration of changing pharmacokinetic parameters that accompany a conventional bioavailability experiment, (2) reduces the time required for the study, (3) reduces the number of samples withdrawn and analyzed, and (4) increases the statistical power of the experiment and allows detection of smaller differences in bioavailability. Some examples of studies that have utilized this approach for determination of absolute and relative bioavailability are listed [35–50].

## C.  Determination of Steady-State Pharmacokinetic Parameters

The determination of differences in pharmacokinetics between single-dose and multiple-dosage regimens is essential for optimizing the therapy of drugs intended for prolonged use. The characterization of autoinduction or autoinhibition phenomena, if present, is important for making suitable dose adjustments for maintenance of requisite plasma levels or reducing adverse effects, respectively. Stable isotope techniques can be used to determine the pharmacokinetics at steady state via a pulse dose that incorporates stable-labeled drug. This pulse dose may be administered in lieu of the normal dose or given intravenously. Mass spectrometry allows concentrations of unlabeled and stable-labeled drug to be determined in parallel and in a single run without having to interrupt the dosing schedule. As mentioned in the preceding section, the considerations regarding identity of pharmacokinetic parameters of the labeled and unlabeled compounds and absence of nonlinear kinetics on pulse dosing apply here as well. The method can be extended to the determination of changes in metabolic profile due to inhibition or induction of drug-metabolizing enzymes [51–58].

## D.  Determination of Differences in Pharmacokinetic Parameters of Enantiomers and Drug-Interaction Studies—Pseudoracemate Approach

It is well accepted that enantiomers can differ from each other in their pharmacological, toxicological, pharmacokinetic, and toxicokinetic profiles [59]. Because drug candidates are initially tested as racemates, a knowledge of the fate of the individual enantiomers is crucial for making appropriate policy decisions regarding development of a chemical entity either as a single enantiomer or a racemate for clinical use. In addition, knowledge of the pharmacokinetics of the individual enantiomers may be required for establishing pharmacokinetic–pharmacodynamic correlations. One method for studying the pharmacokinetics of individual isomers of a racemate involves derivatization of drug present in the sample by using one antipode of a chiral reagent. The resulting diastereomers, due to differences in their physicochemical properties, are then separated by chromatography and quantitated. In another method, chiral stationary phases are used for separation and subsequent quantitation of the individual isomers; derivatization is not required. The major disadvantage of the first method is the requirement of suitable functionalities in the compound that can be derivatized and the limited options with respect to the availability of chiral derivatizing agents. The drawback of using chiral columns is their cost, variable performance, and the limited time of operation. In addition, both methods yield no information regarding the metabolic fate of enantiomers.

A "pseudoracemate" approach, wherein one of the isomers of the racemate is stable labeled and acts as a stereochemical marker, circumvents the issues related to derivatization methods and use of chiral columns for analysis of racemates. Separation and detection by mass spectrometry, using another stable-labeled variant as the internal standard when possible, allows quantitation of the two enantiomers based on their $m/z$ values [60–63]. An alternate application of this approach is the study of drug–drug interactions by monitoring the differential effects on pharmacokinetic parameters of the individual enantiomers in the presence of coadministered drug(s) [64–67]. However, it is important to demonstrate absence of kinetic isotope effects associated with the use of stable-labeled compound and absence of chiral inversion in vivo or during sample work for proper interpretation of the results [68–70].

## E. Tracers for Structure Determination and Quantitation of Metabolites—Ion Cluster Methods

This application is based on the characteristic mass-spectral profiles of halogen-containing compounds due to the high natural abundance of the stable isotopes of chlorine or bromine. A compound containing a single Cl or Br atom will give rise to an isotope peak at $m/z = (M + 2)$, that is, about one-third of or equal in intensity to, respectively, the signal associated with the lighter isotope peak. This is observed as typical ''doublet'' or ''twin ion'' peaks or ''ion clusters'' in the mass spectrum and is also a characteristic of all metabolites derived from the compound that retain the halogen [2–5,71]. Such a signature allows easy detection of metabolites and aids structure elucidation. For compounds lacking halogen substituents, this ''doublet ion'' (or twin ion or ion cluster) signature can be artificially generated by mixing known amounts of unlabeled and stable-labeled analogs (or synthesizing stable-labeled compounds of intermediate isotopic purity) [4,5]. The advantage of this approach is that retention times and the expected ratio of individual ion intensities of an ion cluster can be used both to rule out spurious or noncompound related peaks and for identification of unknown metabolites [4,5,72–76].

A variation of the ''ion cluster'' methodology is useful for identification of metabolites and involves derivatization with a reagent that is composed of a mixture of unlabeled and labeled forms [4,5]. In this case, the ''ion cluster'' signature is generated during sample work-up. Depending on the derivatizing agents and their reactivity toward specific functional groups, important information regarding the presence and number of functional groups (hydroxyl, amino, sulfydryl, aldehyde, etc.) can be obtained, which can assist in the structure elucidation of unknown metabolites [77–80].

## F. Elucidation of Drug Biotransformation Mechanisms

Stable isotopes are an important tool in the investigation of xenobiotic metabolic pathways and mechanisms involved therein. Issues that can be addressed include the source of a new atom or functional group that is introduced into a compound during metabolism, location of the site at which metabolism occurs, molecular rearrangements involved in metabolite generation, mechanism of drug biotransformation, determination of rate-limiting steps in enzyme-catalyzed reactions, and determination of the nature of transition states involved in enzyme reactions. An understanding of these issues can lead to an enhanced knowledge of the topology of the enzyme active site [4,5,8,11,18,81].

Stable isotopes are convenient tools for tracing the origin of new atoms or functional groups that are introduced in course of biotransformations. A classic example of this approach is the use of [18]O-labeled oxygen to show that cytochrome P450-mediated hydroxylations incorporate one atom of oxygen from molecular oxygen into the substrate. An extension of this approach is the use of [18]O-labeled water to clarify the origin of oxygen in the metabolism of unsaturated systems to diol metabolites [82–85]. Likewise, stable-labeled cofactors like S-adenosyl methionine or glutathione may be used for confirmation of metabolic pathways.

On the other hand, studies aimed at the determination of location of metabolic sites can involve deliberate introduction of stable labels at potential sites of metabolism. Subsequent retention or loss of the label is used to explain the formation/existence of specific metabolic pathways and sites of metabolism. For example, oxidative toluene metabolites hydroxylated on the ring or the methyl group are possible. If trideuterated toluene, labeled on the methyl group, is used as a substrate, the benzyl alcohol formed would retain only two deuterium while aromatic hydroxylation would yield trideuterated hydroxytoluenes. Thus, observation of metabolites retaining three or two deuterium atoms provides an immediate indication of the possible nature of metabolites [85–87]. Although this approach finds general utility in determination of the regiospecificity in metabolism [88], aromatic systems have been shown to exhibit anomalous behavior. In aromatic systems, hydroxylation may proceed with retention of the deuterium that was originally present at the site of hydroxylation. This phenomenon, the classic "NIH shift," was in fact discovered as a result of the dubious expectation that loss of label is indicative of site of metabolism [89–91]. The discovery of the "NIH shift" has clarified our understanding of cytochrome P450-mediated aromatic hydroxylations to include the intermediacy of arene oxides and their molecular rearrangement to yield phenolic metabolites [22,82–84,89–96].

An additional important consideration in introducing stable labels at potential sites of metabolism is the possibility of kinetic isotope effects. Kinetic isotope effects are observed as differences in the rate of reaction of a stable-labeled compound relative to the unlabeled counterpart when cleavage of the bond involving stable label is rate limiting or partially rate limiting [18,97–99]. Primary kinetic isotope effects arise due to differences in the zero-point energies of bonds involving stable isotopes (e.g., $C-^2H$) with respect to the unlabeled compound (i.e., $C-H$). This results in the increased stability of the $C-^2H$ bond as compared to the $C-H$ bond and consequently increases the energy required to cleave a $C-^2H$ bond. Kinetic isotope effects of consequence are generally those involving deuterium because the percentage difference in mass of deuterium with respect to its isotopic variant hydrogen is the highest, in contrast to other stable labels ($^{12}C$ vs. $^{13}C$ and $^{16}O$ vs. $^{18}O$) generally used in drug metabolism studies. Kinetic isotope effects are expressed as the ratio of rate constants of the reaction involving unlabeled and labeled compounds, and their values may lie in the range of 1.2 to more than 10 depending on the nature of the reactants, transition state involved, nature of enzyme, and the mechanism of reaction [98]. In general, a primary deuterium kinetic isotope effect of 4 or more is indicative of the C—H bond being rate limiting in a reaction. This is true for several aliphatic hydroxylations and O-dealkylation reactions mediated by cytochrome P450 isozymes [100–105]. In contrast, aromatic hydroxylations show low isotope effects (1.1–1.3) [22,92–94] and N-dealkylation reactions variable isotope effects depending on the substrate and enzyme system involved [106–110]. These findings have opened up new vistas for research aimed at correlation of observed kinetic isotope effects with proposed reaction mechanisms [111–118].

There are three general experimental approaches to the determination of primary kinetic isotope effects based on whether the relative rates of reaction are determined based on intermolecular or intramolecular competition of unlabeled and labeled analogs for enzyme-mediated catalysis [119–121]. The first two methods yield intermolecular isotope effects and the third method provides intramolecular isotope effects. In the first method, "noncompetitive intermolecular isotope effects" are obtained from two experiments, one involving the unlabeled compound and other the stable labeled compound. The relative rates of reaction are then compared based on the relative amount of products formed. Strictly speaking, this method does not necessitate the use of mass spectrometry for analysis. Disadvantages of this approach are that factors pertaining to differences in binding/debinding of the two analogs, the multistep nature of several enzyme reactions, product release, and the fact that two experiments are performed introduce errors in the isotope effect determination and more importantly may attenuate or "mask" the intrinsic isotope effects. The second method yields "competitive intermolecular isotope effects" and involves the use of a 1:1 mixture of labeled (labeled both at the site of metabolism for isotope effect determination and at an alternate site for detection) and unlabeled analogs and subsequent determination of products by mass spectrometry. This method, based on a single experiment, reduces some of the errors of the first method. However, attenuation or "masking" of intrinsic isotope effects may still be observed. In the third approach to determination of isotope effects, the substrate is chosen such that an intramolecular competition is set up between sites that are unlabeled and labeled for the enzyme catalytic site. Examples of substrates amenable to this design are xylenes (where one of methyl groups is trideuterated), dimethoxybenzenes, alkanes, and other symmetrical molecules. In this design, using p-trideutero-methyl toluene, for example, a single experiment with one substrate, followed by the determination of relative amounts of trideuterated and dideuterated *p*-methyl benzyl alcohol formed, directly yields the value for the kinetic isotope effect. The major advantage of this intramolecular isotope effect experimental design is that most factors that contribute to "masking" are eliminated and the observed isotope effects are close to the intrinsic isotope effects. The only requirement for successful determination of isotope effects by this method is that the rate of equilibration of labeled and unlabeled sites of the symmetrical molecule with respect to the catalytic center is faster than the bond breaking reaction, i.e., the catalytic center must have equal access to the two sites at all times [122]. Thus, in special cases, compounds where the unlabeled and labeled sites are far apart may have difficulty in equilibrating within the active site and may show "masking" of isotope effects. This relationship between the distance of the labeled and unlabeled sites and masking of isotope effects has been used to evaluate the active site dimensions of cytochrome P450 enzymes [123].

Unlike primary kinetic isotope effects, secondary kinetic isotope effects arise due to presence of stable isotope at sites close to point of reaction and consequently influence the geometry of the reaction intermediates. Although the absolute values are much lower than those observed for primary effects, secondary kinetic isotope effects yield important information pertaining to the progress of the reaction and transition states involved therein [100,124,125].

The kinetic isotope effect experiments outlined above, along with use of stable isotopes for elucidating the source of oxygen, and observation of the "NIH" shift have been crucial in developing an understanding of cytochromes P450, probably the most important of all the enzyme systems involved in xenobiotic metabolism.

## G.   Drug Metabolism Profiling and Mass Balance Studies

Drug-metabolism profiling and mass-balance studies are essential steps in new drug development and are becoming routing experiments in drug discovery as well (see Chap. 11). These studies are performed with little or no prior knowledge of the metabolic fate of a chemical entity. Thus, compound- or structure-related detection methods involving UV-vis spectroscopy, IR spectroscopy, fluorescence characteristics, electrochemical properties, immunoassays, or mass spectrometry using selected ion monitoring (the method on which almost all applications listed above are based) are of little utility for detection of unknown metabolites. "Ion cluster techniques," although useful for qualitative observation of compound derived peaks, do not ensure complete detection of all possible metabolic products. As such, metabolic profiling and mass-balance studies have traditionally relied on the use of radiolabeled tracers. Radioactivity, dependent only on retention of radiolabel in the compound or its metabolites, allows nonspecific detection (if radiolabel is placed judiciously) to yield drug-metabolism profiles and mass-balance data that are necessary for drug development. Stable-labeled compounds that are devoid of associated radiation hazards, legal issues related to the use of radioactive compounds, and chemical instability associated with radiolysis have also been investigated as potential agents for metabolic profiling and mass-balance studies.

The basis for this application of stable isotopes to these types of investigations is the indirect mass spectrometric determination of the presence of stable labels in a compound or its metabolites. This is achieved by chromatographic separation of the sample and subsequent conversion of the eluate components into carbon dioxide, nitric oxide, or nitrogen (depending on the nature of stable label in the compound). Since carbon dioxide/nitric oxide/nitrogen derived from eluant components that are stable labeled compound related will be enriched in their stable isotope content, they can be easily detected and quantitated after suitable corrections to eliminate background contributions due to unlabeled carbon dioxide/nitric oxide/nitrogen. Although applications of stable isotopes in drug metabolism profiling and mass balance studies have been reported [126–135], issues related to the sensitivity of this method compared to radiation detection, availability of instrumentation, and cost of instrumentation, must be evaluated before this application is used routinely.

## III. CONCLUSION

This chapter has attempted to briefly showcase some of the applications of stable isotopes in pharmacokinetic and drug-metabolism studies. Although many of the examples illustrated here are from the area of drug development, increased application of stable-labeled compounds has occurred in the discovery area. Emphasis has been on the variety of strategies to address research problems, using stable-labeled compounds with mass spectrometry as an aid. However, it must be realized that selection of the stable label and its judicious placement in a molecule (i.e., design of the stable-labeled compounds, synthesis of the stable-labeled compounds, mass spectral analysis, deconvolution of data, and interpretation of the results) necessitates significant specialized skills and intellectual input for successful use of stable-labeled compounds. The scope of investigations with stable-labeled compounds in conjunction with mass spectrometry itself is limited only by the ingenuity of an investigator.

## REFERENCES

1. GD Chase, B Shapiro. Fundamentals of radioisotopes. In: AR Gennaro, ed. Remington: The Science and Practice of Pharmacy. Easton, PA: Mack, 1995, pp 336–391.

2. Mass spectrometry. In: DA Skoog, JJ Leary, eds. Principles of Instrumental Analysis. Orlando, FL: Harcourt Brace College, 1992, pp 420–461.
3. Mass spectrometry. In: RM Silverstein, FX Webster, eds. Spectrometric Identification of Organic Compounds. New York: Wiley, 1998, pp 2–70.
4. LF Chasseaud, DR Hawkins. Use of isotopes in quantitative and qualitative analysis. In: JB Taylor, ed. Comprehensive Medicinal Chemistry: The Rational Design, Mechanistic Study & Therapeutic Application of Chemical Compounds. Oxford, UK: Pergamon, 1990, pp 359–384.
5. TA Baillie. The use of stable isotopes in pharmacological research. Pharmacol Rev 33:81–132, 1981.
6. TA Baillie, AW Rettenmeier, LA Peterson, N Castagnoli Jr. Stable isotopes in metabolism and disposition. In: Annual Reports in Medicinal Chemistry. Vol. 19. New York: Academic Press, 1984, pp 273–282.
7. SD Nelson, LR Pohl. The use of stable isotopes in medicinal chemistry. In: Annual Reports in Medicinal Chemistry. Vol. 12. New York: Academic Press, 1977, pp 319–329.
8. TA Baillie, AW Rettenmeier. Recent advances in the use of stable isotopes in drug metabolism research. J Clin Pharmacol 26:481–484, 1986.
9. U Keller, MJ Rennie. Stable isotopes in clinical research. Eur J Clin Invest 16:97–100, 1986.
10. TR Browne. Stable isotopes in pharmacological studies: present and future. J Clin Pharmacol 26:485–489, 1986.
11. DR Hawkins. The role of stable isotopes in drug metabolism. In: JW Bridges, LF Chasseaud, eds. Progress in Drug Metabolism. Bristol, UK: Wiley, 1977, pp 163–218.
12. TR Browne. Stable isotopes in clinical pharmacokinetic investigations: advantages and disadvantages. Clin Pharmacokinet 18:423–433, 1990.
13. Y Osawa, RJ Highet, LR Pohl. The use of stable isotopes to identify reactive metabolites and target macromolecules associated with toxicities of halogenated hydrocarbon compounds. Xenobiotica 22:1147–1156, 1992.
14. KR Korzekwa, WN Howald, WF Trager. The use of Brauman's least squares approach for the quantification of deuterated chlorophenols. Biomed Environ Mass Spectrom 19:211–217, 1990.
15. WS Caldwell, GD Byrd, JD deBethizy, PA Crooks. Modern instrumental methods for studying mechanisms of toxicity. In: AW Hayes, ed. Principles and Methods of Toxicology. New York: Raven Press, 1994, pp 1335–1390.
16. SCJ Sumner, JP MacNeela, TR Fennel. Characterization and quantitation of urinary metabolites of $(1,2,3,4-^{13}C)$acrylamide in rats and mice using $^{13}C$ nuclear magnetic resonance spectroscopy. Chem Res Toxicol 5:81–89, 1992.
17. K Akira, E Negishi, C Sakuma, T Hashimoto. Direct detection of antipyrine metabolites in rat urine by $^{13}C$ labeling and NMR spectroscopy. Drug Metab Disp 27:1248–1253, 1999.
18. WW Cleland. Use of isotope effects to elucidate enzyme mechanisms. CRC Crit Rev Biochem 13:385–428, 1982.
19. WA Garland, MP Barbalas. Applications to analytical chemistry: an evaluation of stable isotopes in mass spectral drug assays. J Clin Pharmacol 26:412–418, 1986.

20. E Bailey, PB Farmer, JH Lamb. The enantiomer as internal standard for the quantitation of the alkylated amino acid S-methyl-cysteine in haemoglobin by GC-CI-MS with single ion detection. J Chromatogr 200:145–152, 1980.

21. AE Rettie, AC Eddy, LD Heimark, M Gibaldi, WF Trager. Characteristics of warfarin hydroxylation catalyzed by human liver microsomes. Drug Metab Disp 17:265–270, 1990.

22. KR Korzekwa, DC Swinney, WF Trager. Isotopically labelled chlorobenzenes as probes for the mechanism of cytochrome P-450 catalyzed aromatic hydroxylation. Biochemistry 28:9019–9027, 1989.

23. WA Garland, GT Miwa, C Eliahou, W Dairman. The metabolism in mice of two deuterated analogs of diazepam. In: RR Muccino, ed. Synthesis and Applications of Isotopically Labeled Compounds. Proceedings of the Second International Symposium, MO, USA. Amsterdam: Elsevier, 1985, pp 271–276.

24. NJ Haskins, GC Ford, RF Palmer, KA Waddell. A stable isotope dilution assay for disopyramide and its ($^{13}$C, $^{15}$N) labeled analogue in biological fluids. Biomed Mass Spectrom 7:74–79, 1980.

25. JM Neal, WN Howald, KL Kunze, RF Lawrence, WF Trager. Application of negative-ion chemical ionization isotope dilution gas chromatography-mass spectrometry to single dose bioavailability studies of mefloquine. J Chromatogr B Biomed Appl 661:263–269, 1994.

26. KA Regal, WN Howald, RM Peter, CA Gartner, KL Kunze, SD Nelson. Subnanomolar quantification of caffeine's in vitro metabolites by stable isotope dilution gas chromatography-mass spectrometry. J Chromatogr B Biomed Appl 708:75–85, 1998.

27. M Preu, M Petz. Isotope dilution GC-MS of benzylpenicillin residues in bovine muscle. Analyst 123:2785–2788, 1998.

28. SA White, AS Kidd, KS Webb. The determination of lysergide (LSD) in urine by high-performance liquid chromatography-isotope dilution mass spectrometry (IDMS). J Forensic Sci 44:375–379, 1999.

29. HJ Leis, G Fauler, G Raspotnig, W Windischhofer. Quantitative determination of the angiotensin-converting enzyme inhibitor lisinopril in human plasma by stable isotope dilution gas chromatography/negative ion chemical ionization mass spectrometry. Rapid Commun Mass Spectrom 12:1591–1594, 1998.

30. G Singh, V Arora, PT Fenn, B Mets, IA Blair. A validated stable isotope dilution liquid chromatography tandem mass spectrometry assay for trace analysis of cocaine and its major metabolites in plasma. Anal Chem 71:2021–2027, 1999.

31. RJ Scott, J Palmer, IA Lewis, S Pleasance. Determination of a 'GW cocktail' of cytochrome P450 probe substrates and their metabolites in plasma and urine using automated solid phase extraction and fast gradient liquid chromatography tandem mass spectrometry. Rapid Commun Mass Spectrom 13:2305–2319, 1999.

32. FLS Tse, WT Robinson, MG Choc. Study design and the assessment of bioavailability and bioequivalence. In: PG Welling, FLS Tse, SV Dighe, eds. Pharmaceutical Bioequivalence. New York: Marcel Dekker, 1991, pp 17–34.

33. Bioavailability. In: M Gibaldi, ed. Biopharmaceutics and Clinical Pharmacokinetics. Philadelphia: Lea and Febiger, 1991, pp 146–175.

34. Biopharmaceutics. In: RE Notari, ed. Biopharmaceutics and Clinical Pharmacokinetics. New York: Marcel Dekker, 1987, pp 130–220.

35. RL Wolen. The application of stable isotopes to studies of drug bioavailability and bioequivalence. J Clin Pharmacol 26:419–424, 1986.

36. JM Strong, JS Dutcher, W-K Lee, AJ Atkinson Jr. Absolute bioavailability in man of N-acetylprocainamide determined by a novel stable isotope method. Clin Pharmacol Therapeut, 18:613–622, 1975.

37. Y Shinohara, S Baba, Y Kasuya, G Knapp, FR Pelsor, VP Shah, IL Honigberg. Stable-isotope methodology in the bioavailability study of 17alpha-methyltestosterone using gas chromatography-mass spectrometry. J Pharm Sci 75:161–164, 1986.

38. JR Carlin, RW Walker, RO Davies, RT Ferguson, WJA Vandenheuvel. Capillary column GLC-mass spectrometric assay with selected-ion monitoring for timolol and ($^{13}C_3$)timolol in human plasma. J Pharm Sci 69:1111–1115, 1980.

39. M Eichelbaum, A Somogyi, GE von Unruh, HJ Dengler. Simultaneous determination of the i.v. and oral pharmacokinetic parameters of D,L-verapamil using stable labelled verapamil. Eur J Clin Pharmacol 19:133–137, 1981.

40. J Schmid, A Prox, H Zipp, FW Koss. The use of stable isotopes to prove the saturable first-pass-effect of methoxsalen. Biomed Mass Spectrom 7:560–564, 1980.

41. DL Theis, LJ Lucisano, GW Halstead. Use of stable isotopes for evaluation of drug delivery systems: Comparison of ibuprofen release in vivo and in vitro from two biphasic release formulations utilizing different rate-controlling polymers. Pharm Res 11:1069–1076, 1994.

42. SL Preston, GL Drusano, P Glue, J Nash, SK Gupta, P McNamara. Pharmacokinetic and absolute bioavailability of ribavarin in healthy volunteers as determined by stable isotope methodology. Antimicrob Agents Chemother 43:2451–2456, 1999.

43. HJ Pieniaszek Jr, M Mayersohn, MP Adams, RJ Reinhart, JS Barrett. Moricizine bioavailability via simultaneous, dual, stable isotope administration: Bioequivalence implications. J Clin Pharmacol 39:817–825, 1999.

44. KC Yeh, JA Stone, AD Carides, P Rolan, E Woolf, WD Ju. Simultaneous investigation of indinavir nonlinear pharmacokinetics and bioavailability in healthy volunteers using stable isotope labeling technique: study design and model-independent data analysis. J Pharm Sci 88:568–573, 1999.

45. JD Gilbert, TV Olah, MJ Morris, A Bortnick, J Brunner. The use of stable isotope labeling and liquid chromatography-tandem mass spectrometry techniques to simultaneously determine oral and ophthalmic bioavailability of timolol in dogs. J Chromatogr Sci 36:163–168, 1998.

46. JX Sun, AJ Piraino, JM Morgan, JC Joshi, A Cipriano, K Chan, E Redalieu. Comparative pharmacokinetics and bioavailability of nitroglycerine and its metabolites from Transderm-Nitro, Nitodisc, and Nitro-Dur II systems using a stable-isotope technique. J Clin Pharmacol 35:390–397, 1995.

47. J Richard, JM Cardot, J Godbillon. Stable isotope methodology for studying the performance of metoprolol Oros tablets in comparison to conventional and slow release formulations. Eur J Drug Metab Pharmacokinet 19:375–380, 1994.

48. K Hage, K Buhl, C Fischer, NG Knebel. Estimation of the absolute bioavailability of flecainide using stable isotope technique. Eur J Clin Pharmacol 48:51–55, 1994.

49. TR Browne, GK Szabo, C McEntegart, JE Evans, BA Evans, JJ Miceli, C Quon,

CL Dougherty, J Kres, H Davoudi. Bioavailability studies of drugs with non-linear pharmacokinetics. II. absolute bioavailability of intravenous phenytoin prodrug at therapeutic phenytoin serum concentrations determined by double-stable isotope technique. J Clin Pharmacol 33:89–94, 1993.

50. AG Burm, F Haak-van der Lely, JW van Kleef, CJ Jacobs, JG Bovill, AA Vletter, RP van den Heuvel, W Onkenhout. Pharmacokinetics of alfentanil after epidural administration. Investigation of systemic absorption kinetics with stable isotope method. Anesthesiology 81:308–315, 1994.

51. L Bertilsson, B Hojer, G Tybring, J Osterloh, A Rane. Autoinduction of carbamazepine metabolism in children examined by a stable isotope technique. Clin Pharmacol Ther 27:83–88, 1980.

52. IM Kapetanovic, HJ Kupferberg. Stable isotope methodology and gas chromatography mass spectrometry in a pharmacokinetic study of phenobarbital. Biomed Mass Spectrom 7:47–52, 1980.

53. GE von Unruh, BC Jancik, F Hoffmann. Determination of valproic acid kinetics in patients during maintenance therapy using a tetradeuterated form of the drug. Biomed Mass Spectrom 7:164–167, 1980.

54. MH Nelson, AK Birnbaum, PJ Nyhus, RP Remmel. A capillary GC-MS method for analysis of phenytoin and $[^{13}C_3]$-phenytoin from plasma obtained from pulse dose pharmacokinetic studies. J Pharm Biomed Appl 17:1311–1323, 1998.

55. TR Browne, JE Evans, GK Szabo, BA Evans, DJ Greenblatt, GE Schumacher. Studies with stable isotopes. I. changes in phenytoin pharmacokinetics and biotransformation during monotherapy. J Clin Pharmacol 25:43–50, 1985.

56. TR Browne, JE Evans, GK Szabo, BA Evans, DJ Greenblatt. Studies with stable isotopes. II. phenobarbital pharmacokinetics during monotherapy. J Clin Pharmacol 25:51–58, 1985.

57. TR Browne, DJ Greenblatt, JS Harmatz, JE Evans, GK Szabo, BA Evans, GE Schumacher. Studies with stable isotopes. III. pharmacokinetics of tracer doses of drug. J Clin Pharmacol 25:59–63, 1985.

58. JW Hubbard, S Hadad, JP Luo, G McKay, KK Midha. Pharmacokinetics of fluphenazine, a highly lipophilic drug, estimated from a pulse dose of stable isoptomer in dogs at steady state. J Pharm Sci 88:918–921, 1999.

59. B Testa, WF Trager. Racemates versus enantiomers in drug development: dogmatism or pragmatism? Chirality 2:129–133, 1990.

60. H Ehrsson. Simultaneous determination of (−) and (+) propranolol by gas chromatography-mass spectrometry using a deuterium labeling technique. J Pharm Pharmacol 28:662–663, 1976.

61. T Walle, MJ Wilson, UK Walle, SA Bai. Stereochemical composition of propranolol metabolites in the dog using stable isotope labelled pseudoracemates. Drug Metab Disp 11:544–549, 1983.

62. T Walle. Stereochemistry of the in vivo disposition and metabolism of propranolol in dog and man using deuterium-labelled pseudoracemates. Drug Metab Disp 13:279–282, 1985.

63. MJ Kreek, DL Hacye, PD Klein. Stereoselective disposition of methadone in man. Life Sci 24:925–932, 1979.

64. WN Howald, ED Bush, WF Trager, RA O'Reilly, CH Motley. A stable isotope

assay for pseudoracemic warfarin from human plasma samples. Biomed Mass Spectrom 7:35–40, 1980.

65.  RA O'Reilly, WF Trager, CH Motley, WN Howald. Stereoselective interaction of phenylbutazone with ($^{12}$C/$^{13}$C)warfarin pseudoracemates in man. J Clin Invest 65: 746–753, 1980.

66.  RA O'Reilly, WF Trager, CH Motley, WN Howald. Interaction of secobarbital with warfarin pseudoracemates. Clin Pharmacol Therapeut 28:187–195, 1980.

67.  LD Heimark, L Wienkers, KL Kunze, M Gibaldi, AC Eddy, WF Trager, RA O'Reilly, DA Goulart. The mechanism of interaction between amiodarone and warfarin in humans. Clin Pharmacol Therapeut 51:398–407, 1992.

68.  Y Shinohara, K Nagao, N Akutsu, S Baba. Assessment of the metabolic chiral inversion of suprofen in rat by gas chromatography-mass spectrometry combined with a stable isotope technique. J Pharm Sci 83:1521–1523, 1994.

69.  F Trejtnar, V Wsol, B Szotakova, L Skalova, P Pavek, M Kuchar. Stereoselective pharmacokinetics of flobufen in rats. Chirality 11:781–786, 1999.

70.  K Zhang, C Tang, M Rashed, D Cui, F Tombret, H Botte, F Lepage, RH Levy, TA Baillie. Metabolic chiral inversion of stiripentol in the rat. I. mechanistic studies. Drug Metab Disp 22:544–553, 1994.

71.  CG Hammar, B Holmstedt, R Rhyage. Mass fragmentography: identification of chlorpromazine and its metabolites in human blood by a new method. Anal Biochem 25:532–548, 1968.

72.  M Vore, BJ Sweetman, MT Bush. The metabolism of 1-n-butyl-5,5-diethylbarbituric acid (N-n-butyl barbital) in the rat. J Pharmacol Exp Ther 190:384–394, 1974.

73.  TR Burke Jr, RV Branchflower, DE Lees, LR Pohl. Mechanism of defluorination of enfluorane: identification of an organic metabolite in rat and man. Drug Metab Disp 9:19–24, 1981.

74.  WL Nelson, TR Burke Jr. Pathways of propranolol metabolism: use of the stable isotope twin-ion GC-MS technique to examine the conversion of propranolol to propranolol-diol by 9000g rat liver supernatant. Res Commun Chem Pathol Pharmacol 21:77–85, 1978.

75.  F Kasuya, K Igarshi, M Fukui. Metabolism of chlorpheniramine in rat and human by use of stable isotopes. Xenobiotica 21:97–109, 1991.

76.  S Honma, S Iwamura, R Kobayashi, Y Kawabe, K Shibata. The metabolism of roxatidine hydrochloride: liberation of deuterium from the piperidine ring during hydroxylation. Drug Metab Disp 15:551–559, 1987.

77.  DJ Harvey, DB Johnson, MG Horning. Detection by gas phase analytical methods of derivatives of barbiturate epoxides. Anal Lett 5:745–755, 1972.

78.  RE McMahon, HR Sullivan, SL Due, FJ Marshall. The metabolite pattern of d-propoxyphene in man: the use of heavy isotopes in drug disposition studies. Life Sci 12:463–473, 1973.

79.  JL Martin, JW George, LR Pohl. Glutathione-dependent dechlorination of chloramphenicol by cytosol of rat liver. Drug Metab Disp 8:93–97, 1980.

80.  WJ Vandenheuval. Use of stable isotopes in the elucidation of drug biotransformation processes. Xenobiotica 17:397–412, 1987.

81.  WW Cleland. Isotope effects: determination of enzyme transition state structure. Methods Enzymol 249:341–373, 1995.

82. J Holtzman, JR Gillette, GW Milne. The metabolic products of naphthalene in mammalian systems. J Am Chem Soc 89:6341–6344, 1967.

83. DM Jerina, JW Daly, B Witkop. The role of arene oxide-oxepin systems in the metabolism of aromatic substrates. II. synthesis of 3,4-toluene-4-$^2$H oxide and subsequent 'NIH shift' to 4-hydroxytoluene-3-$^2$H. J Am Chem Soc 90:6523–6525, 1968.

84. DM Jerina, JW Daly, B Witkop, P Zaltzaman-Nirenberg, S Udenfriend. The role of arene oxide-oxepin systems in the metabolism of aromatic substrates. III. Formation of 1,2-naphthalene oxide from naphthalene in microsomes. J Am Chem Soc 90:6525–6527, 1968.

85. HA Barton, MA Marletta. Comparison of aniline hydroxylation by hemoglobin and microsomal P450 using stable isotopes. Toxicol Lett 70:147–153, 1994.

86. RP Hanzlik, K Hogberg, CM Judson. Microsomal hydroxylation of specifically deuterated monosubstituted benzenes: evidence for direct aromatic hydroxylation. Biochemistry 23:3048–3055, 1985.

87. JD Baty, RM Lindsay, WR Fox, RG Willis. Stable isotopes as probes for the metabolism of acetanilide in man and the rat. Biomed Environ Mass Spectrom 16:183–189, 1988.

88. LR Pohl, SD Nelson, WA Garland, WF Trager. The rapid identification of a new metabolite of warfarin via chemical ionization mass spectrometry ion doublet technique. Biomed Mass Spectrom 2:23–29, 1975.

89. G Guroff, JW Daly, DM Jerina, J Renson, B Witkop, S Udenfriend. Hydroxylation-induced migration: the NIH shift. Science 157:1524–1530, 1967.

90. G Guroff, M Levitt, JW Daly, S Udenfriend. The production of metatritiotyrosine from para-tritiophenylalanine hydroxylase. Biochem Biophys Res Commun 25:253–259, 1966.

91. G Guroff, CA Riefsnyder, JW Daly. Retention of deuterium in p-tyrosine formed enzymatically from p-deuterophenylalanine. Biochem Biophys Res Commun 24:720–724, 1966.

92. DC Swinney, WN Howald, WF Trager. Intramolecular isotope effects associated with meta-hydroxylation of biphenyl catalyzed by cytochrome P-450. Biochem Biophys Res Commun 118:867–872, 1984.

93. ED Bush, WF Trager. Substrate probes for the mechanism of aromatic hydroxylation catalyzed by cytochrome P-450: selectively deuterated analogs of warfarin. J Med Chem 28:992–996, 1985.

94. JF Darbyshire, KR Iyer, J Grogan, KR Korzekwa, WF Trager. Selectively deuterated warfarin: substrate probe for the mechanism of aromatic hydroxylation catalyzed by cytochrome P450. Drug Metab Disp 24:1038–1045, 1996.

95. LT Burka, TM Plucinski, TL MacDonald. Mechanisms of hydroxylation by cytochrome P-450: metabolism of monohalobenzenes by phenobarbital induced microsomes. Proc Natl Acad Sci USA 80:6680–6684, 1983.

96. JB Tomazewski, DM Jerina, JW Daly. Deuterium isotope effects during formation of phenols by hepatic monooxygenases: evidence for alternative to the arene oxide pathway. Biochemistry 14:2024–2031, 1975.

97. DB Northrop. The expression of isotope effects on enzyme catalyzed reactions. Annu Rev Biochem 50:103–131, 1981.

98. DB Northrop. Steady state analysis of kinetic isotope effects in enzymatic reactions. Biochemistry 14:2644–2651, 1975.

99. DB Northrop. Minimal kinetic mechanism and the general equation for deuterium isotope effects on enzymatic reactions: uncertainty in detecting a rate-limiting step. Biochemistry 20:4056–4061, 1981.

100. JP Jones, WF Trager. The separation of the intramolecular isotope effects for the cytochrome P450 catalyzed oxidation of n-octane into its primary and secondary components. J Am Chem Soc 109:2171–2173, 1987.

101. AB Foster, M Jarman, JD Stevens, P Thomas, JH Westwood. Isotope effects in O- and N-demethylations mediated by rat liver microsomes: An application of direct insertion electron impact mass spectrometry. Chem Biol Interact 9:327–340, 1974.

102. LM Hjelmeland, L Aronow, LR Trudell. Intramolecular determination of primary kinetic isotope effects in hydroxylations catalyzed by cytochrome P450s. Biochem Biophys Res Commun 76:541–549, 1977.

103. K Sugiyama, WF Trager. Prochiral selectivity and intramolecular isotope effects in the cytochrome P-450 catalyzed omega-hydroxylation of cumene. Biochemistry 25:7336–7343, 1986.

104. RP Hanzlik, K Hogberg, JB Moon, CM Judson. Intramolecular kinetic deuterium isotope effects on microsomal and chemical chlorination of toluene-α-d1 and toluene-α, α-d2. J Am Chem Soc 107:7164–7167, 1985.

105. L Higgins, GA Bennett, M Shimoji, JP Jones. Evaluation of cytochrome P450 mechanism and kinetics using kinetic deuterium isotope effects. Biochemistry 37: 7039–7046, 1998.

106. GT Miwa, WA Garland, BJ Hodshon, AYH Lu, DB Northrop. Kinetic isotope effects in cytochrome P-450 catalysed oxidation reactions. Intermolecular and intramolecular deuterium isotope effects on N-demethylation of N,N-dimethylphentermine. J Biol Chem 255:6049–6054, 1980.

107. GT Miwa, JS Walsh, GL Kedderis, PF Hollenberg. The use of intramolecular isotope effects to distinguish between deprotonation and hydrogen atom abstraction mechanisms in cytochrome P-450- and chloroperoxidase- catalyzed N-demethylation reactions. J Biol Chem 258:14445–14449, 1983.

108. JP Dinnocenzo, SB Karki, JP Jones. On isotope effects for the cytochrome P-450 oxidation of substituted N,N-dimethylamines. J Am Chem Soc 115:7111–7116, 1993.

109. SB Karki, JP Dinnocenzo, JP Jones, KR Korzekwa. Mechanism of oxidative amine dealkylation of substituted N,N-dimethylanilines by cytochrome P-450: application of isotope effect profiles. J Am Chem Soc 117:3657–3664, 1995.

110. LR Hall, RP Hanzlik. Kinetic deuterium isotope effects on the N-dealkylation of tertiary amides by cytochrome P-450. J Biol Chem 265:12349–12355, 1990.

111. JP Jones, AE Rettie, WF Trager. Intrinsic isotope effects suggest that the reaction coordinate symmetry for the cytochrome P-450 catalyzed hydroxylation of octane is isozyme independent. J Med Chem 33:1242–1246, 1990.

112. FP Guengerich. Low isotope effects in the hydrogenation of 1,4-dihydro-2,6-di-methyl-4-(2-nitrophenyl)-3,5-pyridinedicarboxylic acid dimethyl ester (nifedipine) by cytochrome P450 enzymes are consistent with an electron/proton/electron transfer mechanism. Chem Res Toxicol 3:21–26, 1990.

113. PF Fitzpatrick. Kinetic isotope effects on hydroxylation of ring-deuterated phenyl-alanines by tyrosine hydroxylase provide evidence against partitioning of an arene oxide intermediate. J Am Chem Soc 116:1133–1134, 1994.

114. FP Guengerich, O Okazaki, Y Seto, TL MacDonald. Radical cation intermediates in N-dealkylation reactions. Xenobiotica 25:689–709, 1995.

115. SB Karki, JP Dinnocenzo. On the mechanism of amine oxidations by P450. Xenobiotica 25:711–724, 1995.

116. Y Seto, FP Guengerich. Partitioning between N-dealkylation and N-oxygenation in the oxidation of N,N-dialkylarylamines catalyzed by cytochrome P450 2B1. J Biol Chem 268:9986–9997, 1993.

117. TJ Carlson, JP Jones, L Peterson, N Castagnoli Jr, KR Iyer, WF Trager. Stereoselectivity and isotope effects associated with cytochrome P450-catalyzed oxidation of (S)-nicotine: the possibility of initial hydrogen atom abstraction in the formation of the delta 1′,5′-nicotinium ion. Drug Metab Disp 23:749–756, 1995.

118. FP Guengerich, CH Yun, TL MacDonald. Evidence for a 1-electron oxidation mechanism in N-dealkylation of N,N-dialkylanilines by cytochrome P450 2BI. Kinetic hydrogen isotope effects, linear free energy relationships, comparisons with horseradish peroxidase, and studies with oxygen surrogates. J Biol Chem 27: 27321–27329, 1996.

119. JR Gillette. Insights into deuterium isotope effects on the metabolism of steroid hormones and foreign compounds by cytochrome P-450 enzymes. Life Sci 49:1–14, 1991.

120. JR Gillette, JF Darbyshire, K Sugiyama. Theory for the observed isotope effect on the formation of multiple products by different mechanisms of cytochrome P450 enzymes. Biochemistry 33:2927–2937, 1994.

121. JP Jones, KR Korzekwa, AE Rettie, WF Trager. Isotopically sensitive branching and its effect on the observed intramolecular isotope effects in cytochrome P-450 catalyzed reactions: a new method for the estimation of intrinsic isotope effects. J Am Chem Soc 108:7074–7078, 1986.

122. KR Iyer, JP Jones, JF Darbyshire, WF Trager. Intramolecular isotope effects for benzylic hydroxylation of isomeric xylenes and 4,4′-dimethylbiphenyl by cytochrome P450: relationship between distance of methyl groups and masking of intrinsic isotope effect. Biochemistry 36:7136–7143, 1997.

123. C Audergon, KR Iyer, JP Jones, JF Darbyshire, WF Trager. Experimental and theoretical study of the effect of active-site constrained substrate motion on the magnitude of the observed intramolecular isotope effect for the P450 101 catalyzed benzylic hydroxylation of isomeric xylenes and 4,4′-dimethylbiphenyl. J Am Chem Soc 121:41–47, 1999.

124. RP Hanzlik, GO Shearer. Secondary isotope effects on olefin hydroxylation by cytochrome P-450. Biochem Pharmacol 27:1441–1444, 1978.

125. WA Garland, SD Nelson, HA Sasame. Primary and beta-secondary deuterium isotope effects in the O-deethylation of phenacetin. Biochem Biophys Res Commun 72:539–545, 1976.

126. TR Browne. Stable isotope techniques in early drug development: an economic evaluation. J Clin Pharmacol 8:213–220, 1998.

127. A Nakagawa, A Kitagawa, M Asami, K Nakamura, DA Schoeller, M Minagawa,

IR Kaplan. Evaluation of isotope ratio mass spectrometry for study of drug metabolism. Biomed Mass Spectrom 12:502–505, 1985.

128. DE Mathews, JM Hayes. Isotope-ratio-monitoring gas chromatography-mass spectrometry. Anal Chem 50:1465–1473, 1978.

129. M Sano, Y Yotsui, H Abe, S Sasaki. A new technique for the detection of metabolites labelled by the isotope $^{13}$C using mass fragmentography. Biomed Mass Spectrom 3:1–3, 1976.

130. CA Goldthwaite Jr, F-Y Hsieh, SW Womble, BJ Nobes, IA Blair, LJ Klunk, RF Mayol. Liquid chromatography/chemical reaction interface mass spectrometry as an alternative to radioisotopes for quantitative drug metabolism studies. Anal Chem 68:2996–3001, 1996.

131. TR Browne, GK Szaba, A Ajami, D Wagner. Performance of human mass balance/metabolite identification studies using stable isotope ($^{13}$C, $^{15}$N) labeling and continuous-flow isotope-ratio mass spectrometry as an alternative to radioactive labeling methods. J Clin Pharmacol 33:246–252, 1993.

132. Y Teffera, FP Abramson. Application of HPLC/CRIMS for the analysis of conjugated metabolites: a demonstration using deuterated acetaminophen. Biomed Mass Spectrom 23:776–783, 1994.

133. FP Abramson, Y Teffera, J Kusmierz, RC Steenwyk, PG Pearson. Replacing $^{14}$C with stable isotopes in drug metabolism studies. Drug Metab Disp 24:697–701, 1996.

134. BL Osborn, FP Abramson. Pharmacokinetic and metabolism studies using uniformly labeled proteins with HPLC/CRIMS detection. Biopharm Drug Disp 19:439–444, 1998.

135. H Song, FP Abramson. Drug metabolism studies using ''intrinsic'' and ''extrinsic'' labels: a demonstration using $^{15}$N vs. Cl in midazolam. Drug Metab Disp 21:868–873, 1993.

# 11
# Cassette Dosing in Drug Discovery

**Jayesh Vora**
*Chiron Corporation, Emeryville, California*

**David T. Rossi and Erick K. Kindt**
*Pfizer Global Research and Development, Ann Arbor, Michigan*

> A man with a new idea is a crank until he succeeds.
>
> —*Mark Twain*

## I. INTRODUCTION

The advent of combinatorial chemistry synthetic programs has resulted in a dramatic increase in the number of new chemical entities (NCEs) for evaluation as possible drug candidates. For discovery programs to be successful, rapid screening of these compounds for optimal biological activity and pharmacokinetic properties is essential [1]. Cassette ("N-in-1" or "cocktail") dosing provides an approach for rapid screening of large number of NCEs for their pharmacokinetic properties in vivo [2].

### A. Cassette Dosing as a Screening Tool in Drug Discovery

A typical cassette-dosing paradigm involves simultaneous administration (oral or intravenous) of 5–10 NCEs from a congeneric chemical series within a discovery program to animals (rats, dogs, etc.) [3–5]. Cassettes consisting of larger numbers of compounds have been reported in the literature [4]. The arrival of liquid chromatography/mass spectrometry (LC/MS) and liquid chromatography/tandem mass spectrometry (LC/MS/MS) technologies (Chapter 3 and 5) has permitted rapid and simultaneous quantitation of these NCEs in plasma samples obtained through cassette dosing [6]. The resulting concentration–time data are analyzed by noncompartmental methods to generate pharmacokinetic parameters (clearance, volume of distribution, area under concentration–time curve, oral bioavailability, and brain penetration). New chemical entities considered "interesting" according to preestablished pharmacokinetic criteria are evaluated further by conventional singlet administration studies [5,7].

## B.  Rationale

Traditional pharmacokinetic evaluation involves administration (orally or intravenously) of a single NCE ("singlet dosing") to animals followed by plasma sampling to generate concentration–time profiles. A typical single oral dose absolute bioavailability study of an NCE (three dogs; oral and intravenous dosing) can result in approximately 30 samples/dosing route/compound. Therefore, screening five compounds for a given discovery program conceivably requires repeating the oral and intravenous arms of the above experiment five times. These studies are typically conducted in a crossover fashion with each dosing conducted once every 2 weeks. Therefore, evaluation of five compounds (oral and intravenous dosing) could easily span 5 months. Further, this approach would result in 600 samples.

Under a cassette-dosing paradigm, five NCEs may be simultaneously administered to three dogs and the oral and intravenous dosing treatment arms completed in a crossover fashion in 1 month. This experimental design generates only 60 samples. The savings in terms of time and resources (number of animals used, number of samples analyzed, etc.) are tremendous. Therefore, cassette dosing has gained increasing acceptance within pharmaceutical research and development.

## C.  Validity of Cassette Pharmacokinetic Data

Implementation of a cassette-dosing paradigm in any discovery effort is invariably preceded by a validation effort. This is usually accomplished by establishing a correlation between cassette pharmacokinetic data with that obtained following singlet administration for a representative group of NCEs. Ideally, the compounds selected for this validation effort should represent the overall range of structures defined by all compounds in a given discovery program. In reality, this is rarely achieved since such validation is usually conducted at the start of the discovery program, when the known structural parameter range is fairly limited. Therefore, it is critical that cassette and singlet pharmacokinetic data are examined at regular intervals during the drug discovery process.

The validity of the cassette dosing approach is well documented [2,4,5,7]. Shaffer et al., demonstrated a linear relationship for clearance, volume of distribution, and elimination $t1/2$ values obtained by cassette or singlet intravenous dosing of 17α-1a antagonists in dogs (Fig. 1) [4]. While there is an obvious trend toward linearity, this relationship shows a fair degree of scatter. This is consistent with the use of cassette dosing as a screening tool that can be useful for rank ordering compounds, but is not an appropriate tool for definitive pharmacokinetic (PK) assessment (see Sec. II).

Similarly, Patel et al. obtained comparable values of oral bioavailability and elimination $t1/2$ following cassette or singlet oral and intravenous administration to rats for a matrix metalloprotease inhibition discovery program [5]. In the first instance, the authors utilized intravenous dosing alone, whereas in the second instance, both intravenous and oral routes were utilized. Further, the animal model in the first example was dogs, while rats were used in the second example. These issues (selection of dosing routes and animal models) represent an interesting dilemma for the discovery pharmacokineticist and are addressed later in the chapter (see Sec. III).

## II. UTILITY OF CASSETTE PHARMACOKINETIC DATA

It is important to distinguish the utility of pharmacokinetic data obtained through cassette dosing compared to that obtained through traditional singlet administration studies. The primary purpose of cassette pharmacokinetic parameters is to prioritize, rank order, or "bin" (categorize) compounds. On the other hand, singlet dosing data characterize the pharmacokinetic properties of a smaller, more select number of NCEs. Therefore, cassette pharmacokinetic data of "interesting" NCEs must be confirmed by singlet administration before selecting drug candidates to be pursued further.

The relatively short turnaround time for generation of cassette pharmacokinetic data allows simultaneous screening of NCEs by in vitro or in vivo biological activity assays. Therefore, compounds and/or structural templates identified through these processes likely incorporate features favorable for biological activity and pharmacokinetic behavior. This is expected to result in selection of drug candidates with fewer development challenges, such as poor bioavailability, rapid clearance, or short elimination half-life [1]. Furthermore, cassette dosing affords an opportunity to create a database of pharmacokinetic parameters within a discovery program. Such databases can be used to develop qualitative and quantitative structure–pharmacokinetic relationships, which can guide future synthetic efforts or permit prediction of pharmacokinetic properties before additional synthesis is undertaken.

The potential for cassette pharmacokinetic data to guide future synthetic effort was demonstrated by Shaffer et al. [4]. Sixty-six NCEs were administered intravenously in several cassettes (5–22 NCEs/cassette) to dogs. The resulting database of pharmacokinetic data, along with the corresponding chemical struc-

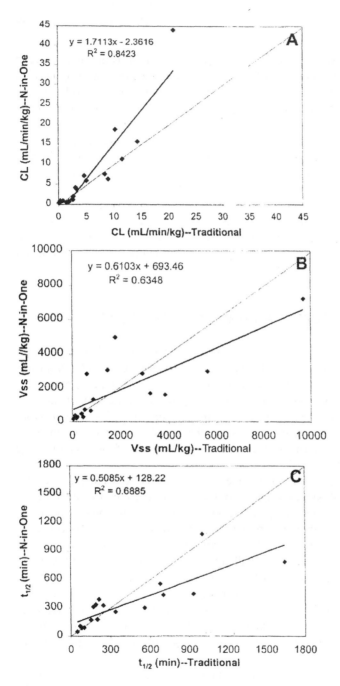

**Figure 1**  Correlation of pharmacokinetic parameters (clearance, panel A; volume of distribution, panel B; and half-life, panel C) of 17α-1a antagonists obtained by cassette (y-axis) or singlet (x-axis) intravenous dosing to rats. (Reproduced with permission from Ref. 4.)

tures, was evaluated for possible qualitative structure–pharmacokinetic relationships. This assessment identified five NCEs that met predetermined criteria for clearance, volume of distribution, and elimination $t1/2$. Further, the data provided clues regarding possible contributions of specific chemical moieties for optimum pharmacokinetic properties. The utility of such structure–pharmacokinetic databases for the potential prediction of pharmacokinetics of NCEs under synthesis has also been demonstrated.

Plasma area under the concentration–time curves (AUCs) of 57 NCEs were determined following oral cassette administration (5–9 NCEs/cassette) to mice. Physicochemical properties [such as, molecular weight, calculated molar refractivity, and calculated lipophilicity (clog$P$)] and molecular descriptors [such as presence or absence of N-methylation, cyclobutyl moiety, or heteroatoms (non-C,H,O,N)] were calculated or estimated for these compounds. This structural data, along with the corresponding pharmacokinetic parameters (primarily AUC), were used to develop artificial neural network models [8]. These models were used to predict the AUCs of compounds under synthesis [10]. This approach demonstrates that predictive models could be developed which potentially predict in vivo pharmacokinetics of NCEs under synthesis. Similar examples have been reported elsewhere [11–13].

## III. IMPLEMENTATION ISSUES AND PITFALLS OF CASSETTE DOSING

Implementation of a cassette-dosing paradigm in discovery screening programs usually requires addressing several issues and acknowledging the pitfalls of this approach. Simultaneous administration of a number of structurally similar NCEs to a single animal can enhance the potential for drug–drug interactions in vivo. For example, if the administered NCEs are metabolized by a limited number of enzyme systems in vivo, the possibility of saturating these metabolic pathways is quite real. This and other possibilities for drug–drug interactions, such as enzyme inhibition and transporter interactions, are discussed in Chapter 9. Similarly, simultaneous administration of a number of NCEs can result in precipitation of the drugs in vivo due to their limited solubility in biological fluids. These factors can be particularly exacerbated during cassette dosing because NCEs selected from a single discovery program likely represent (somewhat) limited structural diversity.

Saturation of metabolic enzyme systems manifests itself as falsely elevated plasma concentrations in vivo. Therefore, when the selected NCEs are evaluated later using singlet administration, their concentrations can be lower than those obtained from cassette dosing (''false positives'') [14]. On the other hand, precipitation of NCEs following cassette administration can artificially result in a reduction in their plasma concentrations when compared to those following singlet administration ("false negatives").

One approach to minimizing these risks is by limiting the overall dose of NCEs administered. Typically, total oral doses of 5–10 mg/kg are considered acceptable. For a cassette composed of five NCEs, this translates to a dose of 1 to 2 mg/kg/compound. For intravenous administration, typical total doses range from 3 to 5 mg/kg, resulting in an individual NCE dose of 0.6 to 1 mg/kg/ compound. However, in some cases, limited analytical assay sensitivity may not permit dose reductions. In such cases, a reduction in the number of NCEs administered per cassette could be considered in lieu of or in addition to a reduction in the total dose administered.

Another complementary approach for managing the overall risk of "false positives or negatives" during cassette dosing involves administration of an *in vivo reference compound* in each cassette [2,4,5,10,14]. It is important that the pharmacokinetics of the in vivo reference compound are adequately characterized by singlet administration before such use. The cassette pharmacokinetic parameters of the reference compound following each cassette experiment are then compared against those established previously following singlet administration. If comparable, the results of the cassette pharmacokinetic data for the unknown NCEs are considered acceptable. Otherwise, drug–drug interaction is suspected [14] and the data are closely examined or discarded or a different combination of NCEs is administered as a separate cassette. The fundamental assumption while employing this approach is that the pharmacokinetics of the in vivo reference compounds changes if drug–drug interaction occurs in cassettes. In reality, it is difficult to verify if this is indeed the case. Furthermore, comparison of the cassette pharmacokinetic parameters of the reference compound with historical singlet dosing values requires a degree of subjective evaluation. Therefore, such comparisons are inherently fraught with the biases of the scientists evaluating the data [2], unless clear unequivocal criteria for such comparisons are preestablished and agreed upon.

As highlighted earlier, the choice of intravenous administration versus oral administration of cassettes represents yet another implementation issue. It has been suggested that cassettes should be administered intravenously alone because the resulting fundamental pharmacokinetic parameters (clearance and volume of distribution) are considered key for decision making [2] and because the saturation of first-pass metabolism following oral dosing is a distinct possibility (see above). It is also likely that at the early drug discovery stage the amount of compound available can be limited, thus intravenous administration can represent the only viable choice because the compound requirement is usually less compared to that of the oral route.

These issues notwithstanding, oral cassette dosing can be a necessity if the solubility of the NCEs is such that an intravenous dosing solution cannot be formulated. Careful dose selection for oral cassette administration can address the issue of saturation of first-pass metabolism because most NCEs are intended for oral use, and information obtained through oral administration of cassettes is relevant for decision making. An alternative to choosing between intravenous or oral dosing of cassettes is to administer cassettes by both routes either in a cross-over fashion (usually dogs) or to separate parallel groups (usually rats) of animals. The advantage of this approach is that comparison of exposures (AUCs) following oral and intravenous administration results in an estimate of oral bioavailability. For most discovery programs, high oral bioavailability and low clearance are highly desirable. Use of both routes of administration provides both these parameters, thus facilitating decision making. This approach results in an increase in the total amount of compound used and a small increase in the time and effort required conducting the study and assaying the samples.

Compound availability also influences the choice of species in which cassette dosing may be carried out. If relatively low amounts of NCEs are available, rats and mice should be considered for cassette dosing. Their low body weight (compared to that of a larger species, such as dogs) results in a lower amount of NCE administered per animal. However, rodents, in general, cannot be used in a crossover fashion. Therefore, the intravenous and oral dosing arms of a cassette screen have to be conducted in parallel resulting in an increase in the number of animals used.

In an effort to reduce the number of animals used for such purposes, intravenous dosing of one cassette (''A'') may be followed immediately by oral dosing of a distinct cassette (''B'') to a group of animals (Group 1). Therefore, analysis of plasma samples drawn at predetermined times results in concentration–time profiles of all NCEs administered. On another occasion, the routes of administration of these two cassettes may be switched or crossed over in a second group of animals (Group 2) to obtain a second set of concentration–time profiles of all NCEs administered, i.e., Cassette ''A'' is administered orally, while Cassette ''B'' is administered intravenously. Comparison of oral concentration–time profiles of a given NCE in Group 1 with the corresponding one following intravenous dosing in Group 2 results in estimates of all relevant pharmacokinetic parameters (such as oral bioavailability, clearance, and volume of distribution). This variation of cassette dosing would reduce the number of animals used by 50% while not increasing turnaround times. However, the potential for drug–drug interactions in vivo increases dramatically since a large number of NCEs are administered simultaneously. This issue can be minimized by administering the two cassettes at slightly different times so that concentrations of NCEs from the first cassette have started to decline when the second cassette is administered. However, there is limit to the total volume of blood that may be withdrawn from an animal. Therefore, even this approach has its limitations. It is apparent from the earlier discussion that several issues must be addressed before implementation of a cassette-dosing paradigm in drug discovery. Furthermore, this approach is not intended to accurately characterize the pharmacokinetics of NCEs. At best, it can assist the drug discovery scientists in prioritizing, rank ordering, or binning compounds so that resources are expended judiciously to thoroughly evaluate "promising" NCEs [1]. Despite lowering the total dose administered (or the number of compounds administered) and the use of in vivo reference compounds, false positives or negatives remain a distinct possibility. Therefore, managing the potential risk of missing a promising drug candidate through good science-based decision making is imperative.

## IV. BIOANALYTICAL ISSUES FOR IN VIVO CASSETTE DOSING

### A. The Impact of Mass Spectrometry on Cassette Dosing

Because of the natural capability for simultaneous multicomponent quantitation, LC/MS with multiple-channel monitoring (MRM) on triple-quadrupole mass spectrometers has made the concept of cassette dosing practical. Although it has been demonstrated that other analytical techniques, such as liquid chromatography with UV absorbance detection, can also be used for quantitative determinations of multiple drug components in plasma [15], the time and effort required to routinely achieve adequate selectivity makes this relatively nonselective approach impractical. Optical (UV, visible, or fluorescence) detection rely heavily on chromatography and sample preparation to achieve assay selectivity for each of the components involved. Typically, development of an appropriately selective HPLC/UV procedure will require a week or longer to develop, depending on the desired quantitation limits.

In contrast, with its intrinsically higher selectivity, tandem mass spectrometry offers a much more effective and practical approach for multicomponent quantitation. This translates into faster method development and, ultimately, rapid and reliable multicomponent quantitation. The routine application of multicomponent quantitation to 5 or 10 compounds in a cassette can be developed in 1 or 2 days, offering typical quantitation range of 5 to 5000 ng/mL. The variability of the assay procedure ranges from 10 to 20%, which is less than the typical intersubject pharmacokinetic variability of 25 to 30%.

### B. Sample Preparation Issues for Cassettes

For a cassette containing structurally similar compounds (Table 1), it is possible to develop and use a high-throughput extraction approach such as deep-well liquid–liquid extraction or a deep-well solid-phase extraction operating under generic wash/elution conditions. These approaches do work for a majority of

**Table 1** Pharmacokinetic Data for Representative Compounds Used in Developing a Structure–Pharmacokinetic Relationship

| Compound | R1 | R2 | R3 | CL (mL/min/kg) | $t_{1/2}$ | Vss |
|----------|----|----|----|----------------|-----------|-----|
| 1 | 4H | O-Me | H | 14 | 59 | 785 |
| 2 | 4H | O-Me | | 2.0 | 144 | 151 |
| 3 | 4H | O-Me | | 9.1 | 144 | 647 |
| 4 | 4H | O-Me | | 10 | 156 | 600 |
| 5 | 4H | O-Me | | 26 | 72 | 2035 |
| 6 | 4H | O-Me | | 6.8 | 72 | 210 |
| 7 | 4H | O-Me | | 11 | 120 | 1102 |
| 8 | 4H | O-Me | | 11 | 114 | 930 |
| 9 | 4H | O-Me | | 22 | 107 | 1241 |
| 10 | 4H | O-Me | | 28 | 123 | 4133 |

**Table 1** Continued

| Compound | R1 | R2 | R3 | CL (mL/min/kg) | $t_{1/2}$ | Vss |
|---|---|---|---|---|---|---|
| 11 | 4H | O-Me | | 18 | 135 | 3175 |
| 12 | 4F | O-Me | | 11 | 432 | 4778 |
| 13 | 4H | O-Me | | 22 | 144 | 3450 |
| 14 | 4H | O-Me | | 25 | 101 | 3004 |
| 15 | 4F | O-Me | | 12 | 204 | 2907 |

[a] *Source*: Ref. 21.

compounds within a class, where small differences in structure will not significantly affect the recovery between compounds. Solvent composition and pH conditions can be quickly fine tuned to obtain acceptable recovery for a given compound class. Cassette work that is geared at obtaining structure/pharmacokinetic relationships is one application of this approach [16].

Aside from investigating structure/pharmacokinetic relationships, cassette dosing can also be used as a high-throughput screen for diverse compounds. These compounds are passed through one or more preliminary screens, such as in vitro efficacy, permeability (Chap. 8), or metabolic screening (Chap. 9). These primary screens typically screen out a significant number of NCEs from consideration. Remaining compounds from diverse therapeutic targets or compound classes are then grouped into cassettes for preliminary in vivo screening. Under this scenario, the compounds in a cassette will look and extract very differently from one another, owing to the presence or absence of ionizing and lipophilic groups. Given the large differences in ionizability and lipophilicity, extractability under a given set of solvent and pH conditions can be highly variable and in many cases unacceptable. It can be impossible or extremely time consuming to find suitable extraction conditions to obtain acceptable recovery for all compounds in a cassette.

For this reason, when a cassette is composed of structurally diverse compounds, it is usually prudent to use protein precipitation as a plasma sample preparation technique. Although the absolute recovery will vary from compound to compound, this approach can be applied with fairly universal success across a wide structural range of compounds and assay development time will be minimal. Generic solid-phase extraction protocols can also be effective, but will experience a higher (10 to 15% of compounds) failure rate. Chapter 6 contains examples of generic extraction protocols that can be adapted for use with cassettes containing diverse NCEs.

## C. The Need for Analytical Separations

As with most LC/MS experiments, the analytical separation is a necessary part of multicomponent quantitation in cassette dosing for three reasons: (i) separations are an important tool for managing collision-cell cross talk, (ii) to minimize possible in-source fragmentation from metabolites, and (iii) to minimize the unique ion-suppression circumstances that can occur in multicomponent quantitation (see Chaps. 3 and 5).

As described in Chapter 3, *cross talk* is observed when the collision cell still holds fragment ions from the previous MRM channel. In cases where two MRM channels have similar fragmentation patterns, the holdover ions from the cell contribute to the signal associated with the next ion channel. Lack of chromatographic separation between compounds increases the potential for cross talk, resulting in likely quantitation errors and a lack of discrimination between compounds. Without an analytical separation, all analytes in a cassette could coelute, thereby increasing the potential for common parent/daughter ion combinations derived from one compound to interfere with those from another due to collision-cell cross talk. The greater the number of NCEs per cassette, the greater the potential for collision-cell cross talk. The most recent generation of triple-quadrupole instruments incorporate hardware features designed to eliminate or avoid collision-cell cross talk by emptying the flow cell of ions prior to manipulating the next set. An example of this feature is the so-called LIN-AC® feature introduced by PE-SCIEX [17]. Although many collision-cell cross-talk problems can be eliminated by intelligent selection of those compounds to be included in a cassette, all possible combinations cannot be predicted. The use of a chromatographic separation to assist in selectivity of a cassette assay has been shown to be an effective way to avoid cross-talk problems [18].

Because cassette dosing is an early-discovery screening tool, little is known about the metabolism of the test compounds. As such, a variety of metabolites can be present in the in vivo sample. Unfortunately, most metabolites (hydroxyls, glucuronides, sulfates, and N-oxides included) are somewhat labile under mass-spectrometer ion-source conditions and fragment to form the parent drug compound. Because these metabolites are more polar than the parent drug and quickly elute from the chromatographic column, they can be readily separated from the unchanged parent drug molecules and it is only necessary to obtain adequate retention to avoid ion suppression. In cassette dosing, as in other pharmacokinetic experiments, it is important to differentiate between the plasma concentrations of intact parent drug and the conjugated or oxidative metabolites. A easy way to minimize this problem is through a chromatographic separation.

Chromatographic separations can also help minimize the unique ion-suppression circumstances that can occur in some multicomponent situations. In cassette dosing, the plasma concentrations observed are dependent on many factors, including the physical-chemical properties of the compounds, the dose, vehicle, route of administration, extent of drug absorption, extent of metabolism, apparent drug distribution volume, and clearance. Therefore, the assay procedure should be designed to have a large dynamic range. This would likely minimize the need for sample dilution and repeat analytical work.

If the chromatography is such that the three compounds coelute (Fig. 2a) then the compounds will enter the ion source at the same time and compete with the residual matrix components and with each other for ionization. Due to this competition, the peak responses of coeluting compounds are generally lower than those of compounds which do not coelute (Fig. 2b). This portion of the mass spectrometry experiment is well controlled and predictable. The factor that varies is the plasma concentration of an NCE in the cassette.

For example, assume that Compounds I and II are poorly absorbed, while Compound III is well absorbed following an oral cassette dose (Fig. 2c). Therefore, the plasma concentrations of Compounds I and II are expected to significantly lower compared to those of Compound III. If these three compounds chromatographically coelute then it is likely that plasma concentrations of Compound III are overestimated due to lack of competition for ionization from the other analytes. This quantitation error can be generally avoided through chromatographic separation of the analytes.

## D. Instrument Sensitivity, Experimental Efficiency, and Cassette Size

There is a loose relationship between $N$, the number of NCEs in a cassette (or the number of channels being monitored by an instrument operating under MRM), and the experimental efficiency (Fig. 3). To a large extent, increased efficiency and throughput may be achieved by increasing "$N$" in a cassette. This

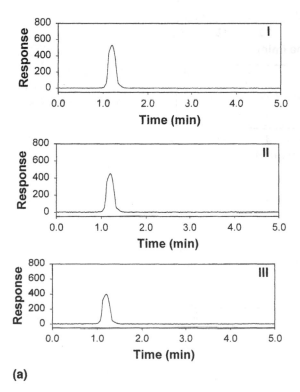

**(a)**

**Figure 2** Multiple-reaction monitoring chromatograms for a three-compound cassette. (a) A standard chromatogram showing the three compounds chromatographically coeluting. In this situation there is a greater chance for ion suppression. (b) A standard chromatogram showing the three compounds chromatographically separated. In this situation there is less chance for ion suppression. (c) A sample chromatogram indicating that Compound I is not absorbed, Compound II is poorly absorbed, and Compound III is well absorbed. Because there is less chance for ion suppression, due to the low abundance of I and II, Compound III is likely to be overestimated when compared to a standard curve containing chromatogram a.

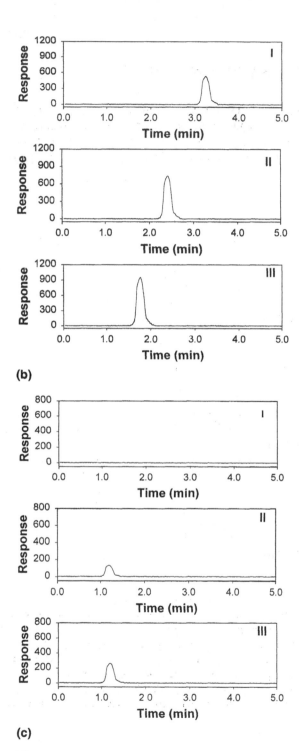

**(b)**

**(c)**

**Figure 2** Continued

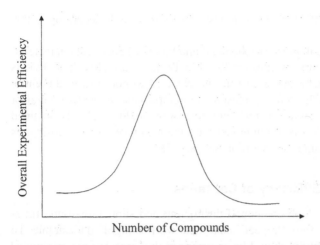

**Figure 3** The approximate relationship between $N$, the number of compounds in a cassette (or the number of channels being monitored by an instrument operating under MRM) and the experimental efficiency.

should result in fewer studies leading to fewer sample analysis. As "$N$" becomes larger, drug–drug interaction issues may become significant (see Sec. III). Also, if $N$ is too large, the analytical method will become complex and unmanageable due to analyte extraction difficulties, potential for interferences (from cross talk or in-source fragmentation), standard preparation, and other practical issues.

From an analytical perspective, the maximum desirable number of compounds in a cassette is less than about 19 compounds. This assumes a minimum chromatographic sampling rate of 1 point/sec for 10-sec chromatographic peaks. With a channel dwell time of 50 msec and an interchannel delay of 3 msec, 19 compounds allows a chromatographic sampling rate of 1.007 data points/sec ($19 \times 0.05 + 19 \times 0.003$). This allows for negligible error introduction from finite chromatographic sampling rates [19]. From a drug-interaction perspective, the optimum number of compounds to include in a cassette is obviously much less. However, larger cassettes have been reported [4].

## E. Compound Selection for Cassettes

Under the best circumstances, compounds are selected for cassettes with their analytical characteristics in mind. In this way, the chances of experimental success are improved. Although it would be useful to choose compounds so as to minimize drug–drug interaction issues, there is generally not enough pharmacological information available to consistently select compounds according to this criterion.

The ideal analytical, physical/chemical and biopharmaceutical characteristics for pooling NCEs into cassettes are listed in Table 2. In reality, it is obvious that no grouping of NCEs into a cassette is expected to conform to the entire list. If a cassette of NCEs offers a reasonable compromise between several of these criteria it can be assumed that useful information could be quickly obtained from the experiment. This is in alignment with the utility of cassette dosing, as a screening and compound ranking tool (see Sec. II).

## F.  Enhancing the Efficiency of Cassettes

During drug discovery, high compound throughput and short turnaround times are important [1]. Therefore, "repeat" analytical work is highly undesirable. To this end, analytical practices should be so arranged that reassays are minimized or involve simple reinjections. To achieve this, the number of compounds included in a cassette should be modest (four to six), the chromatographic separation short and efficient (less than ~5 min) with a near unit resolution between NCEs. The calibration range for each NCE should be as large as possible (say, 5 to 2000 ng/mL) so as to minimize samples that are outside it. Samples near the pharmacokinetic $C_{max}$, which are above the limit of quantitation (ALQ) of the assay, can be anticipated and dealt with a priori by including appropriate dilutions or reinjecting the samples at a lesser volume so that the response for the analyte is below the highest standard in the calibration curve. The response factor (the ratio of analyte to internal standard responses) should then remain unchanged. By avoiding or minimizing reassays, an increase in analytical efficiency for cassette dosing experiments can be attained.

**Table 2**  Potential for Ideal Analytical Characteristics of Cassetted Compounds

| Homologous compounds | Diverse compounds | Ideal characteristic |
| --- | --- | --- |
| − | ++ | Parent and daughter molecular weight differ by >3 AMU |
| ++ | +/− | Same functional groups for consistent extractions and mass spectral response |
| +/− | +/− | Good solubility in dosing vehicle |
| ++ | +/− | Compatibility with a single (either + or −) ionization mode |
| ++ | +/− | Compatibility with either APCI or ESI |
| + | − | Easy and fast LC/MS separation available |

++, very likely; +, likely; +/−, neither likely nor unlikely; and −, unlikely.

As in most bioanalytical work, the use of an analytical internal standard is highly desirable to normalize for intersample differences arising from sample extraction and injection. This analytical internal standard is distinguished from the pharmacokinetic internal standard in that it is used to correct for analytical variability and is not an indicator of drug–drug interactions during the animal experiment. Because there is often little experience with the extraction method for these early discovery stage compounds, analytical internal standards are important in correcting for variability. Stable-label internal standards are almost never available at this stage of drug discovery, therefore structural analog standards will usually be employed as analytical internal standards. Care must be given to selecting an analog internal standard because it should have chromatographic and ionization properties that are similar to each of the analytes, yet not interfere with any of them through cross talk. This task becomes increasingly challenging as the number of compounds in the cassette increases.

## V. CONCLUSIONS

The technological evolution of LC/MS and LC/MS/MS has allowed pharmaceutical scientists to address pharmacokinetics-related questions early in drug discovery. By careful and judicious use of cassette pharmacokinetic data coupled with biological activity data, the overall probability of discovering and developing "drug-like" compounds appears to have increased.

Cassette screening of compounds is now an accepted practice in drug discovery. Because this technique requires in vivo dosing, it is inherently capacity limited. The increase in demand to screen NCEs rapidly and with greater throughput has resulted in in vitro and in silico screening paradigms. Techniques for rapid determination of various characteristics, such as Caco-2 permeability, solubility (Chap. 8), microsomal and/or hepatocyte metabolic stability [20], and Pfizer's "rule of 5" [5], along with in vivo cassette dosing, represent a comprehensive battery of tests to probe the absorption/distribution/metabolism/elimination (ADME)/PK [22] properties of NCEs in drug discovery. The next evolutionary development for rapid ADME assessment may well involve protein binding and renal and biliary clearance. The ultimate goal of these screening efforts must be to efficiently predict human or animal in vivo performance of NCEs following singlet dosing, an area of active research [22–26].

## REFERENCES

1. RA Lipper. How can we optimize selection of drug development candidates from many compounds at the discovery stage? Modern Drug Discov 2(1):55–60, 1999.

2.  LW Frick, KK Adkison, KJ Wells-Knecht, P Woollard, DM Higton. Cassette dosing: rapid in vivo assessment of pharmacokinetics. Pharm Sci Technol 1(1):12–18, 1998.
3.  J Berman, K Halm, K Adkison, J Shaffer. Simultaneous pharmacokinetic screening of a mixture of compounds in dog using API LC/MS/MS anlaysis for increased throughput. J Med Chem 40(6):827–829, 1997.
4.  JE Shaffer, KK Adkison, K Halm, K Hedeen, J Berman. Use of "N-in-one" dosing to create an in vivo pharmacokinetics database for use in developing structure–pharmacokinetic relationship. J Pharm Sci 88(3):313–318, 1999.
5.  VS Patel, ME Dowty, TR Baker, SX Peng, MJ Janusz, YO Taiwo. Utilization of N-in-one and in vivo efficacy data for matrix metalloprotease inhibitors to optimize the compound selection process PharmSci 1, 1998.
6.  AP Watt, D Morrison, DC Evans. Approaches to higher-throughput pharmacokinetics (HTPK) in drug discovery. Drug Discov Today 5(1):17–24, 2000.
7.  MC Allen, TS Shah, WW Day. Rapid determination of oral pharmacokinetics and plasma free fraction using cocktail approaches: methods and application. Pharm Res 15(1):93–97, 1998.
8.  JVS Gobburu, WH Shelver. Quantitative structure–pharmacokinetic relationships (QSPR) of beta blockers derived using neural networks. J Pharm Sci 84:862–865, 1995.
9.  P Veng-Pedersen, NB Modi. Neural networks in pharmacodynamic modeling: is current modeling practice of complex kinetic systems at a dead end? J Pharmacokinet Biopharmacol 20:397–418, 1992.
10. J Vora, C Lathia, S Przybranowski, J Hollembaek, J Blankley, D Leonard, J Kaltenbronn, J Sebolt-Leopold, R Feng, W Li, and W Leopold. Application of neural networks in development of structure–pharmacokinetic relationships. PharmSci 1: S520, 1998.
11. PH Van der Graff, J Nilsson, EA Van Schaick, M Danhof. Multivariate quantitative structure–pharmacokinetic relationships (QSPKR) analysis of adenosine $A_1$ receptor agonists in rat. J Pharm Sci 88(3):306–312, 1999.
12. RA Herman, P Veng-Pedersen. Quantitative structure–pharmacokinetic relationships for systemic drug distribution kinetics not confined to a congeneric series. J Pharm Sci 83(3):423–428, 1994.
13. MD Wessel, PC Jurs, JW Tolan, SM Muskal. Prediction of human intestinal absorption of drug compounds from molecular structure. J Chem Inf Comput Sci 38(4): 726–735, 1998.
14. R Hayes, W Bullen, F Chanoine, T Kulkarni, L Andress, L Radulovic, M Sinz. Identification and characterization of a drug–drug interaction in a cassette dosed pharmacokinetic screen. Proceedings of the 46th ASMS Conference on Mass Spectrometry and Allied Topics, Orlando, FL, 1998 Abstract 1434.
15. BS Quo, T Van Noord, MR Feng, DS Wright. Sample pooling to expedite bioanalysis and pharmacokinetic research. J Pharm Biomed Anal 16:837–846, 1998.
16. JE Shaffer, KK Adkison, K Halm, K Hedeen, J Berman. Use of "N-in-one" dosing to create an in vivo pharmacokinetics database for use in developing structure–pharmacokinetic relationships. J Pharm Sci 88:313–318, 1999.
17. See for example WWW:PEBIO.COM/SCIEX
18. TV Olah, DA McLoughlin, JD Gilbert. The simultaneous determination of mixtures

of drug candidates by liquid chromatography/atmospheric pressure chemical ionization mass spectrometry as an in vivo drug screening procedure. Rapid Commun Mass Spec 11:17–23, 1997.

19. DT Rossi. A simplified method for evaluating sampling error in chromatographic data acquisition. J Chrom Sci 26:101–105, 1988.

20. RE White. Short- and long-term projections about the use of drug metabolism in drug discovery and development. Drug Metab Dispos 26(12):1213–1216, 1998.

21. CA Lipinski et al. Experimental and computational approaches to estimate solubility and permeability in drug discovery and developmental settings. Adv Drug Deliv Rev 23:3–29, 1997.

22. J Kelder, PDJ Grootenhuis, DM Bayada, PC Delbressine, J-P Ploeman. Polar molecular surface as a dominating determinant for oral absorption and brain penetration of drugs. Pharm Res 16(10):1514–1519, 1999.

23. P Stenberg, K Luthman, H Ellens, CP Lee, PL Smith, A Lago, JD Elliot, P Artursson. Prediction of the intestinal absorption of endothelin receptor antagonists using three theoretical methods of increasing complexity. Pharm Res 16(10):1520–1526, 1999.

24. RS Obach. Prediction of human clearance of twenty-nine drugs from hepatic microsomal intrinsic clearance data: an examination of in vitro half-life approach and nonspecific binding to microsomes. J Pharmacol Exp Therapeut 27(11):1350–1359, 1999.

25. G Schneider, P Coassolo, T Lavé. Combining in vitro and in vivo pharmacokinetic data for prediction of hepatic drug clearance in humans by artificial neural networks and multivariate statistical techniques. J Med Chem 1999.

26. T Lavé, P Coassolo. In vitro (hepatocytes) to in vivo scaling of hepatic metabolic clearance in animals and humans (n = 75): Impact of protein binding. Millennial World Congress of Phamaceutical Sciences, San Francisco, CA, April 2000

# 12

# Applications of Microdialysis/Mass Spectrometry to Drug Discovery

**David J. Weiss, Ryan M. Krisko, and Craig E. Lunte**
*University of Kansas, Lawrence, Kansas*

Luck is the residue of design.

*—Branch Rickey*

## I. INTRODUCTION

### A. Basic Principles of Microdialysis

Microdialysis is a chemical sampling technique for biological systems in which site-specific information is provided. This is accomplished by surgically implanting a semipermeable membrane in virtually any tissue of interest (e.g., brain, liver, heart, adipose, and eye). The membrane is perfused with a solution (termed the perfusate) of similar ionic strength and pH as the surrounding fluid or tissue. Mass transport through the membrane is governed by diffusion. If the sampling matrix in the tissue matches that of the extracellular fluid (ECF), there should be no net exchange of ions through the membrane. Where a concentration gradient exists over the membrane, an analyte of interest may diffuse into the perfusion fluid (Fig. 1) and be carried off by the continuous flow for analysis.

The perfusion rate of the probe is typically 0.5 to 5.0 μL/min. At this flow rate there is no net transport of liquid across the membrane. One view of microdialysis is that the probe is like a blood vessel in which mass transport of compounds in and out of the probe is a function of the concentration gradient. If the concentration of a compound is higher inside the probe than in the extracellular fluid some fraction will be delivered to the extracellular fluid. This is termed

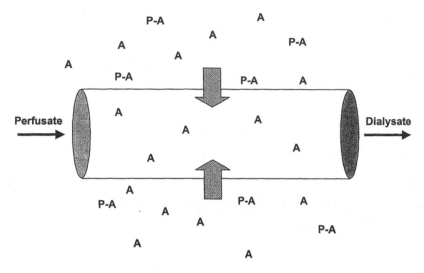

**Figure 1**  Diagram of the microdialysis process. A represents the analyte while P-A indicates a protein-bound analyte.

a delivery experiment. Conversely, if the concentration in the extracellular fluid is greater than that in the perfusate some fraction will be transported into the probe and will be recovered. This is termed a recovery experiment. Several reviews are available which provide more detail on the calculations needed to determine delivery or recovery from the probe [1–3].

## B.  Advantages of Microdialysis Sampling

For traditional blood sampling, the number of blood samples that can be withdrawn is limited, making only a few data points obtainable. Another traditional method used for drug development is tissue homogenization. Degradation can be a serious source of contamination associated with tissue homogenates. For example, short-lived metabolites may not be observed at all with analysis of homogenates [4]. In addition, some experiments require multiple animals to be sacrificed for each data point. A major advantage of microdialysis is that it does not directly remove fluid from the animal. Therefore, no net fluid loss occurs, making continuous in vivo sampling possible. A single animal can also act as its own control decreasing the number of animals needed for experimentation.

Microdialysis is particularly suited for drug transport studies. Multiple probes can be used simultaneously in different parts of the animal. The size of the analyte that can diffuse through the probe is a function of the molecular-weight cutoff of the membrane. This results in large molecules, such as proteins, being excluded and only the non-protein-bound portion of a drug collected. Enzymes are also excluded, resulting in no further enzymatic degradation after dialysis. In addition, simpler separation and detection schemes are possible for complex physiological samples because the membrane provides sample cleanup before the sample undergoes further analysis. As a result, many LC applications require no further cleanup after microdialysis.

The temporal resolution of the sampling is very important in monitoring pharmacokinetics and drug transport. The pharmacokinetics of a drug may be too fast for typical methods used in drug discovery, such as tissue removal and homogenization or multiple blood extractions. Coupling of the analysis modules directly with the continuous sampling of microdialysis allows for on-line in vivo analysis, providing good temporal resolution. Off-line techniques have been performed as well, with samples collected and assayed following microdialysis.

## C. Microdialysis Probe Design

The microdialysis probe body is typically composed of stainless steel needle tubing, fused silica, or a combination of the two. The membrane is often polycarbonate, polyacrylonitrile, or regenerated cellulose. The membranes are hydrophilic and may have a molecular-weight cutoff of 5,000 to 75,000 Da, although membranes of 10–20 kDa are commonly used. The average probe dimensions are 300 μm o.d., with a length of 4–10 mm. These small dimensions result in minimal damage to the tissue. Therefore, awake and freely moving animals can be used in experiments and the probe will maintain the integrity of the internal organs. This allows collection of dialysate for periods of time from hours to weeks.

Probe geometry is dependent on the area in which the probe is implanted [5]. Figure 2 shows the variety of probes that can be used. Microdialysis was originally developed as a tool for sampling the microenvironment of the brain for neurochemical investigations [1,2]. As such, valuable information regarding transport of drugs across the blood–brain barrier (BBB) can be obtained with this technique. The most common probe design for implantation into the rat brain is the rigid cannula probe (Fig. 2a). This probe is commonly used in the brain and consists of an inner and outer length of stainless-steel tubing. The inner cannula extends beyond the outer and is covered with the dialysis membrane. The probe can be cemented to the skull so that it moves with the animal and does not cause further damage to the brain. A guide cannula can be also be used and cemented in place. The probe can be inserted into the guide cannula allowing for another probe to be used later without additional surgery to the animal.

The flexible probe can be implanted into the blood vessel of the rat (Fig. 2b). While the rigid probe can be glued to the animal's skull, the flexible probe

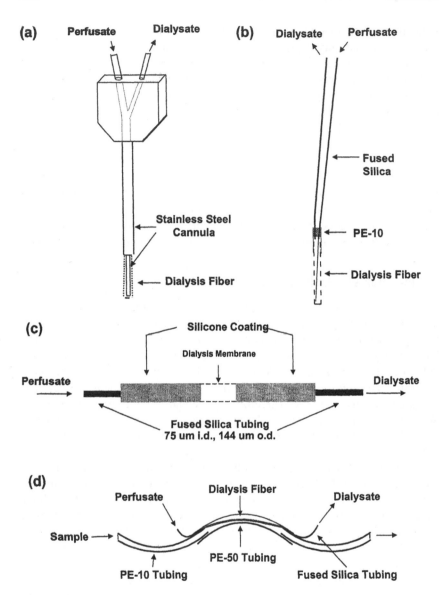

**Figure 2** Typical microdialysis probe geometries. (a) Rigid cannula probe typically used for implantation in the brain. (b) Flexible cannula probe for implantation in the blood vessel of a rat. (c) Linear probe designed for peripheral tissues such as skin, muscle, or liver. (d) Shunt or bypass probe is a linear probe inside a larger tube used for sampling the bile duct of the rat.

needs to be moveable inside the animal without causing additional damage. An advantage of this probe is that it is needlelike and much of it is composed of bendable fused silica. Therefore, it can be easily inserted into the blood vessel without causing extensive damage to the surrounding tissues.

Implantation in skin, muscle, or liver is performed with a linear probe (Fig. 2c). This is the most useful probe design for peripheral tissue. Teflon, PEEK, polyethylene, and fused silica tubing can be used for tubing positioned on opposite ends of the dialysis membrane. The probe can be implanted by pulling it through like a thread. For in vitro or subcutaneous sampling, loop probes (essentially longer linear probes) can be used. The length of the dialysis membrane can be extended because it is more flexible than fused silica. The portion of the membrane not in use is often coated with silicone resin, leaving only a small, uncoated window for sampling.

For sampling from the bile duct, a shunt or bypass probe can be used. The probe can be visualized as a flow-through design, which is essentially a linear probe inside larger tubing, typically made of polyethylene (Fig. 2d). The polyethylene tubing is used as a shunt to transfer biological fluid past the linear probe that is perfused with sampling solution. The rationale for this design is that the shunt allows the solution from the bile duct to flow through without causing blockage and high backpressure.

## D. Analytical Considerations

The flow rates of the microdialysis experiment are such that samples of 1–10 μL are typically obtained. At typical perfusion rates, the perfusate is not at equilibrium with the extracellular fluid. As such, the concentration of sample in the dialysate is some fraction of that in the surrounding tissue. This is termed the extraction efficiency and is a function of the delivery or recovery of the probe. Not only are the sample volumes small but also the concentration in the dialysate may be low, typically ranging from 1 pM to 1 μM. Because the recovery is typically less than 100%, the limit of detection of the method should be lower than the lowest concentration expected in the dialysate. This presents a tremendous challenge for the analyst.

A related issue is that while microdialysis is a continuous process, it is coupled to an analytical separation step that requires discrete sample volumes. Individual samples can be collected off-line with a fraction collector and assayed later (Fig. 3). The temporal resolution is defined by the time interval at which the microdialysis samples are collected. Without the need for further sample cleanup, the temporal resolution for off-line analysis will ultimately be dependent on the perfusion rate and the volume of sample needed for quantitation. If the analytical method does not have sufficient limits of detection, larger sample volumes must be collected, decreasing the temporal resolution of the method.

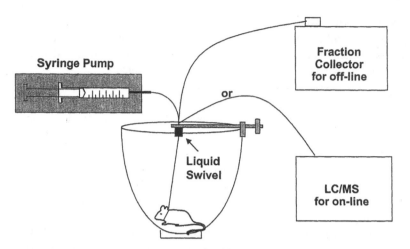

**Figure 3** General diagram of a microdialysis system with a freely moving rat. The dialysate can be collected off-line on a fraction collector or on-line to an LC/MS.

Another approach to handling small sample volumes is to use an on-line approach. This overcomes problems with sample evaporation that can occur with sample manipulation off-line. Although near-real-time data can be obtained, the limit in temporal resolution may be dependent on the separation method. The assay time must be faster than the physiological event in order to provide the best temporal resolution.

When experiments are performed in vivo, previously independent components of the model, the sampling, and the analytical method become interdependent. The conditions that were optimal for each independent component must now be considered in relationship to the other components. For example, if the method requires a large sample volume the flow rate of the perfusate must be increased. If the flow rate is increased, the extraction efficiency will be decreased. The result is that the analytical method will require better limits of detection. On the other hand, if the flow rate is decreased to increase the extraction efficiency, the temporal resolution will be compromised [5]. The increased recovery can also deplete compounds of low molecular mass in the tissue near the probe, thereby perturbing the experimental conditions.

Several characteristics of chromatography impact on the microdialysis experiment. The chromatography process inherently dilutes the sample. If we assume that a typical microdialysis experiment will involve a perfusion rate of 1 µL/min, with sampling for 5 min, 5 µL of sample will be obtained for the assay. A typical analytical column (15 cm × 4.6 mm) with a mobile-phase flow rate of 1 mL/min may have a peak width of 30 sec and would correspond to a 500-

μL sample. This corresponds to a 100-fold dilution factor for the 5-μL sample. In order to decrease the volume of the chromatographic peak, microbore columns (15 cm $\times$ 1 mm) can be used, which typically require 1 μL or less of sample. Using this type of column the mobile-phase flow rate can be reduced from 1 mL/min to 40 μL/min. If the chromatographic peak width is still 30 sec the peak volume is only 20 μL, resulting in a dilution factor of only 4. Capillary electrophoresis is another option because only a few nanoliters of sample are needed. In general, the collection, handling, and injection of microdialysis samples are important parameters to consider when choosing which separation method to use and whether on- or off-line analysis is appropriate.

## E.  Issues Involving Mass Spectrometric Detection

If the ionic strength and pH of the perfusate do not match those of the extracellular environment the physiological conditions will be perturbed. Results from experiments performed under these conditions could then be a product of factors that are not accounted for experimentally. While this can be less important in a blood vessel than in the brain, where all processes are affected by electrolyte levels, the perfusate composition is very important in the reliability of the results. Analysis of microdialysis samples with mass spectrometric detection offers a special challenge because the high salt concentration of the dialysate could affect the sensitivity of the spectrometer.

The remainder of the chapter discusses microdialysis coupled to mass spectrometric detection to both on-line and off-line methods, with an emphasis on analysis of drugs. Applications of microdialysis/mass spectrometry to gas chromatography (GC), liquid chromatography (LC), and flow injection into the mass spectrometer are discussed. In particular, approaches to overcome problems with introduction of high-ionic-strength dialysate into the mass spectrometer are described.

## II.  APPLICATIONS

## A.  Off-Line Methods

### 1.  Microdialysis/GC/MS

One of the early methods for mass spectrometric analysis of microdialysis samples utilized gas chromatography with mass spectrometric detection (GC/MS). Electron ionization is one type of ionization used for GC/MS. It is a "hard" ionization technique, resulting in a high degree of analyte fragmentation. A softer form of ionization, chemical ionization, is also available and often results in less analyte fragmentation. Selected-ion monitoring (SIM) is used frequently with

GC/MS to enhance the selectivity and signal-to-noise ratio of the MS detector [6–8].

Determinations by GC/MS for homovanillic acid [6], cocaine [7], and acetylcholine [8] in microdialysis samples have been previously demonstrated. Even though GC/MS has the capability of separating, identifying, and quantitating individual components of a complex sample, it can be somewhat limited in its application toward microdialysis sampling. For example, microdialysis samples frequently need to be cleaned up before introduction into a gas chromatograph. Sample preparation can include lyophilization [8], extraction of analyte from the dialysate, and/or derivatization of the analyte [6–8]. Such steps are needed to remove nonvolatile species (particularly salt) from the dialysate and to adequately volatilize the analytes of interest. Due to the required sample handling and preparation, an on-line interface for microdialysis/GC/MS has yet to be realized. Furthermore, such sample preparation adds time and complexity to the determination, which is why GC/MS is rarely used in conjunction with microdialysis sampling.

## 2. Microdialysis/LC/MS

Electrospray (ESI) is the most popular ionization method used for microdialysis/ LC/MS. Occasionally dialysate salts can clog the ionization source. To avoid this problem, the dialysate can be collected off-line and extracted with a $C_{18}$ solid-phase extraction (SPE) cartridge [4]. Washing the dialysate with 3% (v/v) aqueous acetic acid and eluting with a methanol–water–acetic acid solution can desalt the sample so that it does not affect mass spectrometer performance. However, even after SPE, direct injection of the dialysate without further chromatographic separation can yield complex spectra by LC/ESI/MS, due to the large number of metabolites formed in vivo. In the characterization of a peptide mixture by LC/MS, Prokai and coauthors found that after SPE followed by LC, three peptides coeluted [4]. Even though complete separation was not observed after chromatographic separation, the parent and metabolite peptides could be confirmed by deconvolution of the mass spectrum to obtain relative molecular weights. A reconstructed ion chromatogram (Fig. 4) shows the metabolites obtained from the primary peptide, Dynorphin A 1-12. Fourier transform ion cyclotron resonance was used to confirm the identities of the peptides in the mixture. Therefore, mass spectrometry can be critical in the identification of components in the presence of poor chromatography.

Instead of SPE, another approach to managing the salt effect on the LC/ MS uses a valve to divert the unretained salts to waste, with the rest of the effluent directed toward the MS [11]. Using this shunt, along with quadrupole ion-trap MS/MS detection, transport of the antipsychotic drug CI-1007 across the BBB was investigated. Dialysis samples were collected every 30 min, frozen, and as-

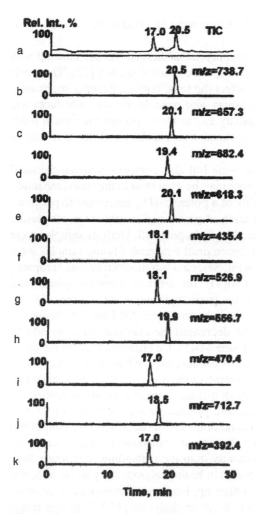

**Figure 4** Total ion current (TIC) (a) and reconstructed ion chromatograms (b–g) from the gradient reverse-phase HPLC/ESI/MS analysis of the microdialysates collected from rat striatum after perfusion of the probe with 100 pmol/µL of Dynorphin A 1-12 (Tyr-Gly-Gly-Phe-Leu-Arg-Arg-Lle-Arg-Pro-Lys-Leu, YGGFLRRIRPLK) at 0.8 µL/min. HPLC conditions: 15 cm × 0.3 mm i.d., $C_{18}$ reverse-phase column, 6.3 µL/min flow-rate, gradient elution from 5 to 95% acetonitrile (0.1 to 0.05% TFA) at 2%/min. Reconstructed ion chromatograms: (b) parent peptide: Dyn A 1-12 ($m/z$ 738.7, doubly charged); metabolites: (c) Dyn A 1-10 ($m/z$ 618.3, doubly charged), (d) Dyn A 1-7 ($m/z$ 435.4, doubly charged), (e) Dyn A 5-12 ($m/z$ 526.9, doubly charged), (f) Dyn A 1-5 ($m/z$ 556.7, singly charged), (g) Dyn A 6-12 ($m/z$ 470.4, doubly charged), (h) Dyn A 1-6 ($m/z$ 712.7, singly charged), (i) Dyn A 7-12 ($m/z$ 392.4, doubly charged), (j) Dyn A 212 ($m/z$ 657.3, doubly charged), and (k) Dyn A 1-11 ($m/z$ 682.3, doubly charged).

sayed within a week. The limits of detection were reported to be 1 ng/mL with this method.

Using capillary LC/microelectrospray no additional sample cleanup was necessary before triple-quadrupole mass spectrometric detection [12]. The small volumes of dialysate injected did not affect the sensitivity of the mass spectrometer. The development of microelectrospray and nanoelectrospray interfaces has made the coupling of micro and capillary LC to mass spectrometry much more feasible. Thus, the low flow rates associated with capillary and micro LC make ESI well suited for this approach.

Not only is sample cleanup an issue but also compensation for chemical noise in the mass spectra of the dialysate may be important in the characterization of drugs [13]. To investigate the ability of a potent 5-HT$_{2a}$ antagonist to penetrate the blood–brain barrier the concentration of parent and metabolite in rat plasma and brain was monitored with LC/triple-quadrupole MS. Dialysis samples were collected over 10-min intervals and frozen until analyzed. Plasma samples were collected from the jugular vein of the rat. Selected-ion monitoring was compared with multiple-reaction monitoring (MRM) for artificial cerebral spinal fluid (aCSF) fortified with parent (I) and the desmethyl metabolite (II) (Fig. 5). Although a large background (Fig. 5d) was observed with SIM for metabolite (II) in aCSF, multiple-reaction monitoring decreased the chemical noise by adding greater selectivity to the experiment. Using MRM, the chemical noise in the blank perfusate was nearly identical for I and II because the tandem mass spectrometry eliminated the high background ion current in the mass range of the metabolite.

Because protein-bound drug cannot diffuse through the dialysis probes, the concentrations of I and II in the brain and plasma are representative of the free drug. Figure 6 is a plot of the plasma and brain extracellular fluid concentrations of I and II on the same time scale. The parent compound is able to cross the blood–brain barrier much more efficiently than the metabolite. The lower limit of quantitation (LLQ) for each compound in brain dialysate and plasma were 500 pg/mL and 1 ng/mL, respectively. Other applications of atmospheric pressure ionization MS have been reported for drugs in dialysate [14,15]. In one recent study, atmospheric pressure chemical ionization (APCI) LC/MS resulted in attomole detection limits with a linear response over 4 decades for anandamide, an endogenous cannabinoid.

Thermospray ionization has also been used with quadrupole mass spectrometry to investigate the novel anticonvulsant topiramate in rat brain [16]. Topiramate crosses the blood–brain barrier and the concentration of the drug remains in the micromolar range for more than 6 hr. Samples were collected every 15 min over this time period, and the authors planned to analyze the dialysate without further sample cleanup. Distilled water was used as the perfusion medium to minimize the effect of dialysis salts on the mass spectrometer. However, it is important to note that microdialysis is best performed with matching ionic

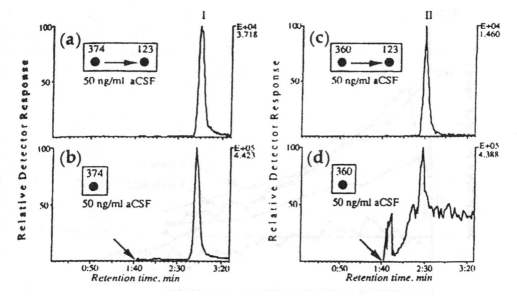

**Figure 5** Chromatographic profiles obtained by following the analysis of CSF fortified with I and II. Both a and c were obtained following SRM analysis of aCSF fortified to 50 ng/mL, whereas b and d were obtained following SIM analysis of aCSF fortified to 50 ng/mL. The arrows in b and d indicate the time where the divert valve was switched, directing the effluent flow to the mass spectrometer.

strength and pH of the perfusate to that of the tissue or fluid being investigated. If the salt concentration in the brain is perturbed with the introduction of water through the probe, the drug levels could be altered. This is particularly important for studies in the brain where electrolyte levels are critical to brain function and where sampling is performed over an extended period of time. Unfortunately, the use of water as a perfusion medium does not take into account the sampling aspects of the experiment. The animal model, the microdialysis system, and the analysis/detection are interdependent even with off-line analysis and the method used should consider all three of these components.

## 3. Microdialysis/Capillary Electrophoresis/MS

Capillary electrophoresis/mass spectrometry (CE/MS) is a useful tool for bioanalysis because it combines a highly efficient separation technique with a sensitive and selective detection system. Because CE offers excellent mass sensitivity but presents challenges when concentration sensitivity is needed, more sensitive detection systems are required. As an example of this, preliminary results for

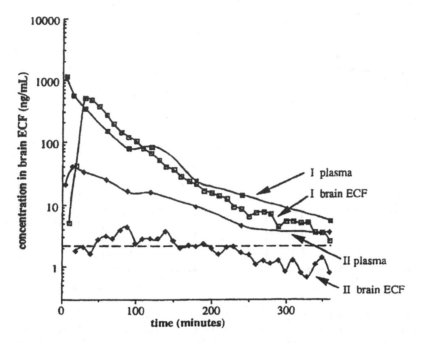

**Figure 6** Semilogarithmic plot of plasma and brain ECF concentration time profiles for I and II following 5-mg/kG iv dose administered to a rat. Note that the brain ECF concentrations represent free concentrations, whereas the plasma values represent the total drug concentration in plasma. The dashed line indicates LLQ of I and II in the brain ECF taking into account the recovery of the microdialysis probe.

analysis of γ-aminobutyric acid (GABA) in the rat brain by microdialysis/CE/MS [17] have been reported. Contaminant ions resulting from poor electrospray ionization made it difficult to detect the protonated GABA. The observed analyte signal was too weak for quantitation and future work is needed to enhance the sensitivity of this method.

More success was obtained using on-line solid phase extraction before off-line analysis by capillary electrophoresis with matrix-assisted laser-desorption ionization (MALDI) MS [18]. Because microdialysis is a method used for analysis of small molecules and peptides, MALDI is not used frequently with microdialysis sampling. Another disadvantage is that MALDI cannot be used on-line with the separation because it is a vacuum ionization technique. However, if the peptide is large enough (≳1000 Da) MALDI can be useful. For the analysis of peptides in dialysate an appropriate separation is important before mass spectrometric detection. In a comparison with direct sampling of dialysate in MALDI, capillary electrophoresis provides the high efficiency separations necessary to resolve all of the peptide fragments in a sample. On-line sample collection can be used to improve recovery (Fig. 7). Because the salt content of dialysate can result in band broadening in the capillary, samples were loaded onto a solid-phase extraction device composed of a short length of electrophoresis capillary. Salts were removed and the salt-free samples eluted onto the separation capillary with organic solvent. After separation, the electrophoretic effluent was deposited in 180-µm-wide spots on a coated cellulose target for MALDI. A nitrogen laser with a spot size of 30 µm was focused on the target with 5 laser shots/spot used for ionization. The system could be automated with computer control of the plate movement, generating an image of the selected ions by simultaneously recording MALDI signal intensities of multiple mass windows. Migration times were reported by knowing the coordinates of the electrophoretic bands with time. Up to 80 metabolites were identified with a limit of detection of $10^{-11}$ M.

## B. On-Line Methods

### 1. Microdialysis/Flow Injection/MS

In some instances, microdialysis sampling can be coupled on-line with mass spectrometry without prior separation or cleanup with techniques such as GC or LC. Continuous flow fast atom bombardment (cfFAB) has been used for the ionization of microdialysis samples without prior sample preparation. Coupling microdialysis directly to mass spectrometry via FAB was more feasible with cfFAB because it provided for a more robust, efficient ionization and prevented source fouling.

Caprioli and Lin first described this method for the characterization of penicillin G in 1990 [19]. A diagram of the microdialysis/cfFAB experimental setup used is shown in Fig. 8. Dialysate was allowed to collect in a 20-µL sampling loop via syringe pump. After collection, the injection valve was switched and a high-performance liquid chromatography pump was used to deliver the microdialysis sample to the mass spectrometer. Mass spectra of penicillin from both a standard and an in vivo microdialysis sample (shown in Fig. 9) compare well.

One requirement of cfFAB is the presence of a matrix, usually glycerol, to assist the ionization process. The matrix can be added to the system through a coaxial sheath flow at the FAB tip [20] or by simply including the matrix in the microdialysis perfusion liquid [19]. A coaxial sheath flow reduces sample dilution and peak broadening that would normally decrease sensitivity. One problem associated with the matrix is that it creates a large abundance of background ions, which can potentially interfere with analyte signal, reducing overall sensitivity. Fortunately, matrix ions are predictable and can be subtracted.

The presence of salt in the dialysate decreases overall sensitivity due to the formation of salt clusters at the FAB tip, decreasing the abundance of analyte

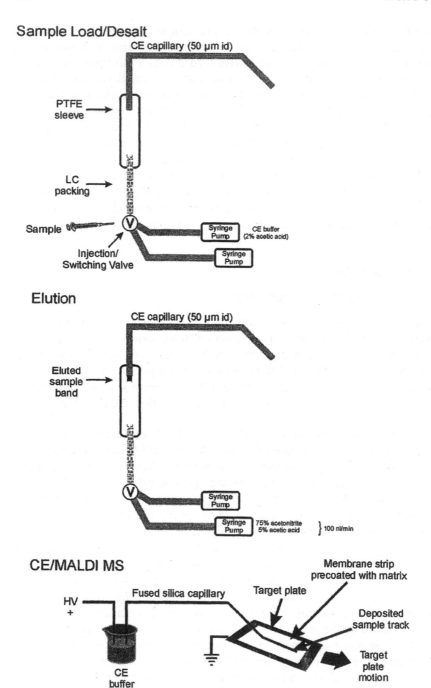

**Figure 7** The ''on-line transfer, off-line analysis'' process.

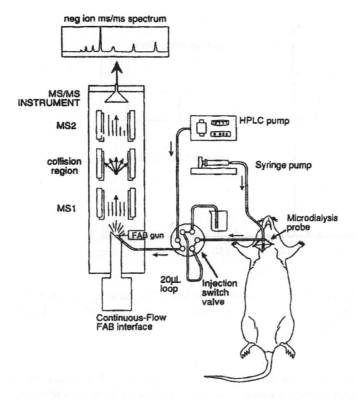

**Figure 8**   General experimental setup for the on-line analysis of drugs in the blood of a live rat. The dialysate from a microdialysis device implanted in the jugular vein of a rat is allowed to flow into the mass spectrometer at the CF–FAB interface.

ions and increasing chemical noise. The flow injection carrier stream provides one solution to this problem because the salt is diluted upon entry into the system [19]. Another method simply uses deionized water as the perfusion liquid instead of a typical microdialysis perfusion liquid such as Ringer's solution [20]. This can be problematic because water creates an osmotic pressure drop across the dialysis membrane, and as a result water can diffuse across the membrane, potentially altering the physiology of the tissue being studied and leading to erroneous data. Single-reaction monitoring (SRM) can also be employed in order to decrease noise and to increase sensitivity.

The coupling of cfFAb to microdialysis can provide valuable pharmacokinetic information. The xenobiotic Tris(2-chloroethyl) phosphate (TRCP) was used to compare the microdialysis/cfFAB technique with a conventional method relying on blood withdrawal [20]. Figure 10 is a direct comparison between the

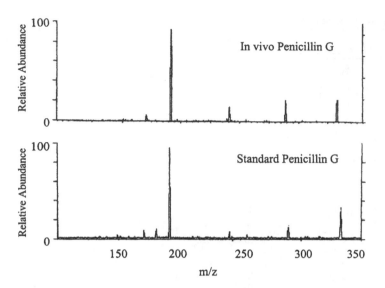

**Figure 9** FAB/collisionally activated dissociation MS/MS negative-ion spectra of penicillin G. (Upper) Spectrum taken from an in vivo analysis as shown in Fig. 2. (Lower) Spectrum of the pure drug.

microdialysis and conventional LC work, with the slopes of the log concentration versus time plots for the two methods showing significant differences from one another. A possible reason for this is that the microdialysis method used anesthetized rats while the other used conscious, freely moving rats. It is possible that the anesthetic used in the microdialysis experiment altered the physiology of the animal, resulting in erroneous results. The use of conscious animals for the microdialysis experiments could eliminate this discrepancy.

A second potential cause for the observed difference between the two methods is related to the ionization of the sample. The presence of ions such as $Na^+$ can form adducts to the analyte, thereby reducing the abundance of the $(M+H)^+$ ion. The presence of $(M+Na)^+$ ions could not be observed because MRM was used.

Third, an in vitro calibration, which does not necessarily provide a good estimate of the recovery in vivo, was used for the microdialysis probe. In vitro calibration is only an approximate calibration technique because it does not always accurately represent what is happening in vivo. In addition, the efficiency of the microdialysis probe can change over time as the biological surroundings adjust to the presence of the probe. A final cause for the discrepancy could be due to the method of sampling for the conventional analysis. Because blood samples were taken from the rat, fewer samples could be taken within a given time.

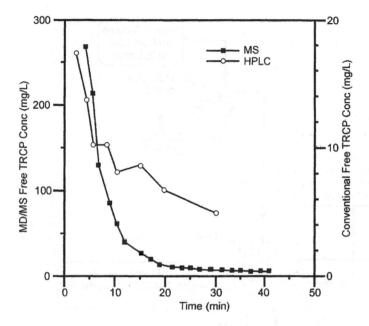

**Figure 10** Free concentration of TRCP in plasma versus time as observed in one animal in the conventional studies (open circles) and one animal in the microdialysis/MS/MS studies (solid squares).

Due to the smaller number of samples taken, the conventional sampling lacks the ability to create an accurate curve of concentration versus time. In addition, the removal of blood can change the physiology of the animal, leading to error.

## 2. Microdialysis/LC/MS

The high ionic strength of the dialysate buffer is a substantial problem for on-line microdialysis/LC/MS, where the samples go directly from the animal through the analytical LC column to the mass spectrometer. A parallel bypass column has been used with a six-port valve, allowing solvent containing the inorganic salts to be voided to waste [21]. A similar setup is described in Chapter 6. Although not required for electrospray experiments, a constant flow can be maintained in the LC column by using the parallel "ballast column." An automated system using this design ran experiments for 8 hr/day for a minimum of 3 weeks without loss of mass spectrometer performance. In addition, a switching valve was used to interface multiple microdialysis probes for simultaneous analysis in both blood and liver.

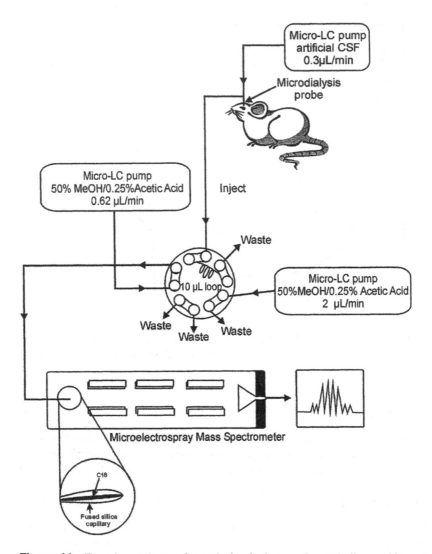

**Figure 11**  Experimental setup for analysis of substance P metabolism and its metabolic fragments. Substance P was infused into the rat striatum through a microdialysis probe at 0.3 μL/min, and the probe was used both to deliver and collect metabolic products. The microdialysate was sampled into a 10-port valve, injected into a fused-silica C$_{18}$ column, and washed with 2% methanol–0.25% acetic acid. The peptides were eluted from the column with 50% methanol–acetic acid at 0.82 μL/min directly into the microelectrospray source on the mass spectrometer by switching the valve back.

Using capillary liquid chromatography with microelectrospray and quadrupole mass spectrometry can result in fewer problems with dialysate salts [22]. Using a perfusion medium of a mock artificial cerebral spinal fluid (KCl, NaCl, $MgCl_2$, $CaCl_2$, and phosphate-buffered saline at pH 7.4) analysis of a peptide in the rat brain was performed with samples collected every 30 min. The perfusate was then directly loaded on-line from the injection valve sample loop to a separation capillary (Fig. 11).

As with off-line microdialysis/LC/MS, APCI can be used for on-line experiments [23]. However, so far experiments reported with APCI LC/MS have used water as the perfusion medium in order to avoid problems with the MS. In general, even though some problems may occur with ion suppression, electrospray is the method of choice for any LC/MS work involving microdialysis.

## III. CONCLUSIONS

Microdialysis coupled to mass spectrometry is a powerful technique for on- and off-line analysis, providing information on pharmacokinetics, drug transport, and metabolite formation. With the widespread availability of LC/ESI/MS instruments there has been a shift toward liquid chromatography and away from gas chromatography and flow injection. Electrospray is the ionization method of choice for most applications and tandem mass spectrometry has grown in popularity. Thermospray and cfFAB applications have been used with microdialysis, but are older, obsolete techniques.

While liquid chromatography, gas chromatography, and flow injection are well-established techniques that are easy to couple to MS detection, development of a simple and reliable sheathless electrospray CE/MS interface is ongoing. As such, coupling microdialysis with CE/MS is a new area with many possible advantages and applications. New mass spectrometric mass analyzers such as the time-of-flight and quadrupole time-of-flight (Q-TOF) instruments allow for the coupling of electrospray. Because the scan rate of a TOF MS is much faster than that of a quadrupole, it will be much more suitable for the narrow peak widths of capillary electrophoresis. The use of new mass spectrometers such as the Q-TOF will likely become more popular because they will offer high scan rates, improved sensitivity, and better mass resolution.

Although mass spectrometry offers major advantages when the chromatography is poor and analytes can be identified by their mass spectra, a problem is that dialysate salts can affect sensitivity and stability of the analyte ionization. The extent of this effect depends on the chromatographic properties and ionization source conditions. The inclusion of a divert valve or extraction and cleanup of the analyte from the dialysate is needed to compensate for this problem. The use of water as a dialysis buffer can minimize ion suppression, but can result in misleading or erroneous results for brain microdialysis because the ionic strength in the brain is critical to cerebral function. While this is less of a problem for blood sampling, care must be taken when developing methods that combine microdialysis and mass spectrometry.

## REFERENCES

1. CE Lunte, DO Scott, PT Kissinger. Sampling living systems using microdialysis probes. Anal Chem 63:774A–780A, 1991.
2. HH Benveniste, Huttemeier. Microdialysis—theory and application. Progr Neurobiol 35:195–215, 1990.
3. PE Andren, S Lin, RN Caprioli. Microdialysis/mass spectrometry. In: DM Desiderio, ed. Mass Spectrometry: Clinical and Biomedical Applications. Vol. 2. New York: Plenum Press, 1992, pp 237–253.
4. L Prokai, H Kim, A Zharikova, J Roboz, L Ma, L Deng, JR Simonick. Electrospray ionization mass spectrometric and liquid chromatographic-mass spectrometric studies on the metabolism of synthetic dynorphin A peptides in brain tissue in vitro and in vivo. J Chromatogr A 800:59–68, 1998.
5. MI Davies, CE Lunte. Microdialysis sampling coupled on-line to microseparation techniques. Chem Soc Rev 26:215–222, 1997.
6. SY Chang, TA Moore, LL Devaud, LCE Taylor, EB Hollingsworth. Analysis of rat brain dialysate by gas chromatography-high-resolution selected-ion monitoring mass spectrometry. J Chromatogr 562:111–118, 1991.
7. LC Nicolaysen, HT Pan, JB Justice Jr. Extracellular cocaine and dopamine concentrations are linearly related in rat striatum. Brain Res 456:317–323, 1988.
8. MR Marien, JW Richard. Drug effects on the release of endogenous acetycholine in vivo: measurement by intracerebral dialysis and gas chromatography-mass spectrometry. J Neurochem 54:2016–2023, 1990.
9. SD Menacherry, JB Justice Jr. In vivo microdialysis and thermospray tandem mass spectrometry of the dopamine uptake blocker 1-[2-[bis(4-fluorophenyl) methoxy]-ethyl]-4-(3-phenylpropyl)-piperazine (GBR-12909). Anal Chem 62:597–601, 1990.
10. PE Andren, MR Emmett, BB DaGue, AF Steulet, P Waldmeier, RM Caprioli. Blood–brain barrier penetration of 3-aminopropyl-n-butylphosphinic acid (CGP 36742) in rat brain by microdialysis/mass spectrometry. J Mass Spectrom 33:281–287, 1998.
11. RN Iyer, MD Davis, PL Juneau, AB Giordani. Brain extracellular levels of the putative antipsychotic CI-1007 and its effects on striatal and nucleus accumbens dopamine overflow in the awake rat. J Pharm Pharmacol 50:1147–1153, 1998.
12. PE Andren, RM Caprioli. Determination of extracellular release of neurotensin in discrete rat brain regions utilizing in vivo microdialysis/electrospray mass spectrometry. Brain Res 845:123–129, 1999.
13. TG Heath, DO Scott. Quantification of a potent 5-HT2a antagonist and an active metabolite in rat plasma and brain microdialysate by liquid chromatography–tandem mass spectrometry. J Am Soc Mass Spectrom 8:371–379, 1997.

14. LJ Petersen, MK Chruch, JP Rihoux, PS Skov. Measurement of interstitial cetirizine concentrations in human skin: correlation of drug levels with inhibition of histamine-induced skin responses. Allergy 54:607–611, 1999.

15. JM Walker, SM Huang, NM Strangman, K Tsou, MC Sanudo-Pena. Pain modulation by release of the endogenous cannabinoid anandamide. Proc Natl Acad Sci USA 96:12198–12203, 1999.

16. JA Masucci, ME Ortegon, WJ Jones, RP Shank, GW Caldwell. In vivo microdialysis and liquid chromatography/thermospray mass spectrometry of the novel anticonvulsant 2,3:4,5-bis-O-(1-methylethylidene)-beta-D-fructopyranose sulfamate (topiramate) in rat brain fluid. J Mass Spectrom 33:85–88, 1998.

17. Y Takeda, M Yoshida, M Sakairi, H Koizumi. Detection of γ-aminobutyric acid in a living rat brain using in vivo microdialysis-capillary electrophoresis/mass spectrometry. Rapid Commun Mass Spectrom 9:895–896, 1995.

18. H Zhang, M Stoeckli, PE Andren, RM Caprioli. Combining solid-phase preconcentration, capillary electrophoresis and off-line matrix-assisted laser desorption/ionization mass spectrometry: intracerebral metabolic processing of peptide E in vivo. J Mass Spectrom 34:377–383, 1999.

19. RM Caprioli, SN Lin. On-line analysis of penicillin blood levels in the live rat by combined microdialysis/fast-atom bombardment mass spectrometry. Proc Natl Acad Sci USA 87:240–243, 1990.

20. LJ Deterding, K Dix, LT Burka, KB Tomer. On-line coupling of in vivo microdialysis with tandem mass spectrometry. Anal Chem 64:2636–2641, 1992.

21. P Michelsen, G Pettersson. An automated liquid chromatography/mass spectrometry system coupled on-line with microdialysis for the in vivo analysis of contrast agents. Rapid Commun Mass Spectrom 8:517–520, 1994.

22. PE Andren, RM Caprioli. In vivo metabolism of substance P in rat striatum utilizing microdialysis/liquid chromatography/micro-electrospray mass spectrometry. J Mass Spectrom 30:817–824, 1995.

23. PSH Wong, K Yoshioka, F Xie, PT Kissinger. In vivo microdialysis/liquid chromatography/tandem mass spectrometry for the on-line monitoring of melatonin in rat. Rapid Commun Mass Spectrom 13:407–411, 1999.

# 13
# Future Directions for Mass Spectrometry in Drug Discovery

**David T. Rossi**
*Pfizer Global Research and Development,*
*Ann Arbor, Michigan*

**Richard A. J. O'Hair**
*University of Melbourne, Victoria, Australia*

> Hokey religions and ancient weapons are no match for a good blaster at your side, kid.
>
> —*Han Solo*

## I. THE FUTURE OF DRUG DISCOVERY REVISITED

Although there has been considerable pressure in the pharmaceutical industry to make the discovery and development of new therapeutic drugs into a routine and predictable venture, drug discovery remains a science and is still subject to the uncertainties of that discipline. There are, however, apparent trends that impact how the science will be conducted, with the aim of making the discovery of drugs more predictable, consistent, and manageable. These trends rely on a fusion of traditional and emerging biological disciplines in an effort to understand disease states (genomics) [1–4], develop molecular targets (proteomics) [5], and evaluate drug candidates (pharmacology, absorption/distribution/metabolism/excretion (ADME), metabolomics, and the use of biochemical molecules as disease-state indicators). These emerging disciplines provide value to drug discovery based on their contributions to the targeting, production, characterization, and selection of the best candidate molecules.

Because it is inefficient, dangerous, or extremely difficult to identify new drugs without them, in vitro and in vivo models will continue to be used as important tools for the selection process [6]. The ability to extract information from drug-selection models is dependent on the availability of analytical techniques to probe them. As a key enabling technology for drug discovery, mass spectrometry is presently unparalleled in its applicability. As described in the proceeding chapters of this book, mass spectrometry provides generally high selectivity and sensitivity and can be either qualitative or quantitative in nature. This powerful and unusual combination of capabilities has led to widespread applicability of the techniques of mass spectrometry for assessments of disease states, drug targets, and drug candidates in various models. For the short and midterms, it is likely that the evolution of drug discovery and mass spectrometry will be closely interlaced, with the needs and capabilities of each area directly impacting the evolution of the other.

## II. TECHNOLOGIES AFFECTING INSTRUMENT USE

In a recent survey on instrument use in drug discovery and development [7], improved software, computing power, and automation were the three technologies cited as having the most impact on the future of analytical instrumentation. The results of this survey, summarized in Fig. 1, also show that other instrumenta-

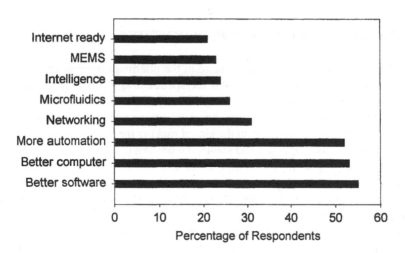

**Figure 1** Results from a recent survey of analytical instrument users [7] shows the relative demand for improved technologies.

tion improvements, such as networking, microfluidics, and Internet technologies, were thought to have less of an impact. Although it represents a composite view of a broad cross section of instrument users, this survey might not accurately predict the future trends of instrument development because it neglects an important aspect of future advances: innovation.

It has been said that incrementalism is the enemy of true innovation [8]. Despite the best attempts to manage it, true innovation does not come in a predictable or steady way. Rather, it comes in spurts and starts, interspersed with dry spells. It is possible, even likely, that the greatest impact on instrumentation advances will come from innovations that have not yet been discovered, developed, or invented or from technologies that do not seem very practical at the present time.

## III. NEW USERS BEWARE: DO NOT FORGET THE FUNDAMENTALS THAT UNDERPIN ALL APPLIED MASS SPECTROMETRY!

Recalling the introductory quote by Stewart (Chap. 4), it is fair to say that mass spectrometry is in a constant state of change. It is worth remembering that mass spectrometry is a science that is over 100 years old and that it was initially the domain of physicists before it moved into the hands of chemists. The extended abstracts of the American Society for Mass Spectrometry [9] for the past 15 years reveal that mass spectrometry has now moved into the hands of a new group of scientists: those working on unraveling the structures and interactions of biological molecules. Recent discussions with ''old school'' organic mass spectrometrists identified a common concern: that these new users would treat mass spectrometers as ''black boxes'' without having (or wanting to have) a deep appreciation of the physicochemical processes that give rise to mass spectra. Why is this an issue? Their answer is that forgotten painful lessons may reappear and new classes of mass spectrometry experiments may turn out to be a ''reinvention of the wheel.''

At present, drug discovery is one of the major vehicles for change in mass spectrometry, with a constant need for faster and cheaper mass spectrometers able to analyze ever-smaller quantities of samples in complex mixtures and matrices. Occasionally, a revolutionary step such as electrospray is made, the impact of which is still being felt as it opens up new avenues for exploration (including fundamental questions as to how and why it works as a method of transferring ions into the gas phase).

A number of obvious areas for growth in mass spectrometry, which have built upon advances in the ''fundamentals'' of mass spectrometry, include the following:

1.  Top-down sequencing of intact biomolecules [10,11], which may re-
    veal interesting high-order structural effects [12]. Note that fundamen-
    tal advances in trapping instruments such as Fourier transform–ion
    cyclotron resonance (FT-ICR) and quadrupole ion traps have allowed
    such experiments to be contemplated. Key challenges are evaluating
    the best ways of dissociating (collision-induced dissociation (CID) or
    electron capture dissociation [10]) the intact protein or DNA ions to
    provide as complete sequence coverage as possible, assigning fragment
    ion structures and charge states, and developing computer programs
    to collect and evaluate all the data. At present, the ultrahigh resolving
    power of FT-ICRs show the most promise, but the novel use of ion
    traps combined with ion–ion reactions offer potential opportunities to
    develop cheap, portable mass spectrometers able to characterize bio-
    molecules.
2.  The observation of noncovalent complexes via electrospray ionization
    has generated much interest. Further fundamental work is required to
    address the question ''do the gas-phase ions observed in the mass spec-
    tra reflect what is going on in solution?'' Although we do not have
    a definite answer, probing the structures and properties of gas phase
    noncovalent complexes is an exciting and emerging area.
3.  Mass spectrometry methods to probe higher order structures of bio-
    molecules is related to item 2 in that we can also ask how secondary
    and tertiary structures might be preserved (or destroyed) on going from
    solution to the gas phase. This is also an emerging area and alternative
    probes to CID (such as ion–molecule reactions) could play an impor-
    tant role.
4.  Robust automation for structural determinations in combinatorial
    chemistry [13]. Two-dimensional mass spectrometry [14] could be re-
    garded as the poor cousin of 2D NMR insofar as it has seen little appli-
    cation to ''real-world problems.'' This could be a technique that is ripe
    for exploiting structural elucidation in complex mixtures.

Perhaps the most exciting opportunities will be afforded when two technologies that have been developed to take advantage of electrospray ionization will
become more readily available. Specifically, these are (1) the new area of ion–
ion chemistry research, which is based on the study of the reactions of high-mass
multiply charged ions with ions of opposite polarity. This exciting area not only
opens up new reaction phenomenology but also offers applications to larger biomolecules [15]. (2) Ion-mobility mass spectrometry, which allows the separation
of ions of the same $m/z$ ratio in terms of their ''shape'' [16,17]. This is achieved
by injecting the ions into a drift tube where they are separated based on differ-

ences in mobility through a buffer gas. This technique is extremely versatile and also has broad implications in both fundamental gas-phase chemistry (e.g., ion–molecule reactions of "shape selected" ions) and applications (e.g., characterization of peptide/protein mixtures and combinatorial libraries) [18,19].

## IV. CHROMATOGRAPHIC SEPARATIONS AND MASS SPECTROMETRY

It has become painfully obvious that most of the excellent approaches and techniques that have been developed for use in liquid chromatography are not applicable to liquid chromatography/mass spectrometry (LC/MS) with atmospheric pressure ionization. Chapter 5 described the reagents and the range of mobile-phase compositions that are compatible with electrospray and atmospheric pressure chemical ionization (APCI), and these are limited to volatile components that do not cause significant ion suppression. Certain problems that are not significant with standard LC separations become difficult to deal with because of the limitations placed on the mobile phase by atmospheric pressure ionization (API) LC/MS.

Because of these restrictions, one significant problem that has arisen for LC/MS is the challenge to gain enough chromatographic retention for ionic compounds so that they are sufficiently separated from early-eluting matrix components. As described in Chapters 3, 5, and 6, ion suppression can be observed when an analyte coelutes with large amounts of salt or other ionizable components. A strategy for avoiding this is to move the analyte peak(s) to a position in the chromatogram where ion suppression is less pronounced. Unfortunately, many common LC/MS mobile phases, such as acidified water, are designed to promote ionization for basic molecules. This ionized state decreases the capacity of reverse-phase LC columns, even with the weakest of mobile phases.

Two classic liquid chromatographic ways to circumvent poor analyte retention are the addition of ion-pair reagents [20] to the mobile phase or the use of ion-exchange separations [21,22]. Neither approach is widely feasible with LC/MS. Ion-pair reagents are either nonvolatile (for example, alkyl sulfonic acids) or cause ion suppression to an unacceptable degree (trialkyl amines). The currently available ion-exchange columns are not designed so that acceptable retention can be achieved with compatible mobile phases (mobile phases limited to $\leq 1\%$ acetic or formic acid, or 1–25 mM ammonium acetate over a 3 to 11 pH range with accompanying organic solvent). To this end, it is expected that the introduction of ion-exchange sorbents (anionic and cationic) and separations that are capable of providing adequate capacity and acceptable chromatography for a wide range of ionic analytes will be forthcoming. Mixed-mode sorbents, containing ion-exchange and partitioning functionalities, are a possibility, but are more difficult to use effectively with the limited liquid chromatography/tandem mass spectrometry (LC/MS/MS) mobile phases.

## V. ALTERNATIVE SEPARATIONS FOR ATMOSPHERIC PRESSURE IONIZATION MASS SPECTROMETRY

Several types of capillary separations, including capillary liquid chromatography [23,24], capillary electrophoresis [25–27], and capillary electrochromatography [27,28], have begun to emerge as alternatives to analytical-scale API LC/MS. Several advantages and disadvantages, relative to analytical-scale separations, should be considered when choosing between separation techniques.

As advantages, capillary separation techniques demonstrate high separation efficiency. On occasion, the number of theoretical plates available from these approaches has exceeded 1 million [29]. Also, very small sample volumes, on the order of 100 to 0.5 nL, are needed for these techniques. This can be an advantage for sample-limited situations, which are often encountered in bioanalysis. High mass sensitivity (the absolute weight of analyte injected) can be achieved, as the narrow capillary concentrates the sample plug and allows less opportunity for band broadening.

Capillary approaches have been shown to be useful for many chiral separations as well as achiral separations. For chiral separations, separation buffer additives containing chirogenic centers (tecoplainin, erythromycin, vancomycin, or cyclodextrans) have facilitated the resolution of enantiomers [26,30,31]. Chiral capillary separations could readily be combined with mass spectrometry because the volume of effluent moving from the separation capillary to the ion source is small and makeup solvent is commonly added by means of an union to stabilize the ion beam. Chiral capillary separations provide an attractive alternative to analytical-scale normal-phase separations when using atmospheric pressure ionization mass spectrometry.

Some of the drawbacks of combining and using capillary separations with mass spectrometry include the nonrobust nature of the capillary relative to larger scale separations. Capillaries are more prone to clogging and can demonstrate poor replication of analyte retention or migration time from injection to injection, especially with dirty biological samples. The physical stability of a capillary is much less than that of a stainless steel column, and there is a great deal of skill necessary to install and maintain them. Nano-flow ion sources, commonly used with capillary separations, are not highly optimized at this time and require greater attention than the more conventional ion sources [32]. If an ion source requires makeup flow, some of the intrinsic sensitivity of the technique is lost because of dilution of the analyte in the effluent flow. Although capillaries provide high-efficiency separations, there are some significant obstacles to their routine use. Because they are currently not robust, workhorse techniques, additional work appears to be needed before these approaches move from investigational to routine use.

## VI. IMMUNOAFFINITY SEPARATIONS WITH MASS SPECTROMETRY

Immunoaffinity separations are widely recognized as having potential to demonstrate more selectivity than any other condensed-phase separation technique [33–35]. These separations are based on biospecific binding interactions between an antibody, often chemically bound to an immobilized support, and a target molecule or antigen in the sample. See Fig. 2 for a representation of this interaction.

The antibody is often an immunoglobulin G (IgG) that has been harvested from animal serum after stimulation of the immune response to the target molecule (antigen). The antisera is harvested from the animal and purified by passing through an immunoaffinity column containing immobilized protein, such as protein A or protein G. To effectively induce an immune response, the target molecule should be of a sufficiently large molecular weight (>2000 Da). Small (<2000) target molecules can be made to better induce the immune response by attaching them to a larger carrier protein such as porcine thyroglobulin or keyhole limpet hemocyanin. After purification, the antibody is covalently immobilized

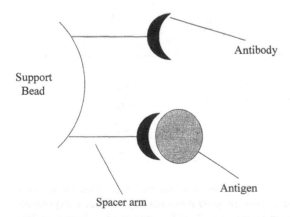

**Figure 2** Conceptualized representation of a simple immunoaffinity separation. In this drawing, the antibody has been covalently immobilized to a support bead.

onto a support that can be used in a simple chromatographic flow arrangement for LC/MS. The immunoaffinity column is conditioned under physiological conditions of pH and ionic strength, the sample is loaded, and the matrix components are removed by washing with buffer, while the target molecules are strongly bound and retained during the washing process. To effect elution, the immunoaffinity column is exposed to a solvent that reversibly denatures the antibody and facilitates release of the target molecules, allowing further separation or detection. A representative instrument configuration for this is displayed in Fig. 3.

When combined with tandem mass spectrometry, capable of selectively detecting a few analytes from the many that could be present, this approach provides for unsurpassed analytical selectivity for difficult chemical problems such as the study of drug–receptor binding [36] or the separation of complex mixtures of proteins or peptides [37]. The detection approach can be implemented in either on- or off-line formats. Alternatively, the purified antibody can be immobilized on a matrix-assisted laser-desorption ionization probe to allow direct application and characterization of a liquid sample containing the target molecule [38].

The immobilized constituent can be a drug receptor protein containing a binding site that emulates the site of pharmacological action. As described in Chapter 7, combinatorial libraries can be screened by allowing drug candidates to bind to the site in a chromatographic flow situation. These candidates are then removed from the site by denaturization and are subsequently identified by a downstream mass-spectrometry experiment [39].

Although not widely used at this time, immunoaffinity separations have also been shown to work with mass spectrometry in other ways as well. For

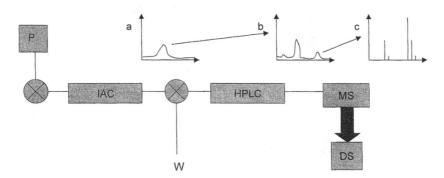

**Figure 3**   Schematic drawing showing the immunoaffinity chromatography/reverse-phase chromatography/atmospheric pressure ionization mass spectrometry experiment. System components are represented as follows: P, HPLC pump; IAC, immunoaffinity column; HPLC, analytical column; MS, mass spectrometer with atmospheric pressure ionization interface; DS, data system; and W, waste. Switching valves are represented as ⊗.

example, it has been proposed that immunoaffinity chromatography/reverse-phase chromatography/tandem mass spectrometry be used to elucidate the metabolism of therapeutic proteins and peptides. An example of this [40] demonstrated that the intact therapeutic peptide and a metabolite could be selectively isolated from rat plasma and identified by mass spectrometry in an on-line configuration. The selectivity of the immobilized antibody was such that all peptides that retained a specific amino acid sequence (epitope) would be bound and those peptides where the epitope has been broken apart or was not present would flow to waste. This property of the antibody allowed the peptide and a major metabolite to be selectively isolated.

Because of the time and effort needed to develop immunoaffinity methodology (4 to 8 weeks typically needed to generate a purified antibody), immunoaffinity separations are usually reserved for the more difficult separation problems in analytical chemistry. The shortened time scale of drug discovery precludes their routine day-to-day use. The power of immunoseparations in combination with mass spectrometry could see wider use as clever solutions to difficult problems are needed.

## VII. FUTURE DIRECTIONS IN AUTOMATED SAMPLE PREPARATION

Until recently, low throughput had been an area of concern in automated sample preparation. In the early days of automated solid-phase extraction, an average chemical analyst could out-produce an automated system and multiple operator interventions were commonplace. Only since the early 1990s have commercial parallel sample processing systems been available, with higher throughput 96-well systems having been available for about half that time. With these systems, the throughput has exploded and the number of operator interventions, while not disappearing across the board, has decreased convincingly. With many semiautomated systems that rely on a vacuum manifold, such as the Tomtec Quadra-96, the Biomek 2000, and the Multiprobe, operator intervention between sorbent washing and analyte elution is still required [41].

Because throughput continues to drive pharmaceutical analytical chemistry, future challenges in the automated solid-phase extraction workstation realm could involve a combination of parallel 96-well liquid handling (as demonstrated by the Tomtec Quadra-96) with fully automated extractions. This total automation could include the method development steps such as standard preparation and sorbent/solvent selection, removal of operator interventions such as positioning of collection tubes, and injection of samples into the chromatographic or electrophoretic system. Along this line, it seems inevitable that a dedicated solid-phase extraction workstation with all of these features will appear by the year 2003.

Beyond this practical and inevitable short-term picture looms the dual-headed phantom of microfluidics and assay miniaturization [42–44]. Although not widely practical at this time, microfluidics and miniaturization hold promise in terms of throughput advantages. Chip-to-mass spectrometry interfaces are being developed in this way to meet the needs of emerging drug discovery paradigms (for example, see http://www.abs-lcms.com). This technology could provide high-throughput sample assays for evaluating and characterizing drug targets and drug candidates by combining microchip-based separation devices with electrospray mass spectrometry. Although some success has been achieved in these areas, enormous technical battles in the area of separation and detection sensitivity need to be waged before this approach can be practical. These battles will be won eventually and the resulting technology will allow automated solid-phase extraction on the nanoliter scale, with thousands of channels being processed in parallel and total assay time shrinking from hours to minutes or seconds.

The current trends indicate that automated sample preparation is now more widely used than in the past. Much of this recent growth stems from increased capabilities of commercially available workstations. In past years, automated systems allowed analytical chemists to redirect time to other tasks while maintaining approximately equal throughput with manual methods. Newer workstations utilize a highly parallel processing approach to achieve greater efficiency and dramatically improved sample throughput. Because they can offer greater sample throughput than either on-line or discrete column approaches, it appears likely that 96-well systems will supplant these other approaches.

Near-term trends for automated sample preparation suggest that more integrated, dedicated systems with decreasing operator intervention are on the horizon. These systems would utilize parallel processing to maintain the high sample throughput of today's most efficient systems. For the longer term, microfluidics and miniaturization could allow parallel processing to take on a whole new meaning, as 96-well processing escalates to hundreds or thousands of samples, as needed. Small sample volumes required by the accompanying miniaturization will place great demands on detection sensitivity. The future of automated sample preparation as an analytical growth area seems secure.

## VIII. STRUCTURAL ELUCIDATION FOR METABOLISM

It would be of considerable value to have metabolite structural elucidation available at the earliest stages of drug discovery. This would allow a greater understanding of clearance mechanisms of novel drug candidates and allow synthetic chemists the opportunity to build greater stability into rapidly metabolized structures. To facilitate this, a fair amount of effort has recently gone into automated mass spectral interpretation software [45]. These simple pattern recognition programs make use of full-scan spectral data collected from LC/MS experiments and make approximate structural assignments (indication of hydroxylation, glucuronide, or sulfate conjugation, etc.), based on the observed $m/z$ ratios for unknowns. Comparisons to blank samples are also made to build additional reliability into the assignment.

A simple example of this process is for a chromatographic peak corresponding to a drug with a molecular weight of 350. In the positive ion spectrum, a base peak of 351 would be observed. Observation of a chromatographic peak with a mass-to-charge ratio of 367 suggests a hydroxylation reaction. A follow-up experiment would be to perform CID as a check that the fragmentation for the compound is consistent with the hydroxylated metabolite. Additional confidence can be built into this set of experiments by using the exact mass capabilities of time-of-flight (TOF) instrumentation, such as the quadrupole-TOF configuration.

This quick and approximate qualitative approach can readily be automated. It does not, however, compensate for the possibility that the ionizable groups can be lost from the molecule during metabolism. Some poorly ionized compounds can be completely missed, thereby confounding metabolic structural elucidation. In addition, unless the structure is very favorable, positional isomers cannot be distinguished.

A stronger approach to this process is to combine LC/MS with another universal and information-rich detection technique such as proton NMR. The combination technique of LC/NMR/MS has a higher information content than either technique alone. Some of the sensitivity and flow-related problems associated with flow stream NMR techniques are now starting to be addressed [46].

## IX. ASSAYS FOR BIOCHEMICAL INDICATORS OF DISEASE STATES

A difficult challenge in discovering and developing human pharmaceuticals is to ascertain their efficacy. Traditionally, this question has only been answered very late in preclinical development through the use of animal efficacy models or in costly clinical trials that are very demanding of time and other resources. More recently, it has been found that the monitoring of appropriate molecular entities that are biochemically indicative of a disease state has been valuable in determining the ability of a potential drug to impact a disease [47]. Historically, the most significant impediment to providing timely efficacy evaluation from a biomarker was the time needed to develop the bioanalytical method. Therefore, biomarker characterization was typically relegated to end-stage discovery and utilized to select which of two or three potential lead candidates was apparently most efficacious. Clearly, if efficacy information is to be available for early discovery decisions, such as compound and dose selection, then more readily available biomarkers and analytical methodologies for them are necessary.

Because of the high intrinsic sensitivity and selectivity of LC/MS/MS, analytical method development time for small molecule biochemical markers has been shortened from weeks or months to hours or days. Quantitation limits also have been lowered by 1 to 2 orders of magnitude, in some cases, relative to other established methodology, such as high-performance liquid chromatography (HPLC) with absorbance detection. With the capability of distinguishing differences in the characteristic mass of each analyte, mass spectrometry imparts a degree of specificity that classic chromatographic detection or enzyme assays cannot normally achieve.

An example application of LC/MS/MS has recently been reported for two compounds, pyridinoline (Pyd) and deoxypyridinoline (Dpd), that have been proposed as potential biomarkers for drug efficacy in collagen degradation models [48,49]. These pyridinium cross-links, formed during maturation of extracellular collagen fibrils, have been reported to offer a quantitative approach for estimating bone degradation as well as tissue collagen content. While Pyd is present in bone, cartilage, and soft tissues, Dpd is primarily restricted to bone collagen. Pyridinoline and deoxypyridinoline from bone collagen are excreted into urine as free ($\sim$40%) and peptide-bound ($\sim$60%) forms. The availability of a facile bioanalytical method is allowing new chemical entities to be evaluated as possible treatments for osteodegeneration. As they are administered in an efficacy study, the Pyd and Dpd levels are determined and can give a preliminary indication of how well the drug candidates work at an unusually early stage.

Atmospheric pressure ionization LC/MS has several distinct advantages over other more established quantitative approaches for these biochemical markers. The HPLC/fluorescence methods that have been reported require involved sample preparation, analyte extraction, and long chromatographic run times and are prone to interferences. The reported enzyme assays are less selective in that they cannot differentiate between the two compounds in a single run. Pyridinoline must therefore be quantified by difference. The API LC/MS/MS approach has allowed for a decrease in chromatographic run time, a streamlining of the sample preparation approach (dilute/inject as opposed to extraction), and a decrease in the analytical run time, while allowing simultaneous determination of both components.

Biochemical markers are gathering interest as early indicators of drug efficacy. Mass spectrometry techniques will continue to be applied to the quantitation of those discrete small-molecule chemical entities that can indicate disease or treatment states. The science of developing suitable surrogates is developing rapidly and mass spectrometry will be increasingly challenged by the imposing sensitivity requirements needed to make these advances possible.

## X. CONCLUSION

Our ability to study the physical and biological world has become closely linked to the capabilities of our instrumentation. As instrumentation improves, so do our abilities to make observations. The evolution of mass spectrometry instrumentation and drug discovery have become intertwined, so that many of the advances in the instrumentation have arisen because of a demand for throughput, sensitivity, or a need to perform chemical analysis with respect to a given pharmacological problem. This symbiosis appears destined to continue and should provide many fruitful benefits for both mass spectrometry instrumentation and drug discovery. Although many evolutionary changes have been forecast, it is possible, perhaps even likely, that the most important innovations have yet to be recognized or devised.

## REFERENCES

1. MD Garrett, P Workman. Discovering novel chemotherapeutic drugs for the third millennium. Eur J Cancer 35(14):2010–2030, 1999.
2. WE Evans, MV Relling. Pharmacogenomics: translating functional genomics into rational therapeutics. Science 286(5439):487–491, 1999.
3. S Paul. CNS drug discovery in the 21st century: from genomics to combinatorial chemistry and back. Br J Psychiat 174(37):23–25, 1999.
4. SR Wiley. Genomics in the real world. Curr Phar Des 4(5):417–422, 1998.
5. JH Wang, RM Hewick. Proteomics in drug discovery. Drug Discov Today 4(3): 129–133, 1999.
6. SL Schreiber. Target-oriented and diversity-oriented organic synthesis in drug discovery. Science 287(5460):1964–1969, 2000.
7. G Karet, J Boguslavsky. Buoyant demand for analytical instruments. Drug Discov Dev Jan/Feb:45–48, 2000.
8. K Kelly. New Rules for the New Economy. New York: Penguin Putnam Inc., 1998.
9. The most recent abstracts can be downloaded from the web page: http://www.asms.org/.
10. FW McLafferty, EK Fridriksson, DM Horn, MA Lewis, RA Zubarev. Biochemistry—biomolecule mass spectrometry. Science 284:1289–1290, 1999.
11. DP Little, DJ Aaserud, FW McLafferty. Sequence information from 42–108-mer DNAs (complete for a 50-mer) by tandem mass spectrometry. J Am Chem Soc 118: 9352, 1996.
12. JA Loo, JX He, WL Cody. Higher order structure in the gas phase reflects solution structure. J Am Chem Soc 120:4542–4543, 1998.
13. RD Sussmuth, G Jung. Impact of mass spectrometry on combinatorial chemistry. J Chromatogr B 725:49, 1999.

14. CW Ross, S Guan, PB Grosshans, TL Ricca, AG Marshall. Two dimensional Fourier transform ion cyclotron resonance mass spectrometry/mass spectrometry with stored-waveform ion radius modulation. J Am Chem Soc 115:7854–7861, 1993.
15. SA McLuckey, JL Stephenson. Ion ion chemistry of high-mass multiply charged ions. Mass Spectrom Rev 17:369–407, 1998.
16. DE Clemmer, MF Jarrold. Ion mobility measurements and their applications to clusters and biomolecules. J Mass Spectrom 32:577–592, 1997.
17. YS Liu, SJ Valentine, AE Counterman, CS Hoaglund, DE Clemmer. Injected-ion mobility analysis of biomolecules. Anal Chem 69:A728–A735, 1997.
18. SC Henderson, SJ Valentine, AE Counterman, DE Clemmer. ESI/ion trap/ion mobility/time-of-flight mass spectrometry for rapid and sensitive analysis of biomolecular mixtures. Anal Chem 71:291–301, 1999.
19. CA Srebalus, JW Li, WS Marshall, DE Clemmer. Determining synthetic failures in combinatorial libraries by hybrid gas-phase separation methods. J Am Soc Mass Spectrom 11:352–355, 2000.
20. LR Snyder, JJ Kirkland, JL Glajch. Practical HPLC Method Development. 2nd ed. New York: Wiley, 1997.
21. M Waksmundzka-Hajnos. Chromatographic separations of aromatic carboxylic acids. J Chromatogr B Biomed Sci Appl 717(1–2):93–118, 1998.
22. PL Buldini, S Cavalli, A Trifiro. State-of-the-art ion chromatographic determination of inorganic ions in food. J Chromatogr A 789(1–2):529–548, 1997.
23. RS Plumb, GJ Dear, D Mallett, IJ Fraser, J Ayrton, C Ioannou. The application of fast gradient capillary liquid chromatography/mass spectrometry to the analysis of pharmaceuticals in biofluids. Rapid Commun Mass Spectrom 13(10):865–872, 1999.
24. A Ducret, N Bartone, PA Haynes, A Blanchard, R Aebersold. A simplified gradient solvent delivery system for capillary liquid chromatography-electrospray ionization mass spectrometry. Anal Biochem 265(1):129–138, 1998.
25. CL Flurer. Analysis of antibiotics by capillary electrophoresis. Electrophoresis 20(15–16):3269–3279, 1999.
26. SH Chen, YH Chen. Pharmacokinetic applications of capillary electrophoresis. Electrophoresis 20(15–16):3259–3268, 1999.
27. JF Banks. Recent advances in capillary electrophoresis/electrospray/mass spectrometry. Electrophoresis 18(12–13):2255–2266, 1997.
28. JP Quirino, S Terabe. Electrokinetic chromatography. J Chromatogr A 856(1–2):465–482, 1999.
29. JP Quirino, S Terabe, X Shao, Y Shen, K O'Neill, ML Lee. Capillary electrophoresis using diol-bonded fused-silica capillaries. J Chromatogr A 830(2):415–422, 1999.
30. H Nishi. Capillary electrophoresis of drugs: current status in the analysis of pharmaceuticals. Electrophoresis 20(15–16):3237–3258, 1999.
31. R Vespalec, P Bocek. Chiral separations in capillary electrophoresis. Electrophoresis 20(13):2579–2591, 1999.
32. T Wachs, RL Sheppard, J Henion. Design and applications of a self-aligning liquid junction-electrospray interface for capillary electrophoresis-mass spectrometry. J Chromatogr B Biomed Appl 685(2):335–342, 1996.

33. DS Hage. Affinity chromatography: a review of clinical applications. Clin Chem 45(5):593–615, 1999.
34. DS Hage. Survey of recent advances in analytical applications of immunoaffinity chromatography. J Chromatogr B Biomed Appl 715(1):3–28, 1998.
35. SE Katz, M Siewierski. Drug residue analysis using immunoaffinity chromatography. J Chromatogr 624(1–2):403–409, 1992.
36. AK Mehta, MK Ticku. An update on GABAA receptors. Brain Res Rev 29(2–3): 196–217, 1999.
37. D Petsch, FB Anspach. Endotoxin removal from protein solutions. J Biotechnol 76(2–3):97–119, 2000.
38. X Liang, D Lubman, DT Rossi, GD Nordblom, CM Barksdale. On-probe immuno-extraction by matrix-assisted laser desorption/ionization mass spectrometry. Anal Chem 70:498–503, 1998.
39. JE Bruce, GA Anderson, R Chen, X Cheng, DC Gale, SA Hofstadler, BL Schwartz, RD Smith. Bio-affinity characterization mass spectrometry. Rapid Commun Mass Spectrom 9(8):644–650, 1995.
40. K Zheng, X Liang, DT Rossi, GD Nordblom, CM Barksdale, D Lubman. Elucidation of peptide metabolism by on-line immunoaffinity liquid chromatography mass spectrometry. Rapid Commun Mass Spectrom 14:261–269, 2000.
41. DT Rossi, N Zhang. Automated solid-phase extraction: current aspects and future prospects. J Chromatogr 885:97–113, 2000.
42. D Figeys, SP Gygi, G McKinnon, R Abersold. An integrated microfluidics-tandem mass spectrometry system for automated protein analysis. Anal Chem 70:3728–3734, 1998.
43. N Zhang, H Tan, ES Yeung. Automated and integrated system for high-throughput DNA genotyping directly from blood. Anal Chem 71:1138–1145, 1999.
44. L Silverman, R Campbell, JR Broach. New assay technologies for high-throughput screening. Curr Opin Chem Biol 2:397–403, 1998.
45. X Yu, D Cui, MR Davis. Identification of in vitro metabolites by intelligent automation LC/MS/MS (INTAMS) using the Finnegan TSQ 7000 in Abstract #19, 46th American Society of Mass Spectrometry National Meeting, Orlando, Fl (1998).
46. RS Plumb, J Ayrton, GJ Dear, BC Sweatman, IM Ismail. The use of preparative high performance liquid chromatography with tandem mass spectrometric directed fraction collection for the isolation and characterisation of drug metabolites in urine by nuclear magnetic resonance spectroscopy and liquid chromatography/sequential mass spectrometry. Rapid Commun Mass Spectrom 13(10):845–854, 1999.
47. FP Perera, JB Weinstein. Molecular epidemiology: recent advances and future directions. Carcinogenesis 21(3):517–524, 2000.
48. PD Delmas. Biochemical markers of bone turnover. 1. theoretical considerations and clinical use in osteoporosis. Am J Med 95:5A–16S, 1993.
49. M Saito, K. Marumo, K Fujii, N Ishioka. Single-column high-performance liquid chromatographic-fluorescence detection of immature, mature, and senescent cross-links of collagen. Anal Biochem 253:26–32, 1997.

# Index

T - #0076 - 101024 - C0 - 234/156/23 [25] - CB - 9780824706074 - Gloss Lamination